U0327713

Details in Architecture

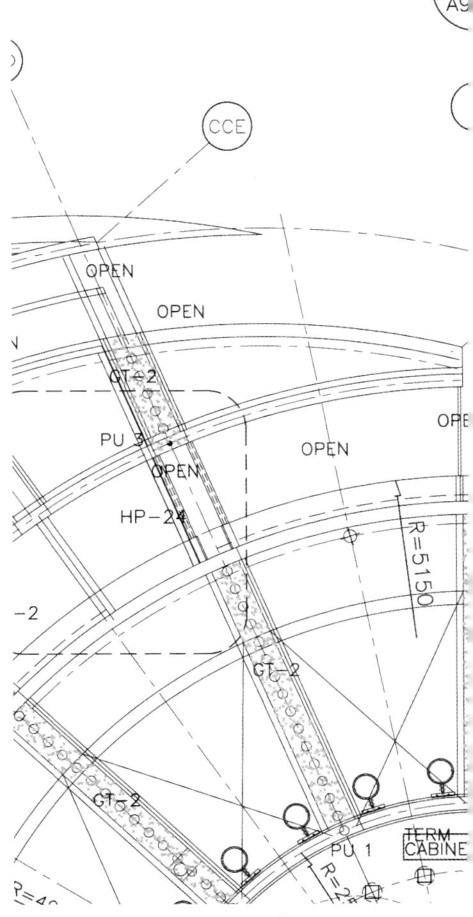

DA 建筑名家细部设计创意 5

——加拿大专集

王 真 编

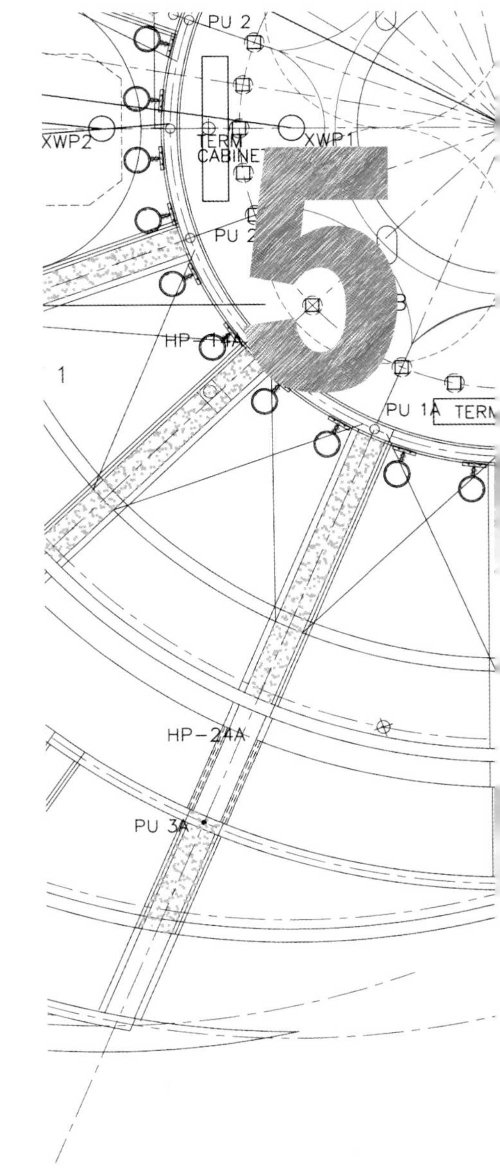

中国建筑工业出版社

图书在版编目（CIP）数据

DA 建筑名家细部设计创意 5——加拿大专集 / 王真编. —北京：中国建筑工业出版社，2005
ISBN 7-112-07440-1

Ⅰ. D… Ⅱ. 王… Ⅲ. 建筑结构－细部设计－加拿大－图集
Ⅳ. TU22-64

中国版本图书馆 CIP 数据核字（2005）第 050845 号

责任编辑：张　建
责任设计：郑秋菊
责任校对：李志立　李志瑛

***DA* 建筑名家细部设计创意 5**
——加拿大专集
王真　编
*
中国建筑工业出版社出版、发行（北京西郊百万庄）
新 华 书 店 经 销
伊诺丽杰设计室制版
北京中科印刷有限公司印刷
*
开本：787 × 1092 毫米　1/10　印张：30⅘　字数：500 千字
2005 年 12 月第一版　2005 年 12 月第一次印刷
印数：1 — 1,500 册　定价：220.00 元
ISBN 7 - 112 - 07440 - 1
(13394)

本社网址：http: //www.cabp.com.cn
网上书店：http: //www.china-building.com.cn

致编者

The Architectural Institute of British Columbia (AIBC) is very pleased that a book, *Leading Canadian Architectural Firms and Their Projects*, is being produced for the Chinese market. We believe that the book demonstrates that Canadian designers have made an important contribution and the work of British Columbia's Architects add a rich architectural history in the urban landscape and built environment of Western Canada.

In particular, we are delighted that a publication will showcase their work.

我们非常荣幸地看到《加拿大专辑》即将在中国出版面世。我们相信这本书展示出了加拿大建筑师所作的重大贡献，以及他们通过这些作品在加拿大西部的城市环境中谱写的丰富瑰丽的建筑历史。

看到本书展示出了加拿大建筑师的众多优秀作品使我们感到特别高兴。

Stuart Howard
AIBC President

不列颠哥伦比亚建筑师协会主席

Dorothy Barkley
AIBC Executive Director

不列颠哥伦比亚建筑师协会总裁

The AIBC is a self-governing body dedicated to excellence in the profession of architecture for the benefit of its membership, society and the environment. Through our many community initiatives, professional programs and partnerships with other professional organizations throughout the province and across the country, we strive to be effective and successful ambassadors on behalf of the profession, its practitioners and British Columbia.

前　言

《DA建筑名家细部设计创意5——加拿大专集》是第一本专门介绍加拿大建筑名家作品的书籍。

书中所有的照片和图纸都是首次在中国发表，绝大部分照片、图纸甚至与在北美出版的建筑书籍同步。

本书收录了加拿大近期完工的建筑佳作和获奖作品，全部图纸直接来自各个事务所的施工图。书中所列建筑都精选了一些效果独特的建筑细部照片和其施工图纸，展示出了建筑师大胆而新奇设计的创意。

此外，本书还收录了对于建筑设计事务所的介绍，以便读者对加拿大建筑名家有更直观的认识。与其他建筑书籍相比，本书所展示的内容更全面、详细，力求给读者一个全面了解加拿大建筑的机会。

本书收录的加拿大著名建筑师及其事务所有：亚瑟·埃里克森、谭秉荣建筑设计事务所、Patkau建筑设计事务所，以及第一家在北美上市的Stantec建筑设计事务所。

本书在加拿大住宅抵押贷款公司的协助之下完成，在此表示感谢。

Contents

目　录

Arthur Erickson with Nick Milkovich Architects Inc.

亚瑟·埃里克森和Nick Milkovich建筑设计事务所

Nick Milkovich Architects Inc. was established in November 1991 as a result of the reorganization of Arthur Erickson Architects Inc., and is dedicated to continuing the approach to design, which has gained the former firm a worldwide reputation for design excellence.

The original group of five architects had, until this date, been the custom design team for primarily small to medium scale projects within the larger Arthur Erickson office. With its total of over sixty years experience under the mentorship of Arthur Erickson on a wide range of local and international projects, Nick Milkovich Architects Inc. has grown to a staff of 11 and continues to offer professional services in architecture, interior design, urban design, planning and programming to all levels of government, institutions, corporations, developers, and private clients. Arthur Erickson works with NMA as a design consultant on a wide range of projects.

Arthur Erickson with Nick Milkovich Architects Inc. undertakes projects of any type, which lend themselves to creative and precedent setting solutions. A thorough understanding of the aspirations of the clients and user groups, as well as their schedule and budget concerns is the starting point of the approach to every commission. The firm seeks to establish the role the project plays in relation to its environmental, historical, social, and economic context. Our objectives are to let the project demands speak to us and for us to interpret and design without preconceptions in order to arrive at the most appropriate solution.

To encourage the most appropriate and creative use of the talents of the firm a flexible management approach has been adopted. While each team leader has a clearly defined role, the team structure and assignment of responsibilities are tailored to suit the specific needs of each project.

Arthur Erickson with Nick Milkovich Architects Inc. calls on the expertise of the finest outside consulting services for prime engineering disciplines, structural, mechanical, and electrical, and for cost control estimating. If the project demands, more specialized consultants are retained for disciplines such as landscape, lighting, and acoustics.

Nick Milkovich建筑设计事务所成立于1991年9月，是在亚瑟·埃里克森建筑设计事务所的基础上重新组合而成的，延续了与其国际化声誉相匹配的设计素质。

直至今日，最初的5个建筑师组成的设计团队依然在大的亚瑟·埃里克森建筑设计事务所的名义下完成中小型建筑的设计。Nick Milkovich建筑设计事务所继承了亚瑟·埃里克森建筑设计事务所的六十年本土的和国际的建筑设计经验。目前事务所拥有11位建筑师，提供广泛的建筑设计服务。其主要业务范围有：建筑设计、室内设计、城市设计、城市规划和不同规模的政府工程、文化教育建筑、地产开发和私人工程。在很多工程项目设计中亚瑟·埃里克森都是Nick Milkovich建筑设计事务所的顾问。

Nick Milkovich建筑设计事务所有创新化的思维和丰富的经验，承接不同类型的设计工程。全面理解业主和用户的意图、工程工期与预算是建筑设计开始的第一步。Nick Milkovich建筑设计事务所会根据建筑物的环境、历史、社会和经济条件的要求寻求最佳的解决方案。其设计目标是突出建筑项目本身的要求，无偏见地把这些条件以最适宜的方式付诸于建筑实践。

为了充分发挥每一个设计师的才能，Nick Milkovich建筑设计事务所采用灵活的管理方式。每一个设计主管都有清晰的目标任务，而设计团队和其工作范围将为其更好的完成任务而量身定做。

Nick Milkovich建筑设计事务所与最适宜的结构、设备和预算顾问公司合作。因工程的需要，其他专业顾问公司，如：景观、采光和声效工程等，也可以成为工程总承包的一部分。

MUSEUM OF GLASS
玻璃博物馆

地　　点：华盛顿州 塔科马市
规　　模：约7 000m²
完成时间：2002年
合作建筑师：托马斯·库克（Thomas Cook）、里德·赖因瓦尔德（Reed Reinvald）
摄 影 师：尼克·勒乌（Nic Lehoux）

The Museum of Glass is sited near Tacoma's cultural and historical corridor. The Washington State History Museum and the Tacoma Art Museum stand across a highway and railway, parallel to the new museum.

Many visitors to the Museum of Glass approach it by crossing from the history museum over the Chihuly Bridge of Glass. The challenge was to design a meaningful pathway for visitors to travel from the footbridge, which connects to the rooftop plaza of the museum, to the museum entrance two levels below. The entire museum rooftop is a stepped plaza. Ramps offer one means of descent from the rooftop to the museum's entrance at the water's edge. They take visitors by the reflecting pools designed to mirror art installations and the distant landscape. Alternatively, a grand staircase that circles the distinctive cone will provide an intriguing and faster route down. The composition of descending forms reflects the demands of the location and gives the shape of the building a landscaped rather than structural presence. The building is clad in subtly patterned pre-cast concrete panels, the cone is sheathed with diamond-shaped stainless steel scales and glass is used where possible to allow natural light to fill the museum interior.

The 75,000 square foot Museum of Glass contains glass workshops and an artist studio, exhibition spaces, a permanent collection display area, a retail area for the sale of handmade glass art and books, a restaurant, a library, a theater and classrooms.

The key attraction is the authentic "hot shop" amphitheatre - a giant cone, 90-feet tall and 100-feet diameter and one of the Museum's most striking architectural features. Inside the cone the public can watch the dynamic process of glass art creation unfold.

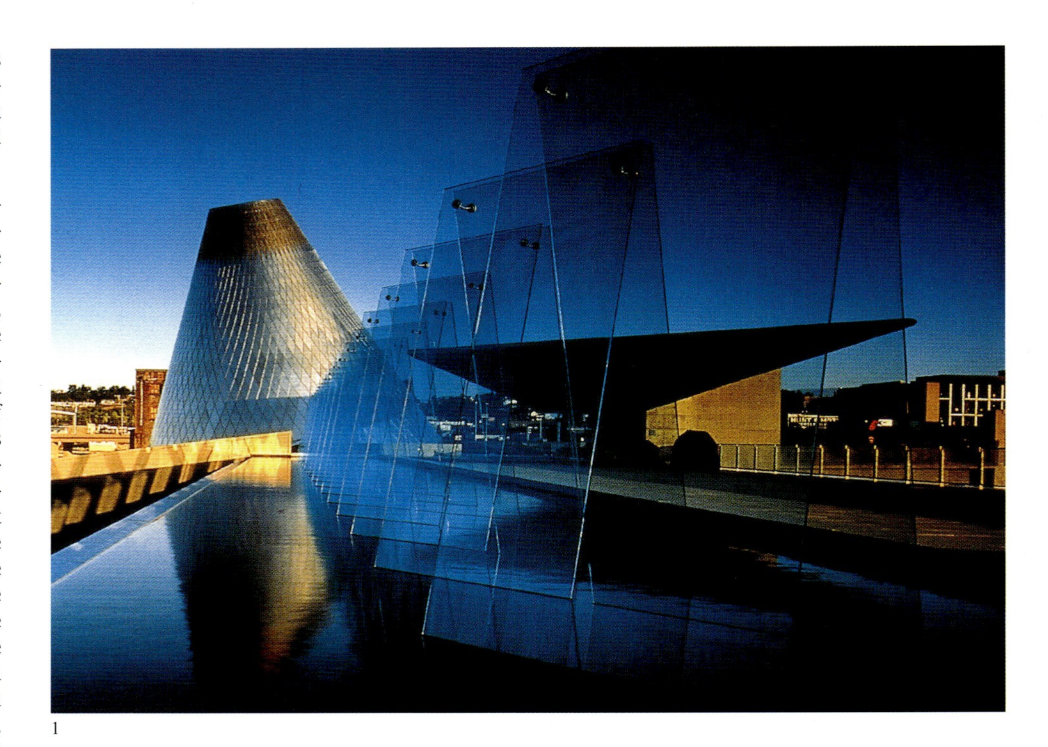

1

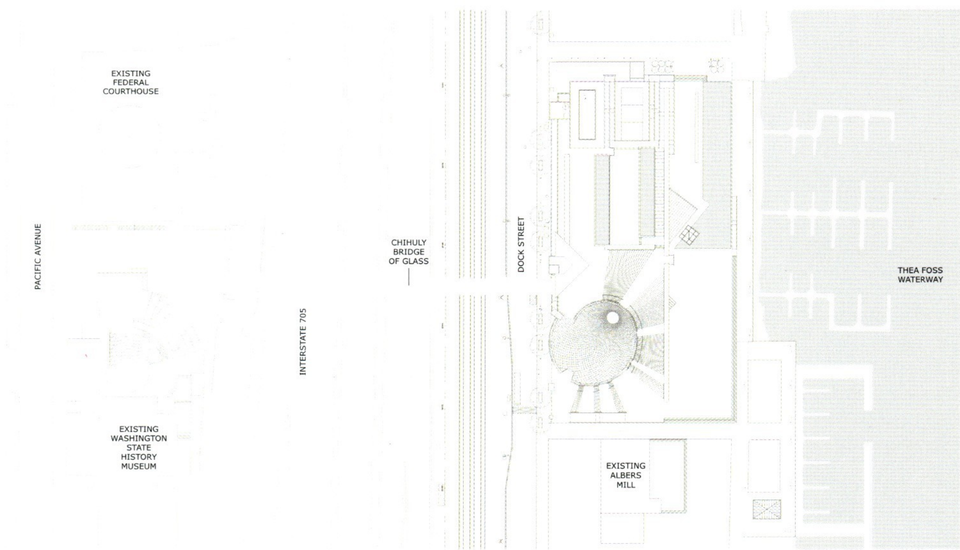

2

玻璃博物馆位于塔科马市文化和历史带附近，与华盛顿州历史博物馆隔高速公路相望，和塔科马艺术博物馆仅隔一条铁路。

大多数的参观者是从历史博物馆的方向穿过Chihuly玻璃桥进入玻璃博物馆。一个步行桥通向博物馆的屋顶广场。玻璃桥设计为一个象征意义的入口，方便参观者从步行桥到博物馆下两层的入口。整个博物馆的屋顶是阶梯状的广场。坡道一直由屋顶延伸到博物馆入口的水池边。反射水池、镜面艺术品和远处的景观吸引着参观者的注意力。一个位于醒目的桶状物内的楼梯提供另一个快速向下的通道。

这种下沉式的设计反映出用地的特殊性，其建筑形式与其说是建筑构造的要求，不如说是景观的需求。建筑物以精巧的混凝土预制板为外墙，筒状体包裹着钻石形的不锈钢片。在自然光可以照入室内的地方均采用玻璃。

75 000m^2的玻璃博物馆内包括：玻璃工厂、艺术工作室、展示空间、永久收藏品展示区、一个手工玻璃器皿和书籍零售空间、一家餐馆、一座图书馆、一个剧场和教室。

最主要的焦点是环幕电影厅，一个巨大的筒状体高27m，直径30m。在厅内，观众可以观看具有动感的玻璃艺术品。

3

4

1　入口水池和玻璃装饰
2　平面图
3　建筑模型
4　屋顶延伸到博物馆入口水池的坡道

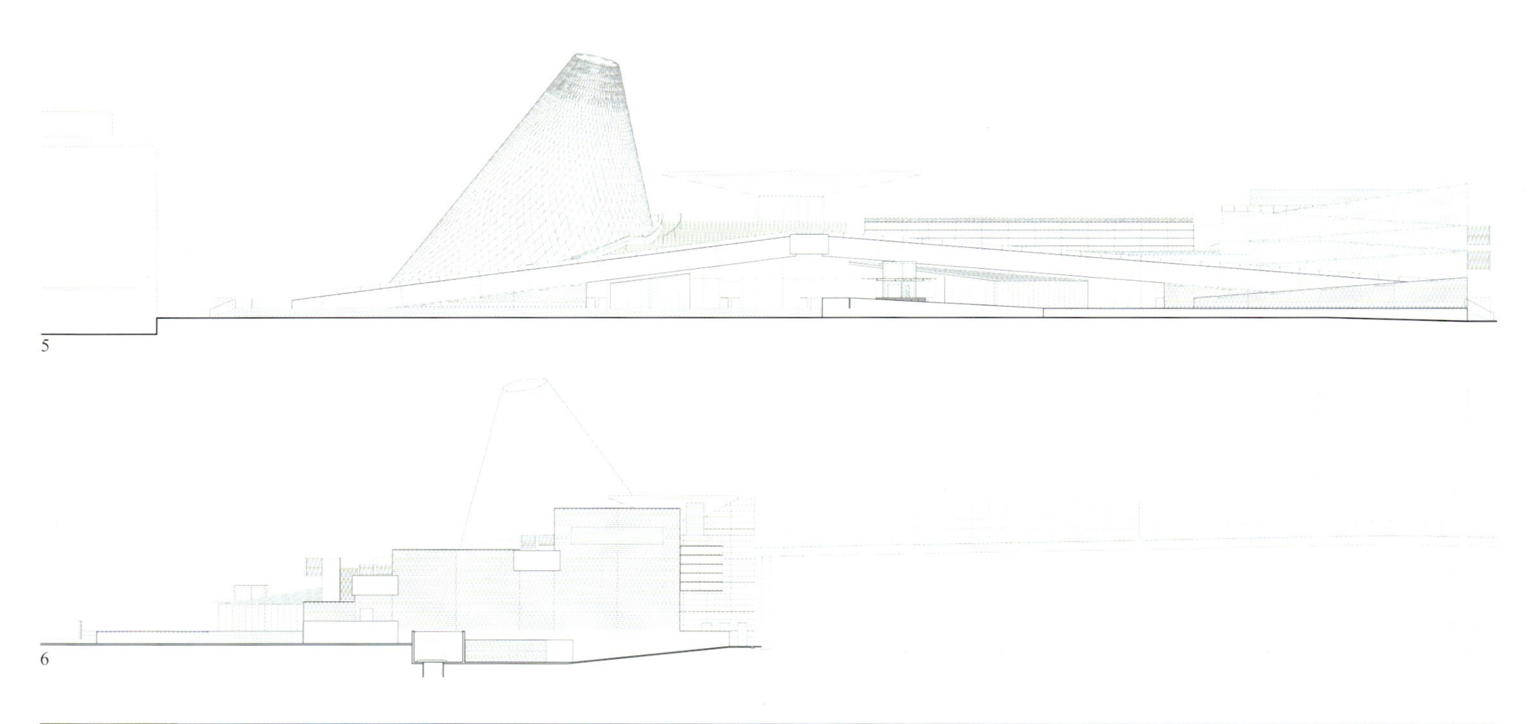

5

6

7

5 东立面图
6 北立面图
7 通向博物馆屋顶广场的步行桥
8 南立面图
9 西立面图
10 二层平面图

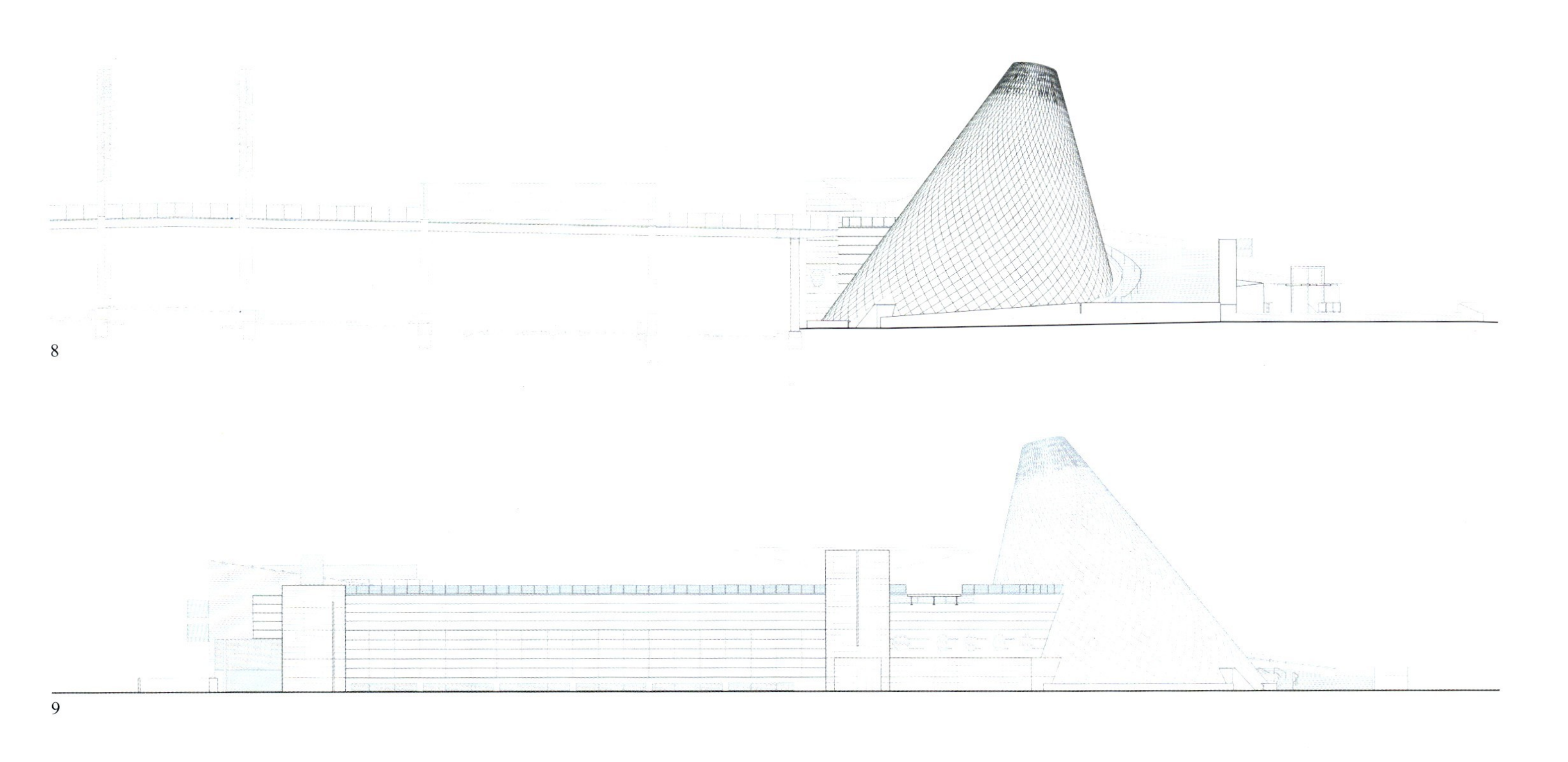

8

9

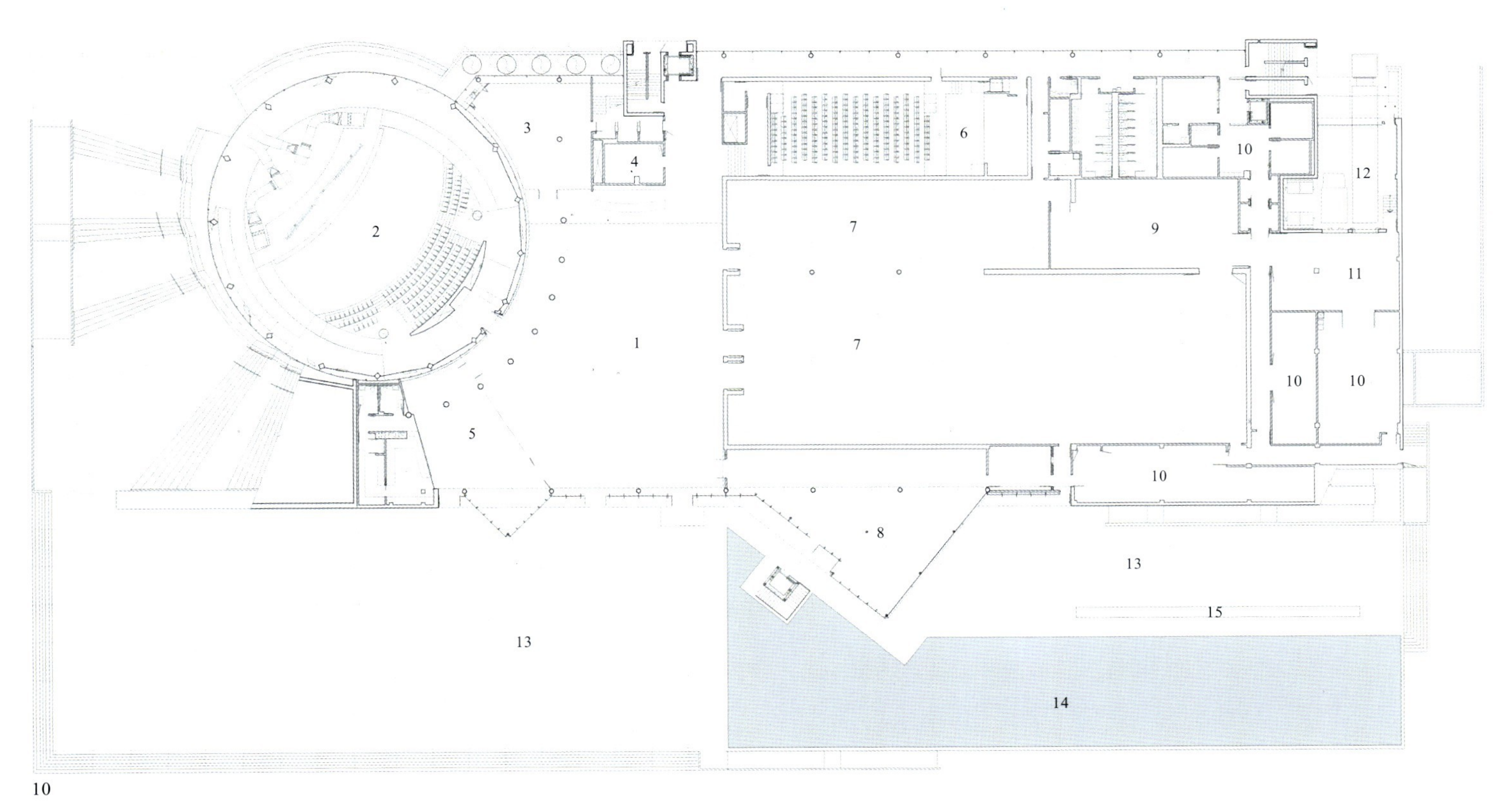

10

1 Grand Hall
2 Hot Shop
3 Education Centre
4 Coat Room
5 Cafe
6 Theatre
7 Exhibit Room
8 Museum Store
9 Preparation Staging Area
10 Back of House
11 Shipping/Receiving
12 Loading Bay
13 Plaza
14 Reflecting Pool
15 Planter

11

8 7 6 5 4 3 2 1

25'-6" 29'-6" 25'-6" 29'-6" 25'-6" 6'-0" 8'-9"

CANOPY
N.I.C.

ELEVATOR NO.1

308
MEZZANNINE

315
LOUNGE

GLASS PARTITIONS
AND DOORS N.I.C.

208
CAFE (BEYOND)

GRAND HALL

247
MEMBER'S DESK

218
LOBBY

PACKAGE TWO

PACKAGE ONE

7'-6" MIN. AT BEAM

120
C.O.T. PARKING

10'-6"

115
MECHANICAL ROOM

118
LOBBY

13'-11 1/4"

15'-11 1/4"

12'-6 3/4"

17'-0"

5'-1 3/4"

12

11　环幕电影厅
12　剖面图
13　电影厅剖面图

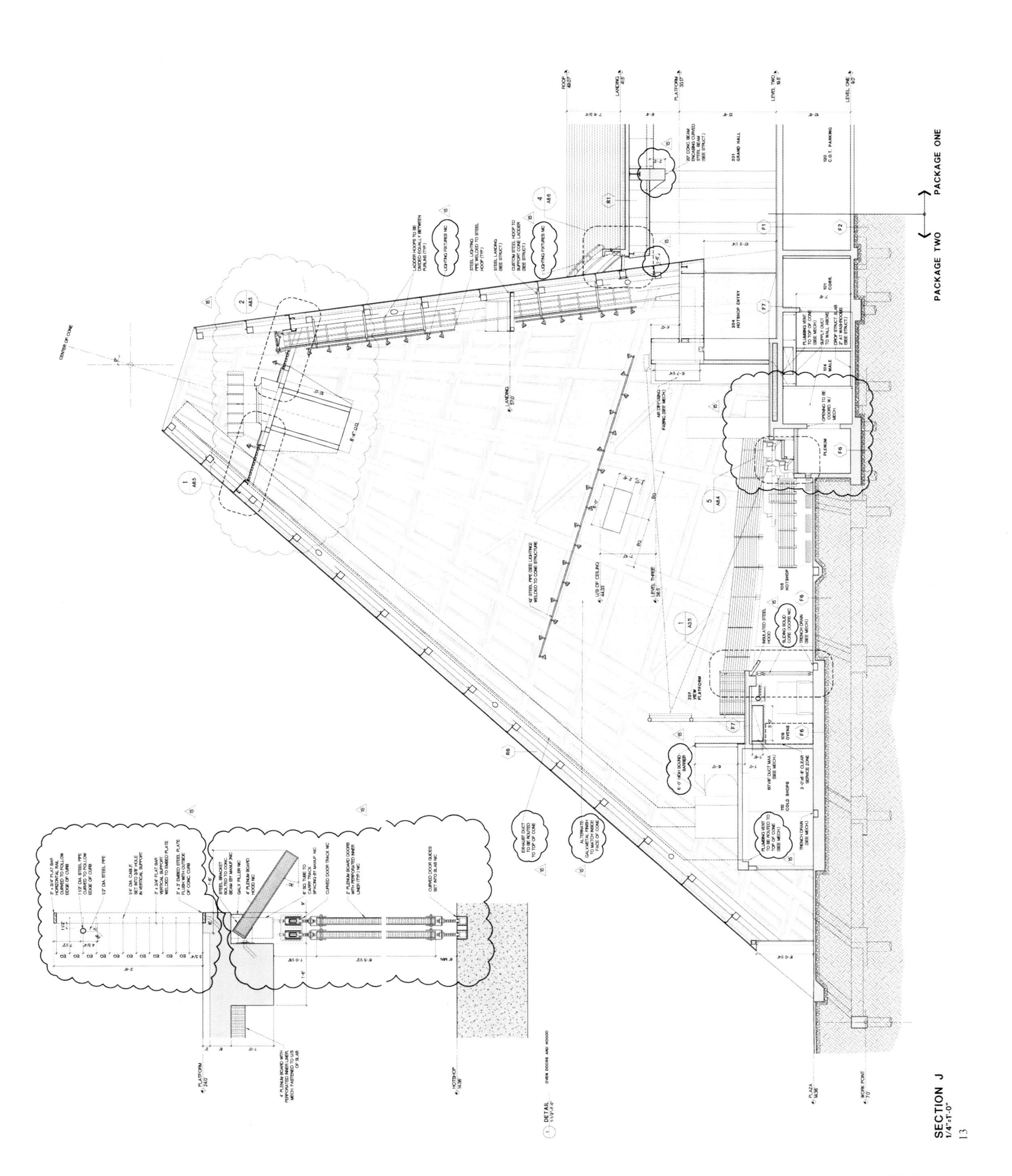

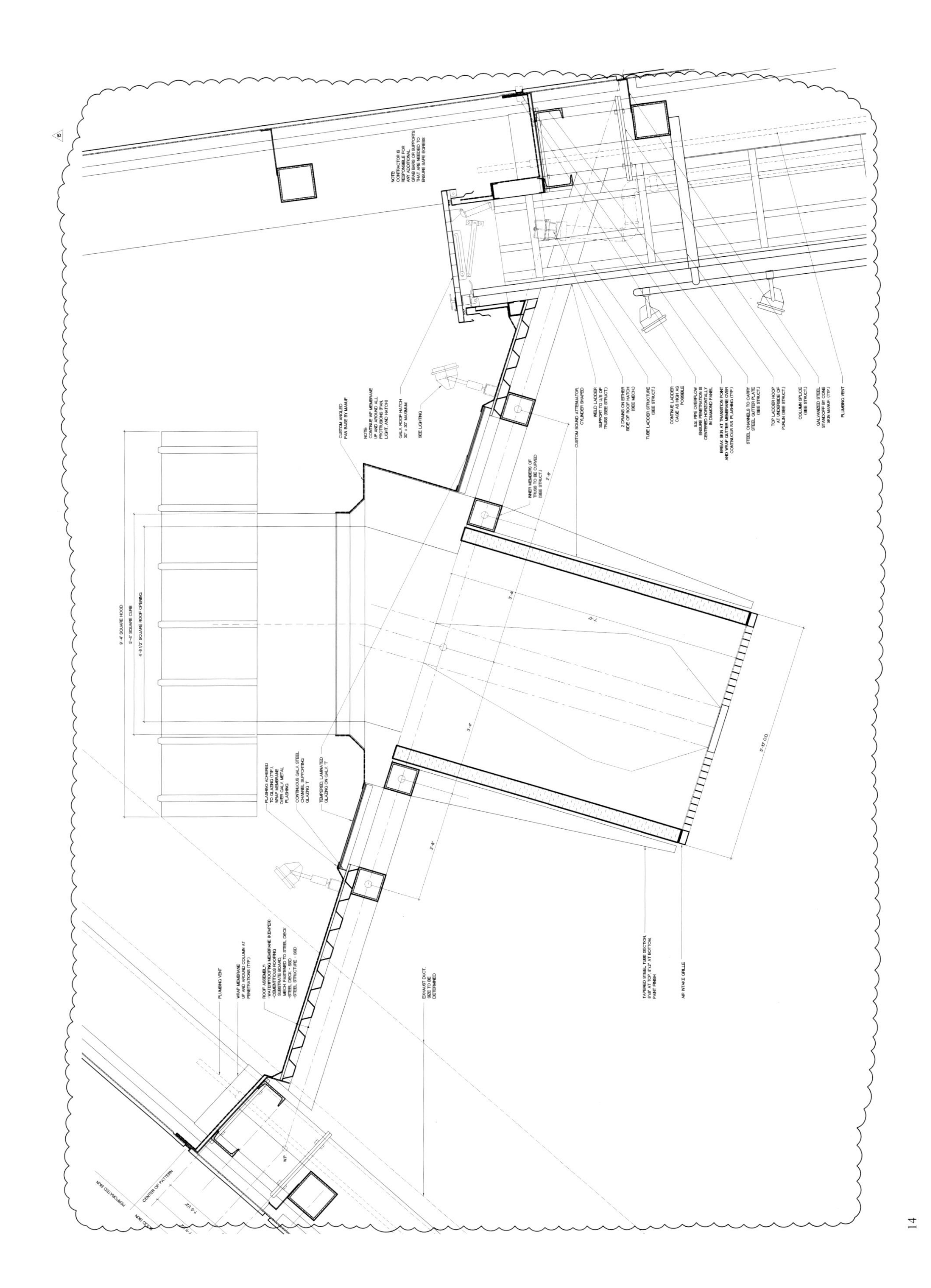

14 环幕电影厅天窗开口详图
15 锥形体详图

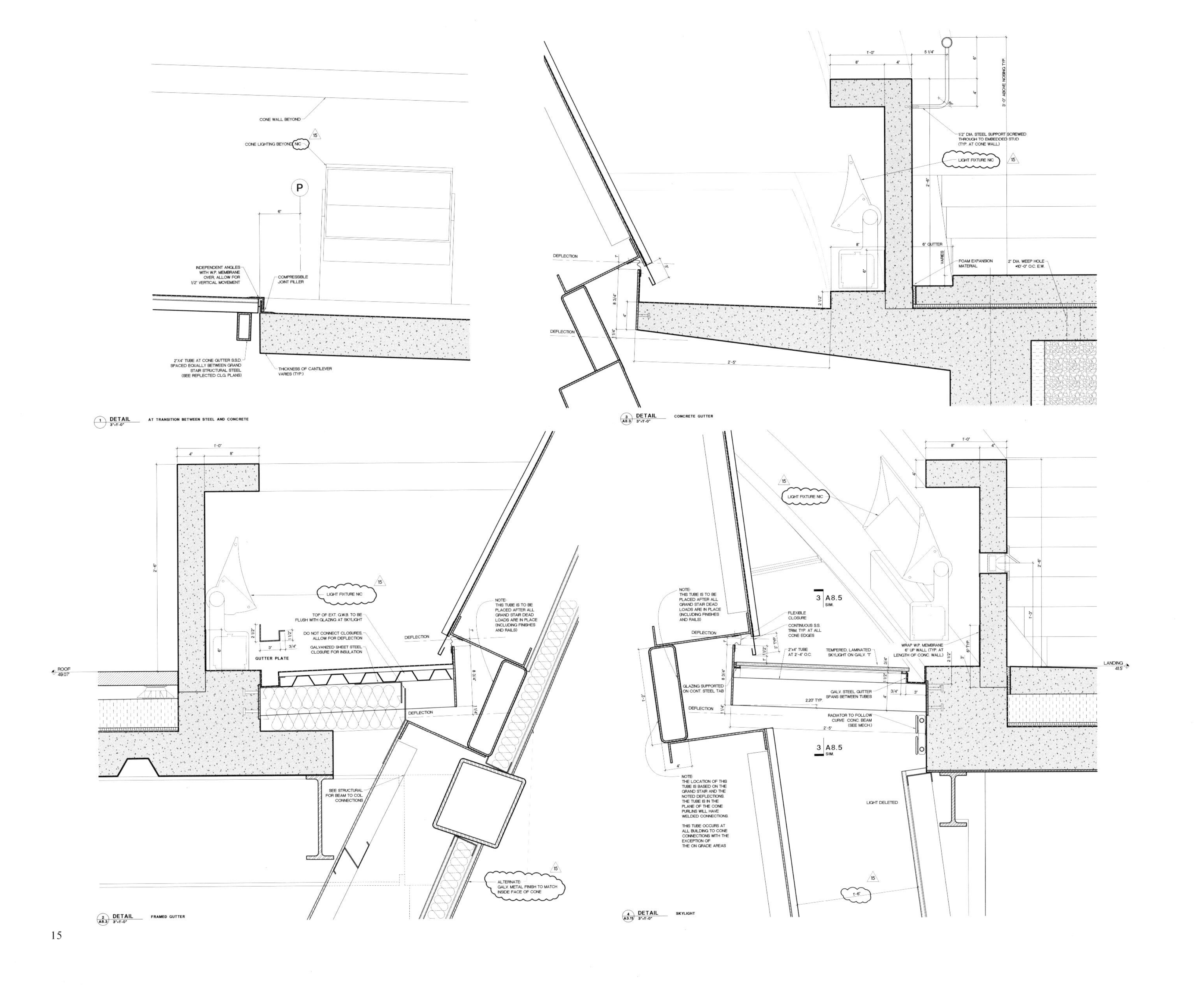
CONE WALL BEYOND
CONE LIGHTING BEYOND NIC
P
INDEPENDENT ANGLES WITH W.P. MEMBRANE OVER. ALLOW FOR 1/2" VERTICAL MOVEMENT
COMPRESSIBLE JOINT FILLER
2"X4" TUBE AT CONE GUTTER S.S.D. SPACED EQUALLY BETWEEN GRAND STAIR STRUCTURAL STEEL (SEE REFLECTED CLG. PLANS)
THICKNESS OF CANTILEVER VARIES (TYP.)
DETAIL
3"=1'-0"
AT TRANSITION BETWEEN STEEL AND CONCRETE
1/2" DIA. STEEL SUPPORT SCREWED THROUGH TO EMBEDDED STUD (TYP. AT CONE WALL)
LIGHT FIXTURE NIC
3'-0" ABOVE NOSING TYP.
6" GUTTER
VARIES
FOAM EXPANSION MATERIAL
2" DIA. WEEP HOLE @10'-0" O.C. E.W.
DEFLECTION
DETAIL
3"=1'-0"
CONCRETE GUTTER
LIGHT FIXTURE NIC
TOP OF EXT. G.W.B. TO BE FLUSH WITH GLAZING AT SKYLIGHT
DO NOT CONNECT CLOSURES. ALLOW FOR DEFLECTION
GALVANIZED SHEET STEEL CLOSURE FOR INSULATION
GUTTER PLATE
NOTE: THIS TUBE IS TO BE PLACED AFTER ALL GRAND STAIR DEAD LOADS ARE IN PLACE (INCLUDING FINISHES AND RAILS)
ROOF 49.07
SEE STRUCTURAL FOR BEAM TO COL. CONNECTIONS
ALTERNATE: GALV. METAL FINISH TO MATCH INSIDE FACE OF CONE
DETAIL
3"=1'-0"
FRAMED GUTTER
LIGHT FIXTURE NIC
NOTE: THIS TUBE IS TO BE PLACED AFTER ALL GRAND STAIR DEAD LOADS ARE IN PLACE (INCLUDING FINISHES AND RAILS)
FLEXIBLE CLOSURE
CONTINUOUS S.S. TRIM TYP. AT ALL CONE EDGES
2"x4" TUBE AT 2'-4" O.C.
TEMPERED, LAMINATED SKYLIGHT ON GALV. "T"
WRAP W.P. MEMBRANE 6" UP WALL (TYP. AT LENGTH OF CONC. WALL)
LANDING 41.5
GLAZING SUPPORTED ON CONT. STEEL TAB
GALV. STEEL GUTTER SPANS BETWEEN TUBES
RADIATOR TO FOLLOW CURVE CONC. BEAM (SEE MECH.)
3 A8.5 SIM.
NOTE: THE LOCATION OF THIS TUBE IS BASED ON THE GRAND STAIR AND THE NOTED DEFLECTIONS. THE TUBE IS IN THE PLANE OF THE CONE PURLINS WILL HAVE WELDED CONNECTIONS.
THIS TUBE OCCURS AT ALL BUILDING TO CONE CONNECTIONS WITH THE EXCEPTION OF THE ON GRADE AREAS
LIGHT DELETED
DETAIL
3"=1'-0"
SKYLIGHT
15

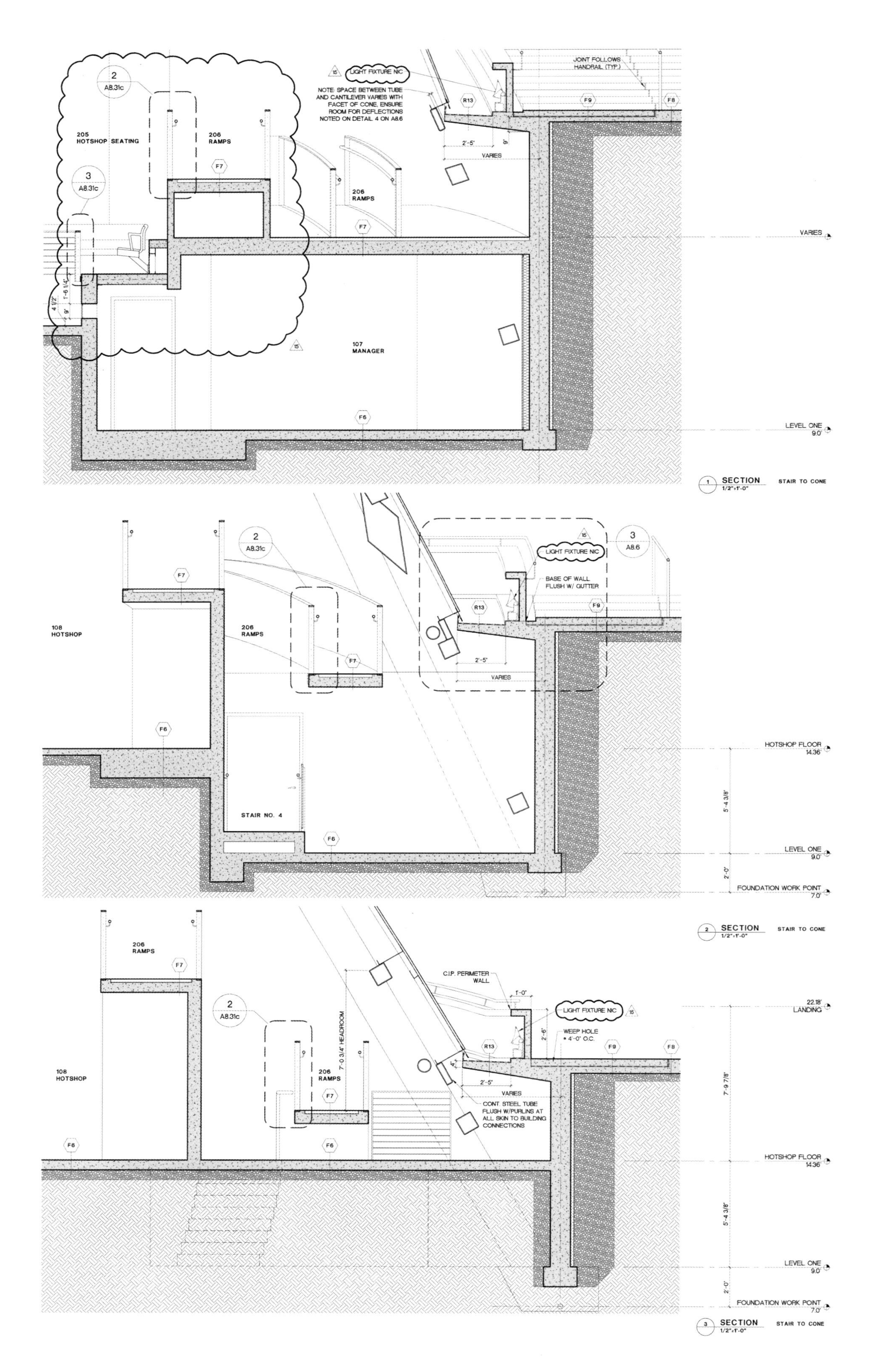
205
HOTSHOP SEATING
206
RAMPS
LIGHT FIXTURE NIC
NOTE: SPACE BETWEEN TUBE AND CANTILEVER VARIES WITH FACET OF CONE, ENSURE ROOM FOR DEFLECTIONS NOTED ON DETAIL 4 ON A8.6
JOINT FOLLOWS HANDRAIL (TYP.)
VARIES
107
MANAGER
LEVEL ONE
9.0'
SECTION
STAIR TO CONE
108
HOTSHOP
BASE OF WALL FLUSH W/ GUTTER
STAIR NO. 4
HOTSHOP FLOOR
14.36'
FOUNDATION WORK POINT
7.0'
C.I.P. PERIMETER WALL
WEEP HOLE @ 4'-0" O.C.
7'-0 3/4" HEADROOM
CONT. STEEL TUBE FLUSH W/PURLINS AT ALL SKIN TO BUILDING CONNECTIONS
22.18'
LANDING

16

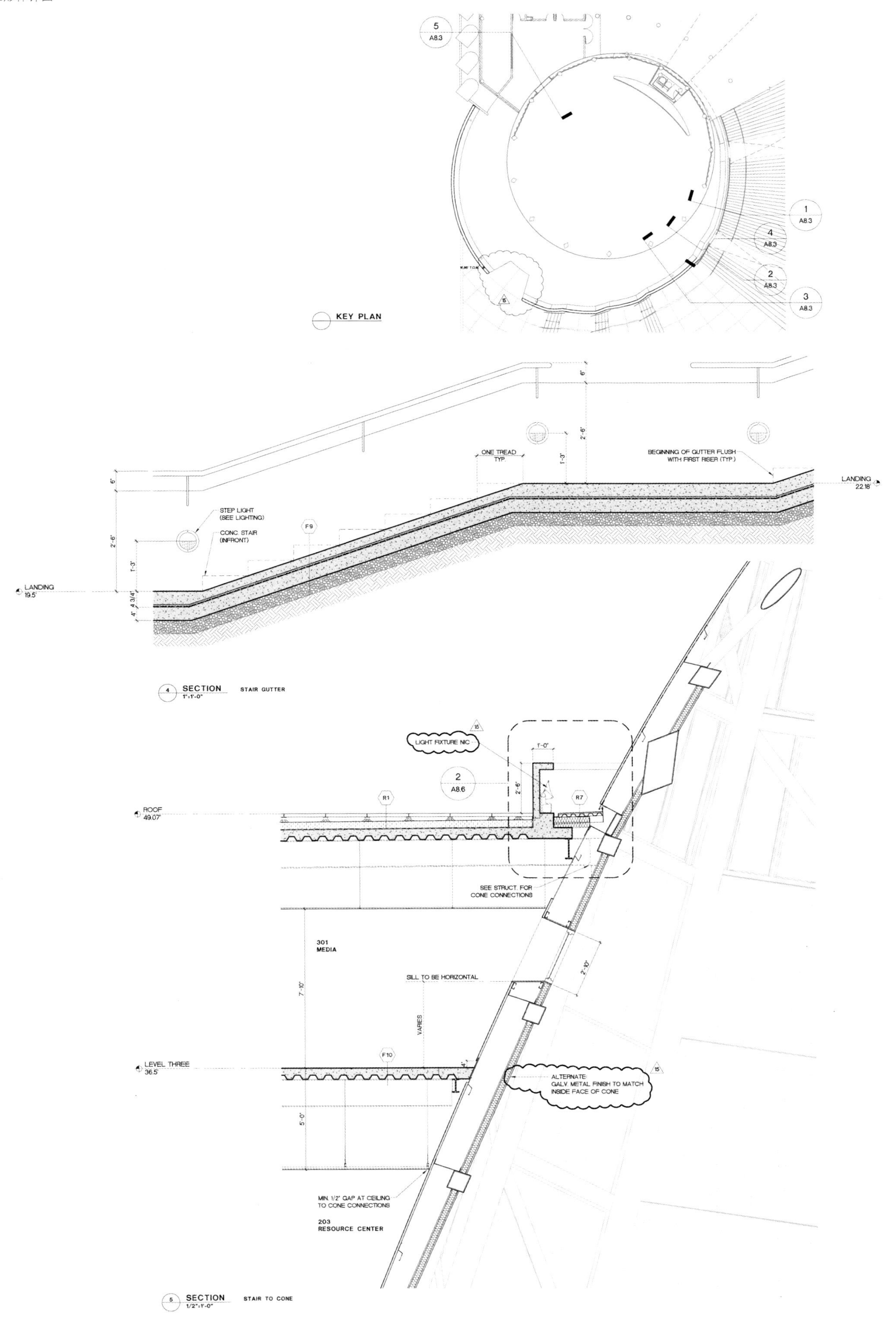
KEY PLAN
SECTION STAIR GUTTER
SECTION STAIR TO CONE
LANDING 19.5'
LANDING 22.18'
STEP LIGHT (SEE LIGHTING)
CONC. STAIR (INFRONT)
ONE TREAD TYP.
BEGINNING OF GUTTER FLUSH WITH FIRST RISER (TYP.)
LIGHT FIXTURE NIC
ROOF 49.07'
SEE STRUCT. FOR CONE CONNECTIONS
301 MEDIA
SILL TO BE HORIZONTAL
LEVEL THREE 36.5'
ALTERNATE: GALV. METAL FINISH TO MATCH INSIDE FACE OF CONE
MIN. 1/2" GAP AT CEILING TO CONE CONNECTIONS
203 RESOURCE CENTER

17

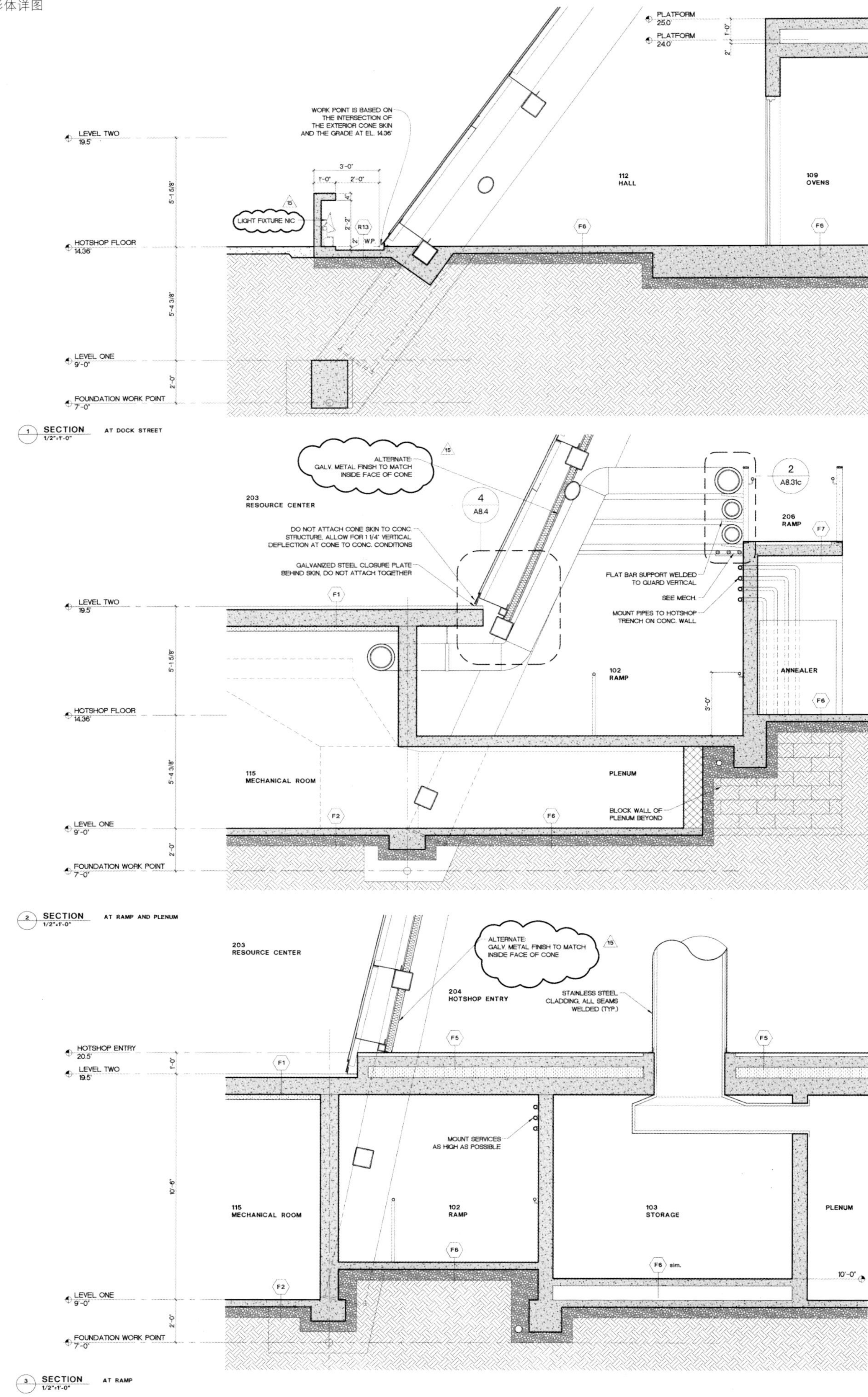
PLATFORM
25.0'
PLATFORM
24.0'
WORK POINT IS BASED ON THE INTERSECTION OF THE EXTERIOR CONE SKIN AND THE GRADE AT EL. 14.36'
LEVEL TWO
19.5'
112
HALL
109
OVENS
LIGHT FIXTURE NIC
HOTSHOP FLOOR
14.36'
LEVEL ONE
9'-0"
FOUNDATION WORK POINT
7'-0"
SECTION
1/2"=1'-0"
AT DOCK STREET
ALTERNATE: GALV. METAL FINISH TO MATCH INSIDE FACE OF CONE
203
RESOURCE CENTER
DO NOT ATTACH CONE SKIN TO CONC. STRUCTURE. ALLOW FOR 1 1/4" VERTICAL DEFLECTION AT CONE TO CONC. CONDITIONS
GALVANIZED STEEL CLOSURE PLATE BEHIND SKIN. DO NOT ATTACH TOGETHER
206
RAMP
FLAT BAR SUPPORT WELDED TO GUARD VERTICAL
SEE MECH.
MOUNT PIPES TO HOTSHOP TRENCH ON CONC. WALL
102
RAMP
ANNEALER
115
MECHANICAL ROOM
PLENUM
BLOCK WALL OF PLENUM BEYOND
SECTION
1/2"=1'-0"
AT RAMP AND PLENUM
204
HOTSHOP ENTRY
STAINLESS STEEL CLADDING, ALL SEAMS WELDED (TYP.)
HOTSHOP ENTRY
20.5'
MOUNT SERVICES AS HIGH AS POSSIBLE
103
STORAGE
SECTION
1/2"=1'-0"
AT RAMP

18

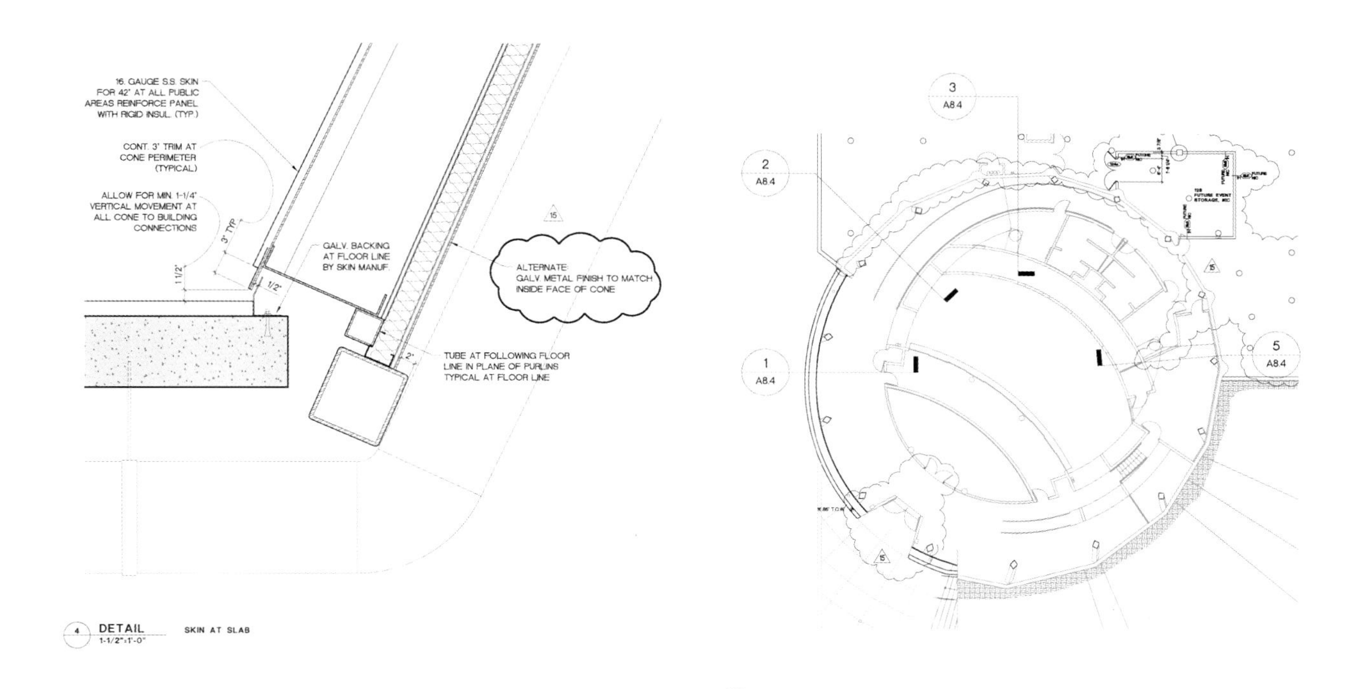
16 GAUGE S.S. SKIN FOR 42" AT ALL PUBLIC AREAS REINFORCE PANEL WITH RIGID INSUL. (TYP.)
CONT. 3" TRIM AT CONE PERIMETER (TYPICAL)
ALLOW FOR MIN. 1-1/4" VERTICAL MOVEMENT AT ALL CONE TO BUILDING CONNECTIONS
GALV. BACKING AT FLOOR LINE BY SKIN MANUF.
ALTERNATE: GALV. METAL FINISH TO MATCH INSIDE FACE OF CONE
TUBE AT FOLLOWING FLOOR LINE IN PLANE OF PURLINS TYPICAL AT FLOOR LINE
4 DETAIL SKIN AT SLAB
1-1/2"=1'-0"
3 A8.4
2 A8.4
1 A8.4
5 A8.4
6 KEY PLAN

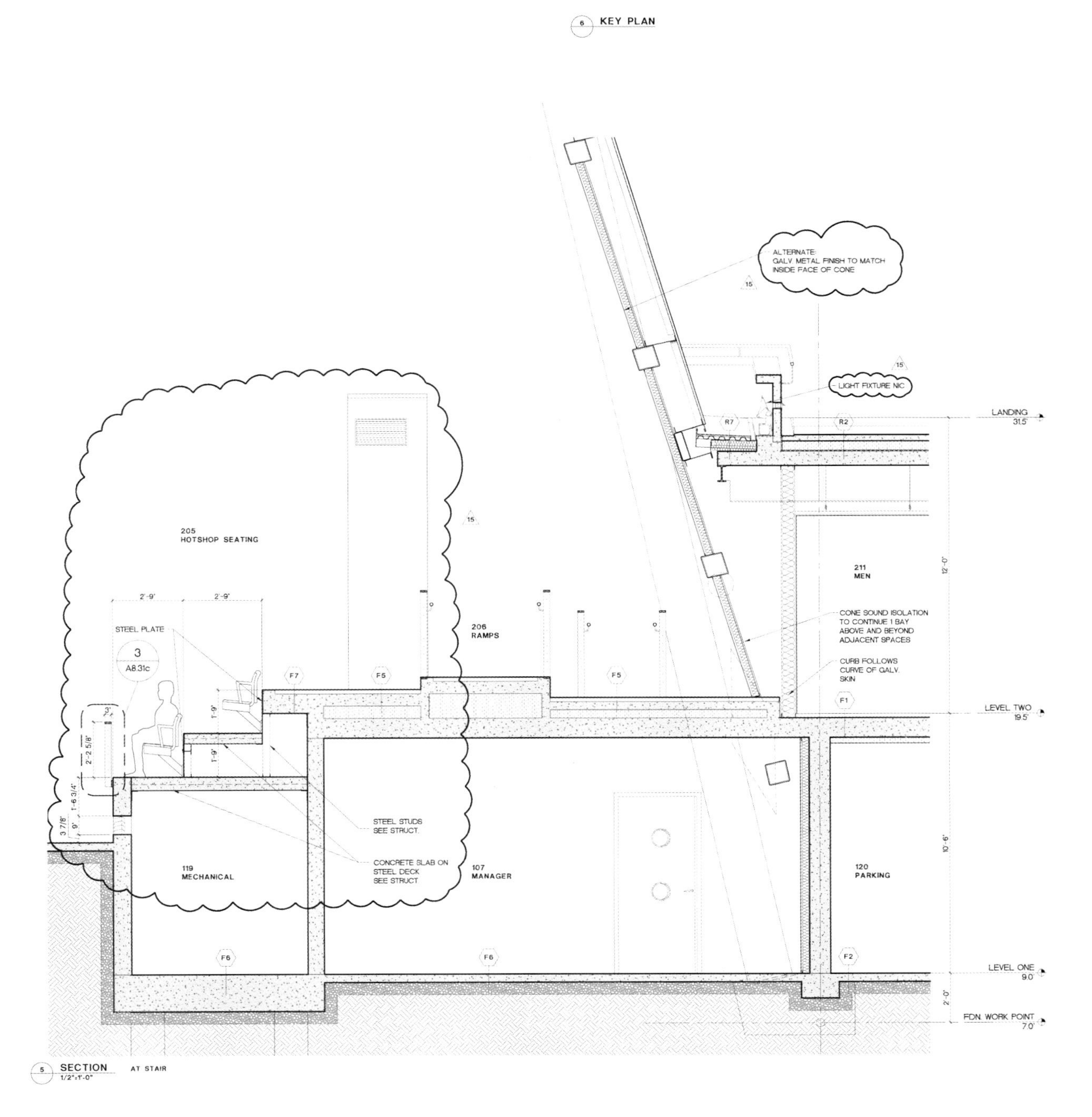
ALTERNATE: GALV. METAL FINISH TO MATCH INSIDE FACE OF CONE
LIGHT FIXTURE NIC
LANDING 31.5'
205 HOTSHOP SEATING
211 MEN
206 RAMPS
STEEL PLATE
3 A8.31c
CONE SOUND ISOLATION TO CONTINUE 1 BAY ABOVE AND BEYOND ADJACENT SPACES
CURB FOLLOWS CURVE OF GALV. SKIN
LEVEL TWO 19.5'
STEEL STUDS SEE STRUCT.
CONCRETE SLAB ON STEEL DECK SEE STRUCT.
119 MECHANICAL
107 MANAGER
120 PARKING
LEVEL ONE 9.0'
FDN. WORK POINT 7.0'
5 SECTION AT STAIR
1/2"=1'-0"

WATERFALL BUILDING ARTIST LIVE / WORK STUDIOS
水幕建筑——艺术家工作室

地　　点：温哥华
规　　模：约 6 000 m^2
完成时间：2001 年

The concept of the scheme was to provide simple, elegant spaces that provide maximum natural light and a "blank palette" for the tenants to customize. The client wanted to engender a "community spirit", which is why all the units are grouped around a south facing inner courtyard that is private, but connected to the street.

The development comprises Artist Live/Work Studios and commercial retail space, with two levels of parking below grade. The studios are based on an interlocking unit plan that allows every studio to have a 16-foot high clear space and exposure in two directions, giving every studio natural through-ventilation. The 16-foot high section of the studio is fully glazed with sliding doors that open to French balconies. The construction of the studios is sandblasted concrete that is left exposed on the interiors. The finishes are robust: galvanized steel, stainless steel, steel mesh and concrete. The floors of the studios have radiant heating installed under a polished concrete topping.

The studios are broken into four blocks that take into account the slope of the site and help to define the inner courtyard. There is a large 65-foot opening between the street and the inner courtyard. The underside of the opening is curved and a 40-foot long curtain of water flows from the centre of the curve into a large reflecting pool underneath that helps bounce light up into the opening. Two glass elevators reach the landscaped roof terraces providing a dynamic experience of the inner courtyard and city views to the north.

Directly across from the opening, there is a large glazed wedge-shaped art gallery that is landscaped with white roses tumbling down the sides. The courtyard is simply landscaped with deer ferns, moss gardens, cherry trees and grass planted between concrete pavers.

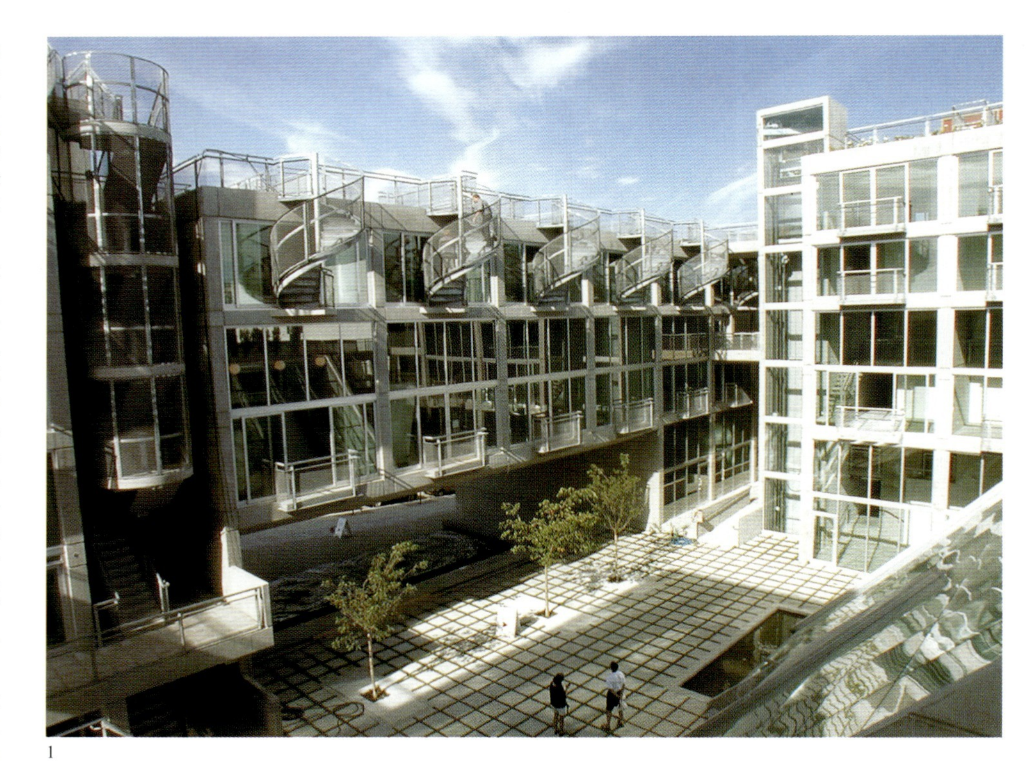

1

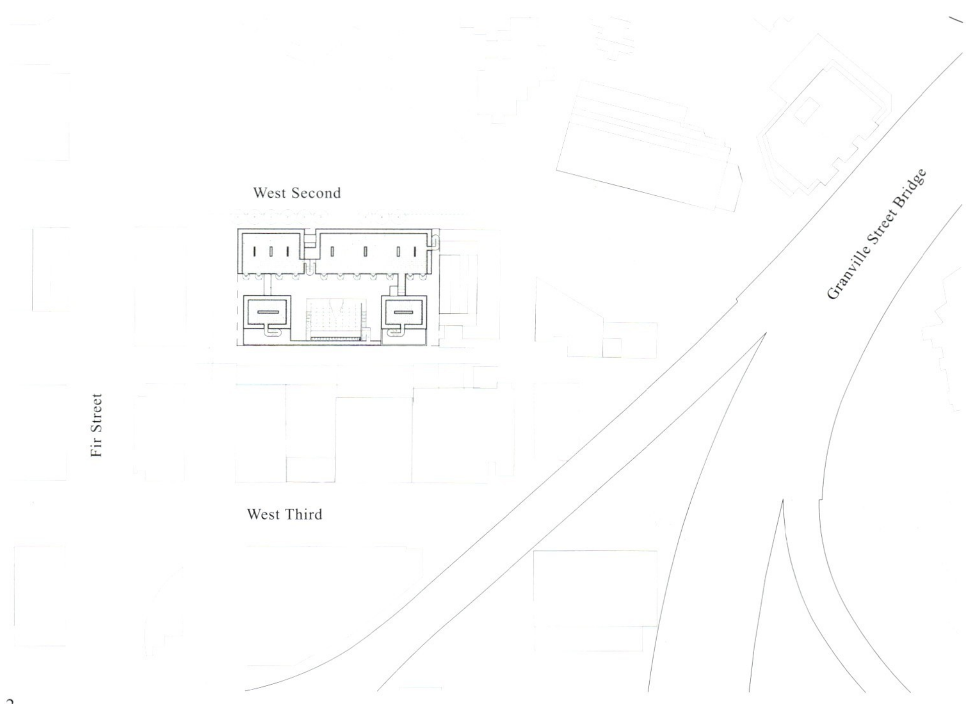

2

1　从大玻璃楔状艺术品展示厅上方俯视庭院
摄影师：斯蒂芬·海因斯（Stephen Hynes）
2　总平面图
3　在艺术品展示厅室内看庭院
摄影师：斯蒂芬·海因斯
4　沿街立面

水幕建筑的主题是提供简单和优雅的空间，一个可以有最大化的自然光和让使用者可以根据自身需要而设计的“空白区域”。业主希望创造一种“社区精神”。这正是所有居住单位围绕这个通向街道的南向庭院的原因。

水幕建筑包括艺术家工作室，零售空间和地下两层车库。工作室的错层设计使每一个单位有4.9m高的室内空间和穿堂风。工作室由推拉玻璃门通向法式阳台，室内结构部分裸露，建筑完成面都很粗糙：电镀钢材、不锈钢，钢网和混凝土。抛光混凝土地面下安装有供暖系统。

地面的斜坡将工作室分为四块，并以这四部分为庭院的界线。庭院和街道之间有20m的开口。在弧形开口的中间是12m宽的水幕。水聚集在下方的一个反射水池内，水池的设置有利于开口处的光线组合。两台玻璃电梯可到达屋顶花园，这提供了中央庭院和北面城市景观的动态联系。

开口的正对面是一个大玻璃楔状艺术品展示厅，白色玫瑰花由两侧垂挂下来，庭院铺地由混凝土、蕨类植物、苔类、樱桃树和草坪组成。

3

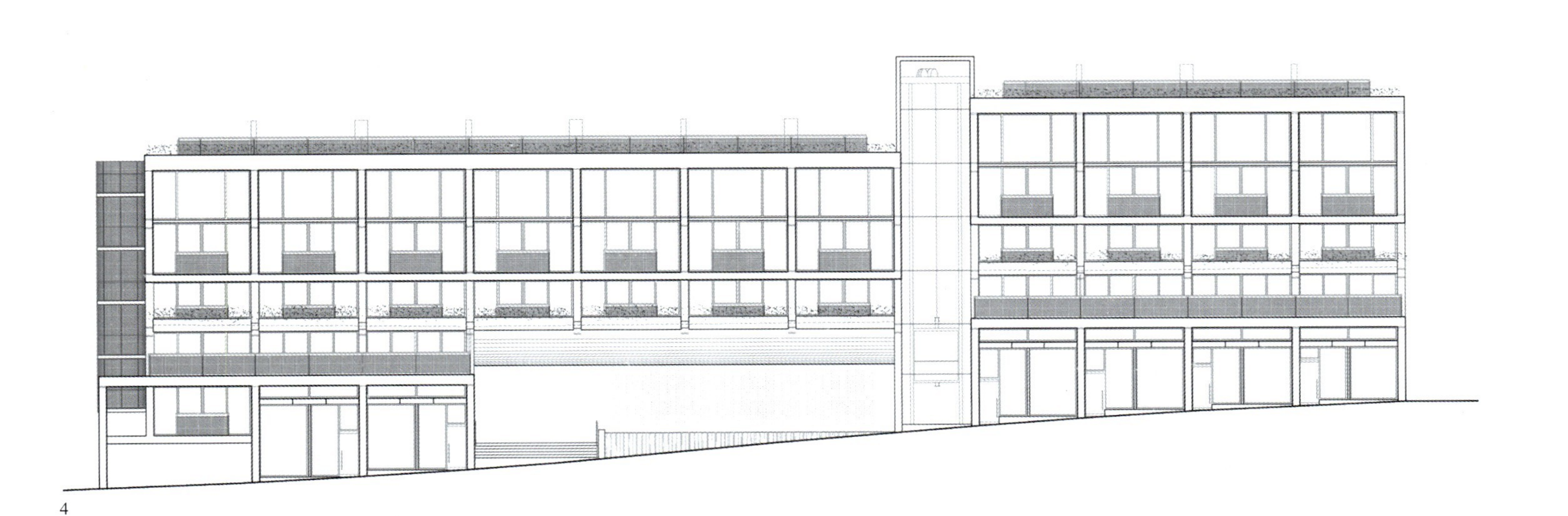
4

5　平面图
6　剖面图
7　工作室之间的旋转梯
摄影师：Wyn Bielaska
8　工作室之间的旋转梯
摄影师：斯蒂芬·海因斯
9　大玻璃楔状艺术品展示厅
摄影师：斯蒂芬·海因斯

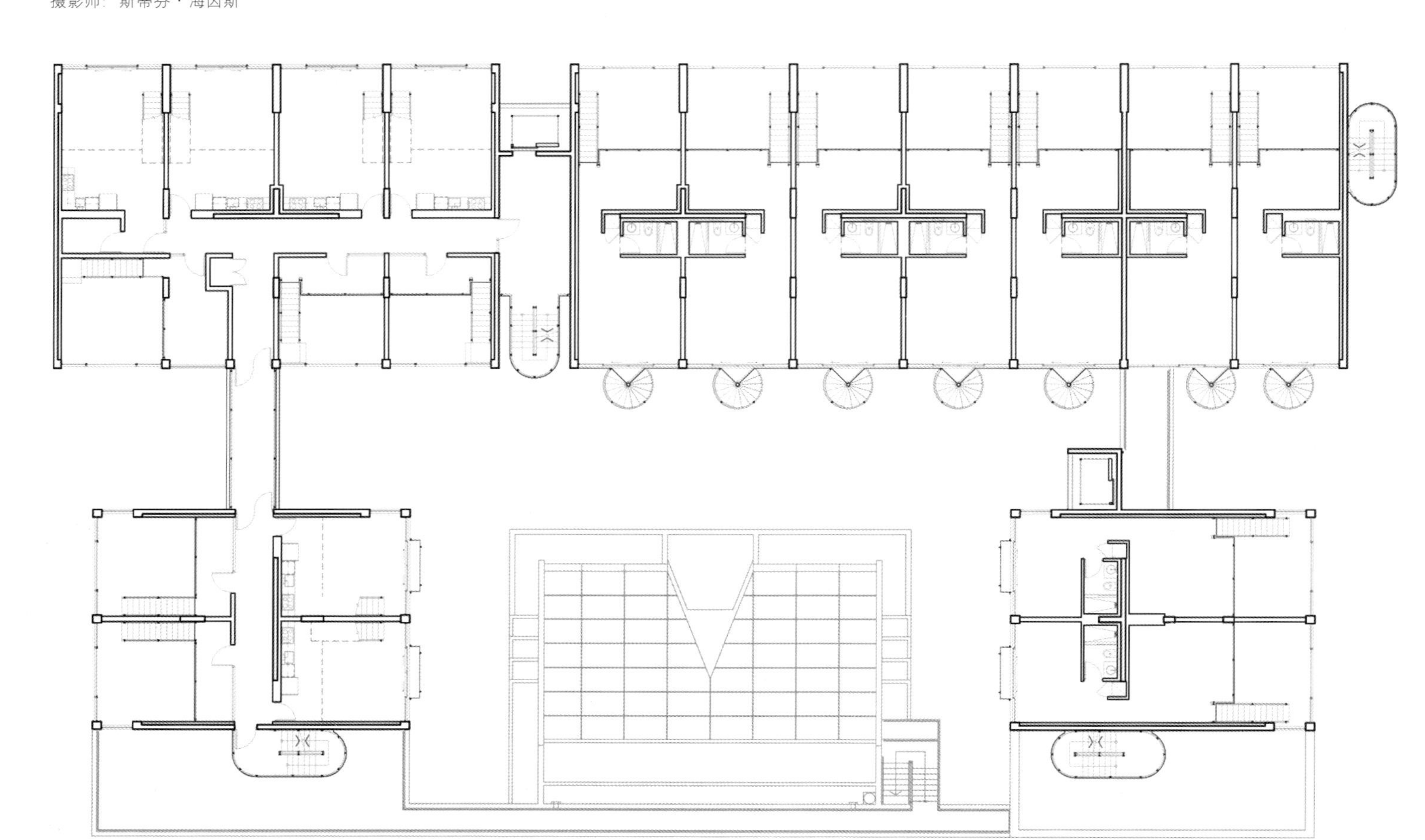
5

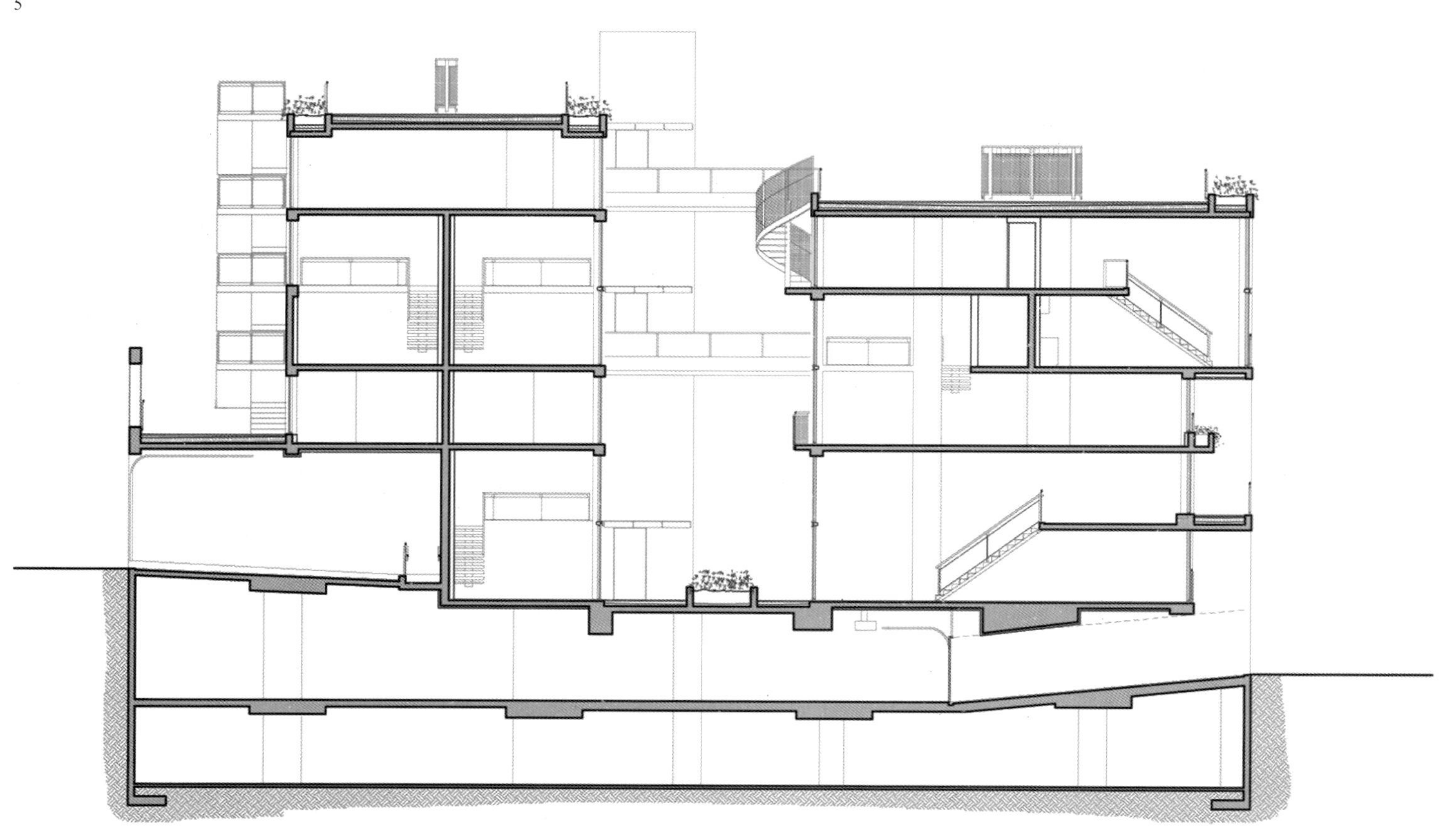
6

7

8

9

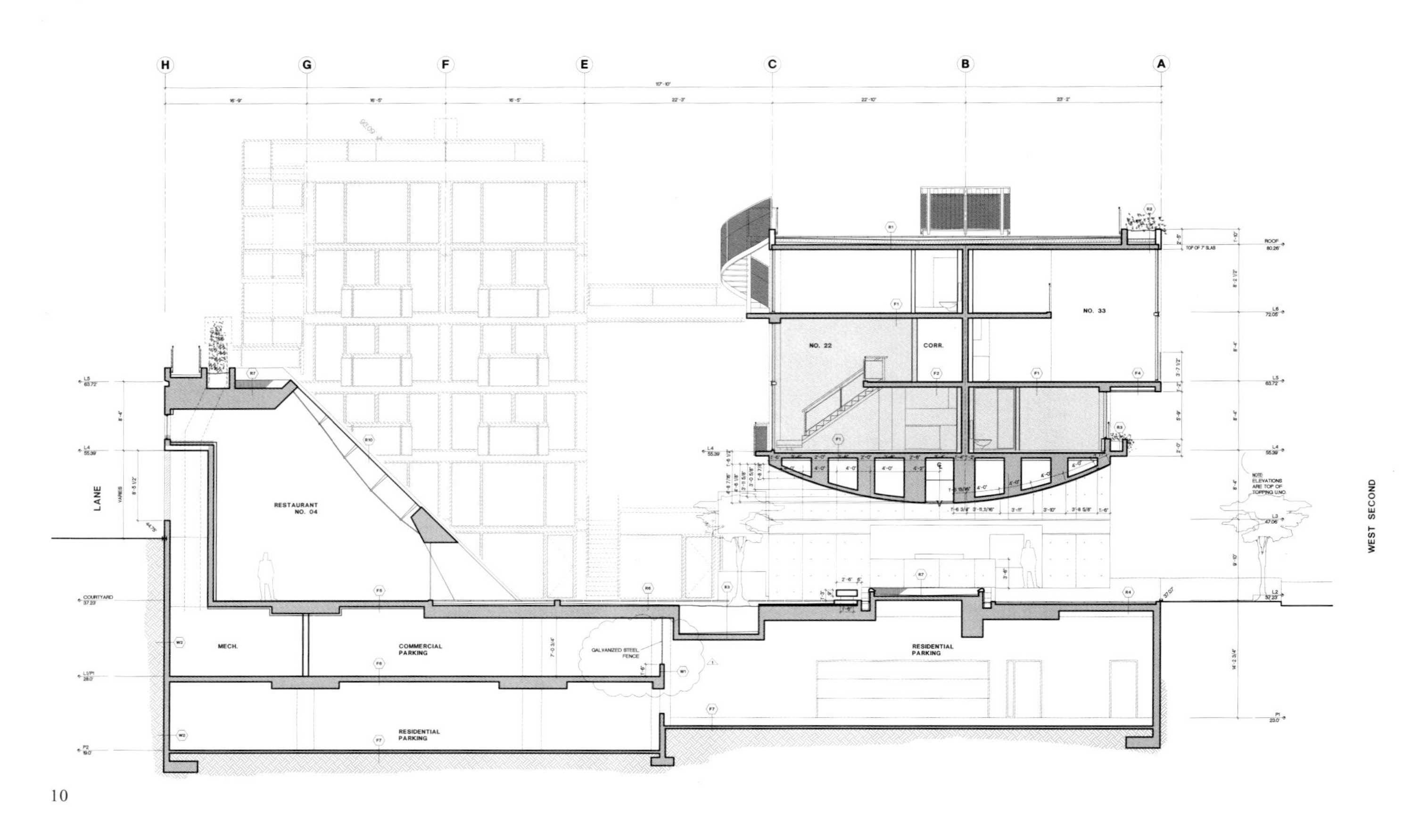
H
G
F
E
C
B
A
NO. 33
NO. 22
CORR.
RESTAURANT
NO. 04
LANE
WEST SECOND
MECH.
COMMERCIAL
PARKING
GALVANIZED STEEL
FENCE
RESIDENTIAL
PARKING
RESIDENTIAL
PARKING

10

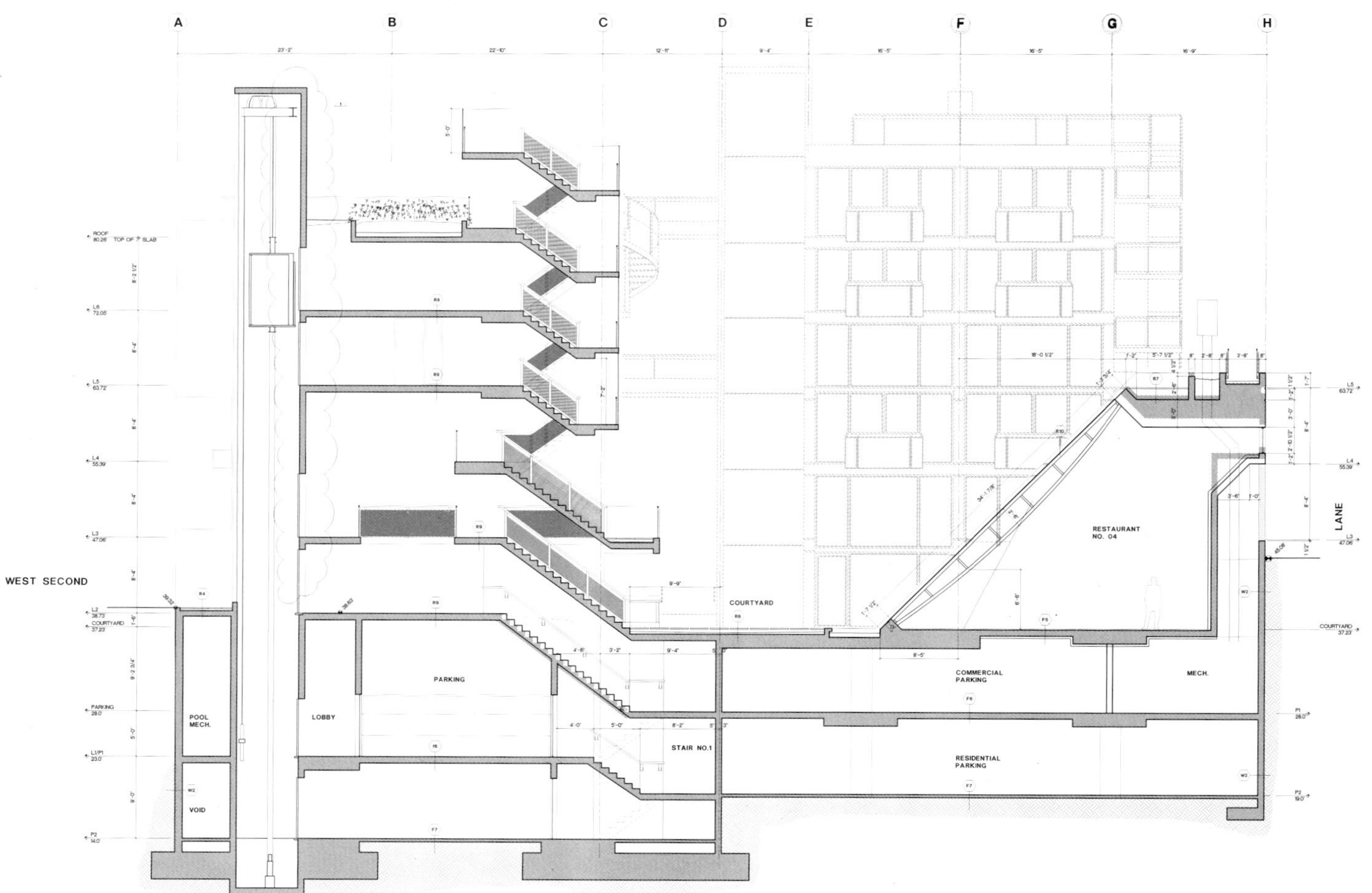
A
B
C
D
E
F
G
H
WEST SECOND
COURTYARD
RESTAURANT
NO. 04
LANE
PARKING
COMMERCIAL
PARKING
MECH.
POOL
MECH.
LOBBY
STAIR NO.1
RESIDENTIAL
PARKING
VOID

11

10~11 纵剖面图
12 开口水幕
摄影师：斯蒂芬·海因斯
13 开口水幕细部
摄影师：Ricardo Castro
14 水幕节点详图

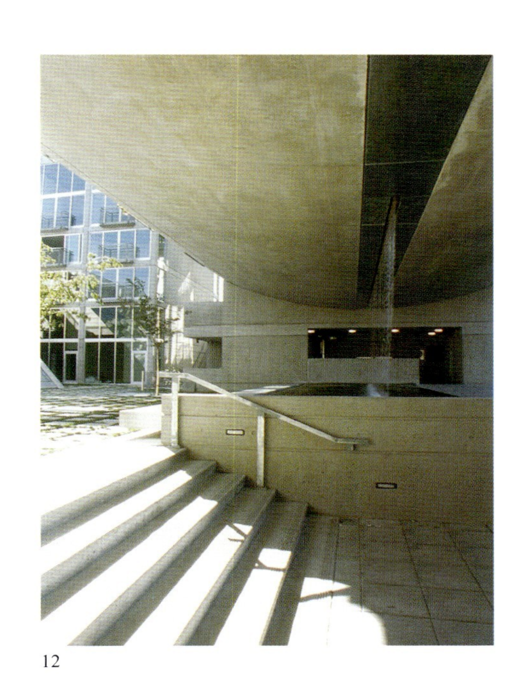

12

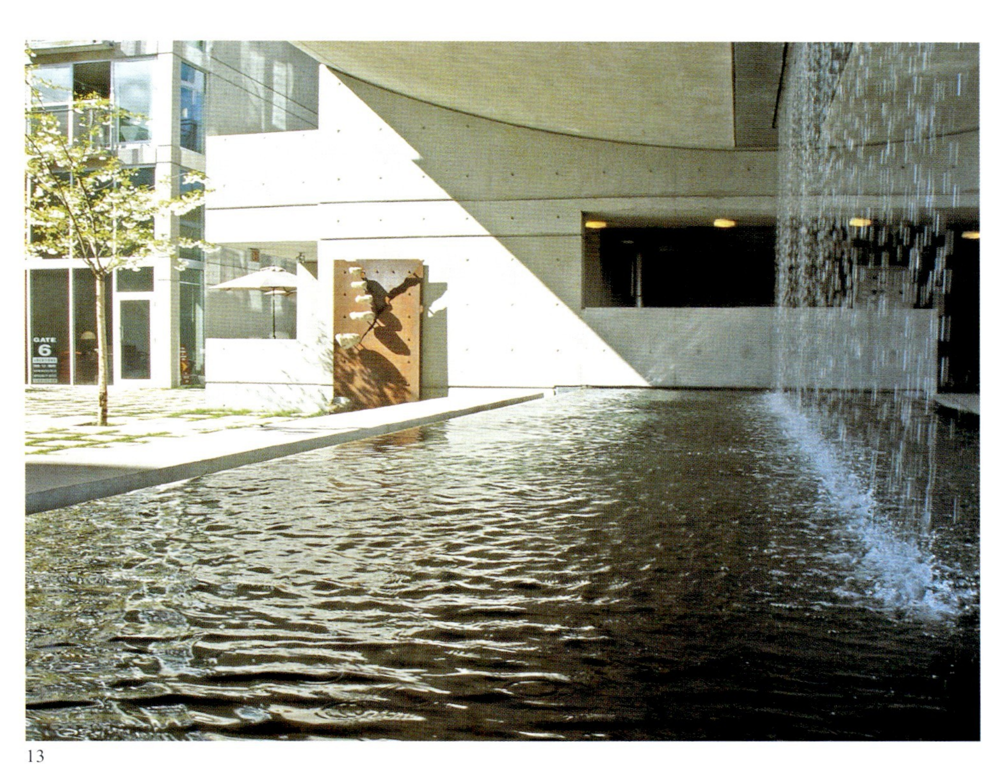

13

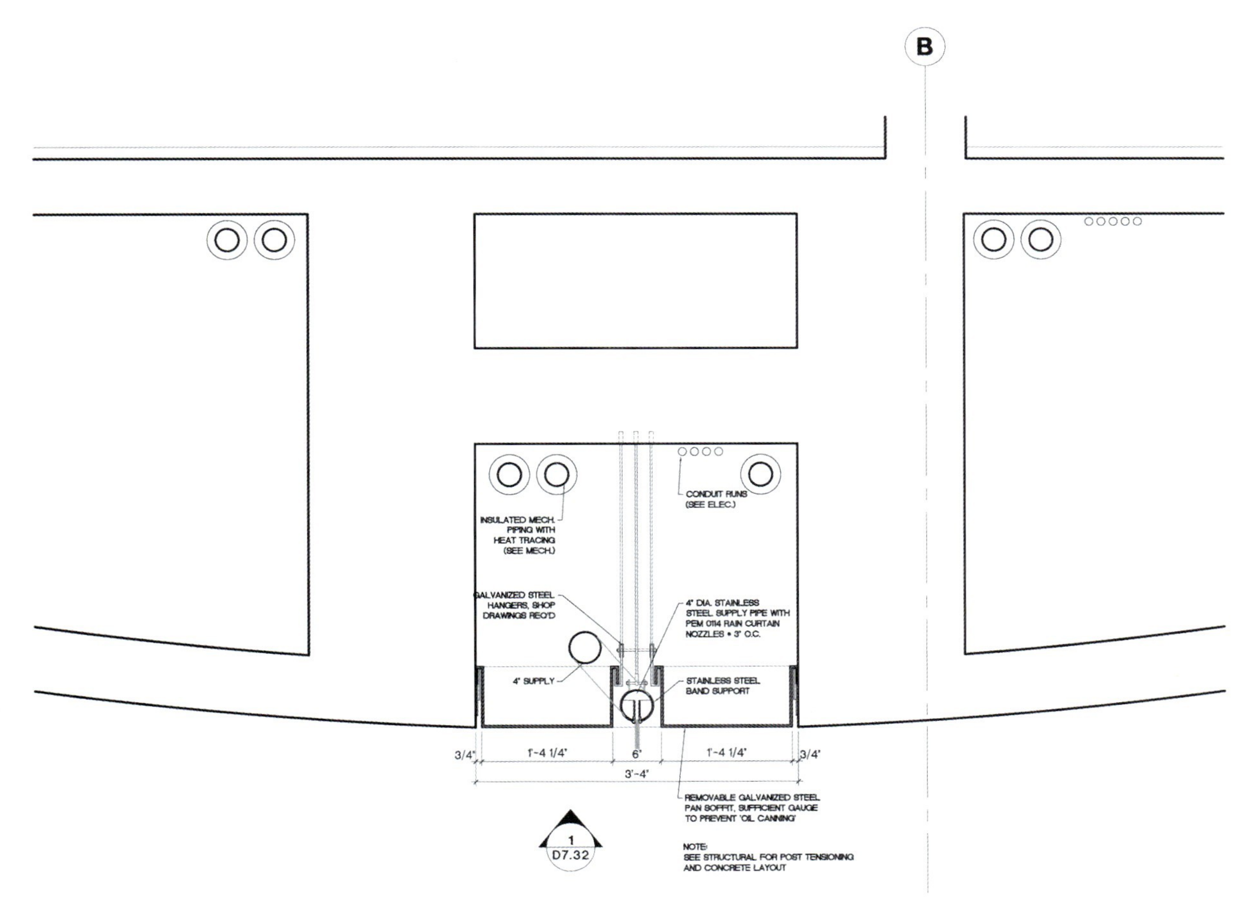

1 A3.4 **DETAIL** CURVED SOFFIT - WATER CURTAIN
1"=1'-0"

14

LANYON PHILLIPS COMMUNICATIONS INC.

LANYON PHILLIPS 广告公司

业　　主：LANYON PHILLIPS COMMUNICATIONS INC.
地　　点：温哥华
规　　模：880 m²
完成时间：1999年2月

1

Lanyon Phillips Communications is a non-traditional advertising agency that takes a more holistic approach to solving client's problems. The objective for their new office space was to build a stimulating interactive working environment, which would promote creative thinking. They wanted their clients and visitors to sense who they are when they experience the space.

The 9600 square foot space, 48 feet wide and 200 feet long, is located in the World Trade Centre at Canada Place in Vancouver. Full height glass walls give access to terraces on the south and north ends of the space, while the east side has a full length strip of glazing overlooking sweeping views of the harbor and the eastern slopes of Vancouver.

The program required 38 individual work stations for four distinct work groupings, 3 executive offices, two conference rooms, two workrooms and various back of house functions. The main entry and reception are situated at the south end of the space and overlooks the deck and the city view. The enclosed spaces treated as islands define the various working areas and give form to the progression through the space. The experience is that of a meandering path from the reception through work areas to openings, or meeting places, ending with the staff common area, which opens to the north terrace overlooking the harbor and the mountains.

All the principal walls and partitions in the north/south direction are constructed of polygol polycarbonate set in aluminum frames anchored to round aluminum columns, which also act as door frames for frameless glass doors. The polygol gives visual and acoustic privacy to conference rooms and private offices while allowing light to penetrate to the working areas behind. The curved translucent walls here arranged to offer visual connection between the various working areas thus promoting an interactive working atmosphere.

Two joined enclosed circular rooms used for intensive group work sessions are housed within solid walls of drywall with sound absorbing pin-up paneling on the interior. The doors are frameless glass offering visual relief and connection to the rest of the office.

The floor and ceilings are treated as two seamless planes giving continuity to movement through the space. The ceiling is steel wire woven mesh backed by acoustical panels set in the T-bar frames painted to match the clear finish steel mesh. The floor is cobalt blue, coloured vinyl tile highly polished to give a seamless appearance and to enliven the inner spaces with reflected light. The custom designed workstations with five foot high enclosures allow privacy but also offer the ability to overlook other areas of the office. They are finished in silver automotive paint.

2

LANYON PHILLIPS广告公司不是传统意义上的广告代理公司，公司提供全方位的服务以解决业主的问题。新的办公室设计目标是创造一个可以产生更多灵感的空间。业主的要求是：顾客和访问者可以从办公室本身了解业主的素质。

办公室占地880m²，14.6m × 60m，位于温哥华世贸中心。通高的落地玻璃幕墙可以看到南面和北面的花园，西面可以望见海湾码头全景和温哥华西坡的景色。

工程包括4大组，共38个工作间、3个经理室、2个会议室、2个工作室和其他设备辅助间。主出入口和接待前台位于南端，可以俯瞰城市景观。封闭空间为岛屿形。一个由接待前台到会议室，再到职员休息室的交通流线贯穿整个办公室。通道尽端职员休息室的北部花园可以俯瞰海湾和山景。

所有南北向的墙和隔断是由一种塑料（polygol polycarbonate）固定在铝合金框架上。铝合金框同时也是无框玻璃门的代用门框。独特的材质可以提供办公空间的私密性，却允许光线穿透。为了营造互动的工作氛围，半透明的弧形墙在各办公区域之间提供了视线的交流与联系。

两个集中工作室的墙壁为木隔墙，并设有隔声板。无框的玻璃门提供视觉上的轻盈感和与其他部分的联系。

地板和顶棚的无缝设计，使空间的流动性更强。顶棚是T型龙骨钢丝网上加隔声板，龙骨的颜色与钢丝网一致。地板是深蓝色乙烯板。高抛光的地板感觉上无缝，加之反射光线的设计使空间更有活力。量身定制的工作站隔断高1.5m，提供私密性的同时，也不阻碍视觉的交流。地板和顶棚的完成面是银色汽车漆。

3

4

1 接待前台
2 来访者等候区的三角形坐椅
3 交通流线
4 职员休息室

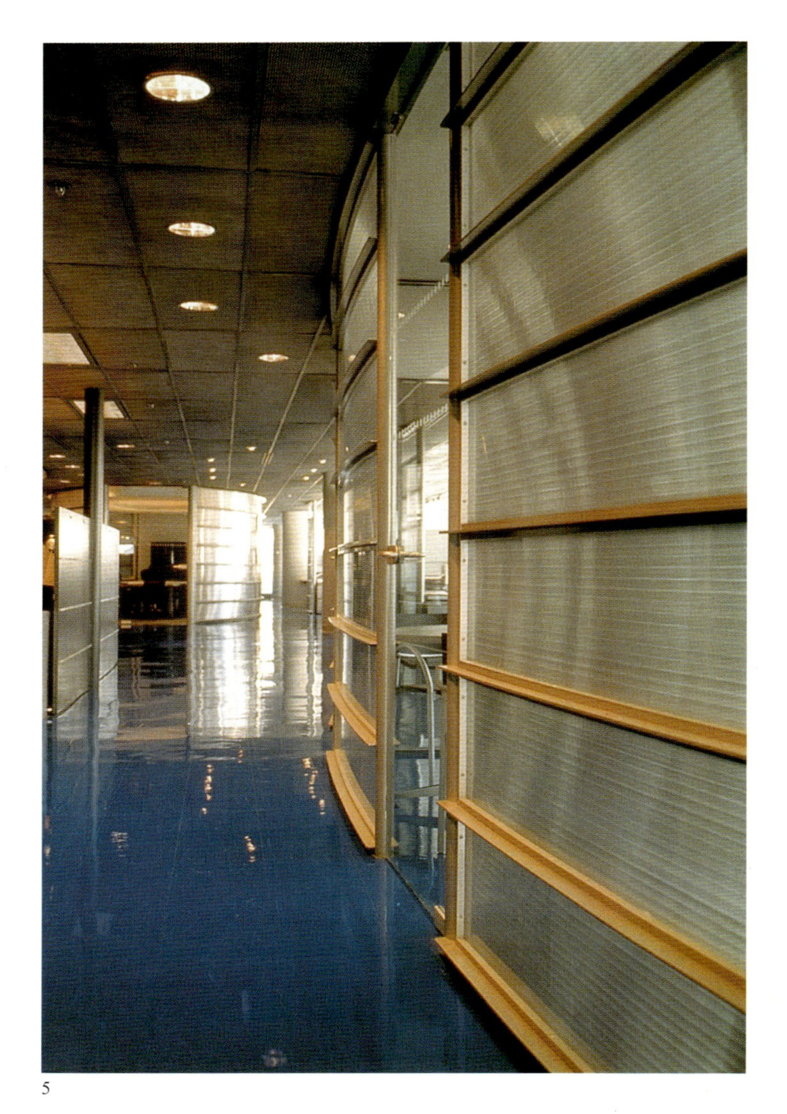
5

6

7

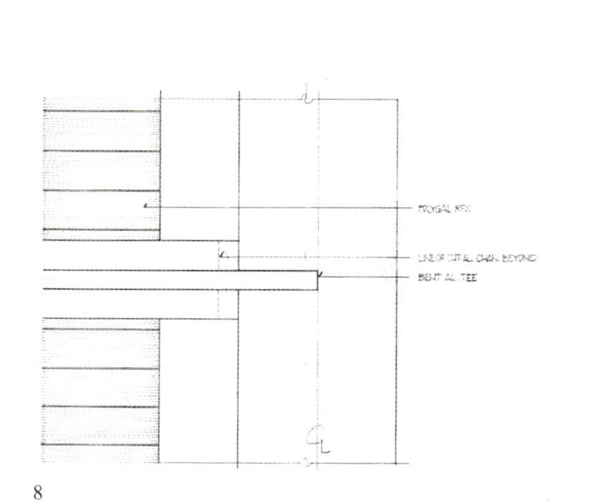
8

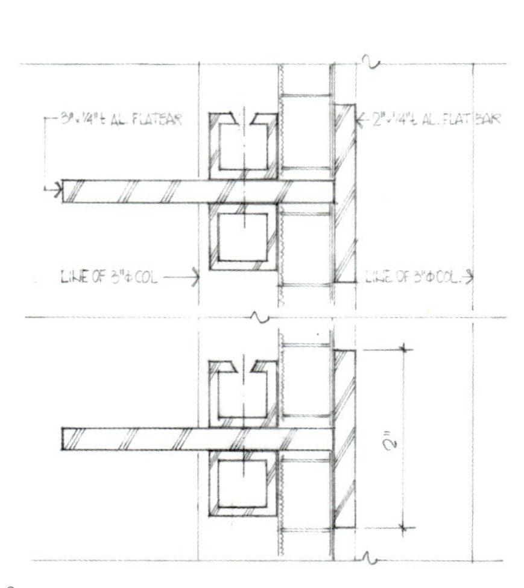
9

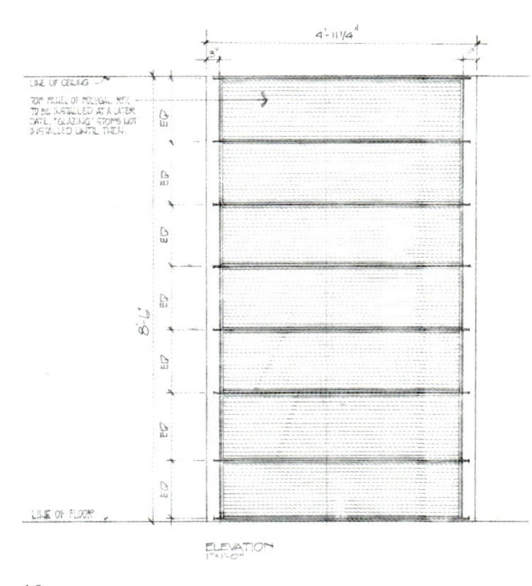
10

5~7,13　墙和隔断
8~9,11　墙和隔断详图
10　墙和隔断立面详图
12　扶手详图
14　等候区一角
15　扶手细部

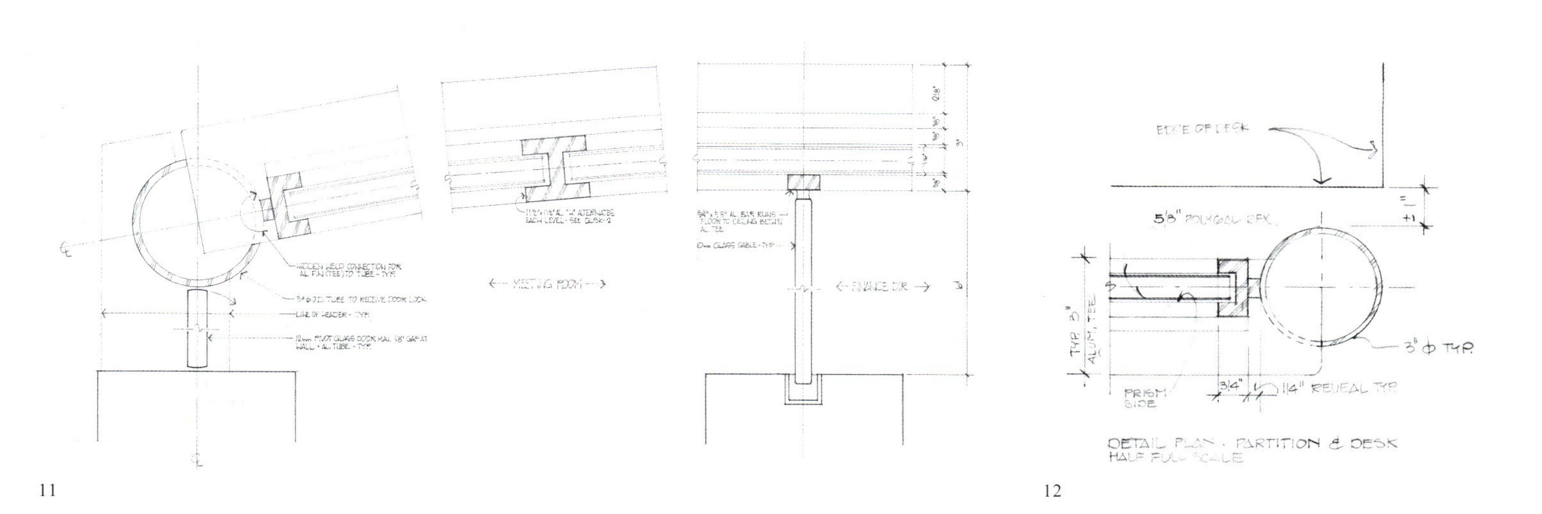

11　　12

13

14

15

Stantec Architecture Ltd.

Stantec 建筑设计事务所

One Team- Infinite Solution

Stantec, founded in 1954, provides professional design and consulting services in planning, engineering, architecture, surveying, and project management. They support public and private sector clients at every stage, from initial concept and financial feasibility to project completion and beyond. In simple terms, "the world of Stantec is the water we drink, the roadways we travel, the buildings we visit, the industries in which we work, and the neighborhoods we call home."

Stantec's services are offered through more than 4,000 employees operating out of over 50 locations in North America and the Caribbean. Stantec trades on the Toronto Stock Exchange under the symbol STN.

Stantec = Infinite Solutions

In 50 short years, Stantec has evolved from a one-person consulting engineering business in Edmonton, Alberta, to a more than 4,000-person multidiscipline design firm operating from over 50 offices across North America.

The Stantec vision, and driver for their growth plan, is to be a top 10 global design firm by 2008—attaining annual revenues approaching $1 billion and employing 10,000 people. Stantec will achieve this goal through excellence in design and projectde livery and by following an orderly growth plan that builds on our solid foundation. Stantec aims to work with the best clients on the best projects, and to provide the best services.

Architecture

We create designs that are timeless, intelligent, and sustainable.

At Stantec they are passionate about the design quality of the built environment. Their practice extends from macro to micro-from master planning to architectural and interior design-throughout the life cycle of a project.Offering services in architecture, interior design, and planning, Stantec's team enjoys a reputation for creating high-performance, fiscally responsible, and, award-winning buildings and interiors.

Stantec offers a multidisciplined team of experienced professionals-architects, project managers, interior designers, planners, and technologists-to provide creative and integrated solutions for clients. With a diverse client base, Stantec specializes in airports; attractions; commercial and residential buildings; health care and research facilities; community industrial, and transportation projects; educational environments; and hospitality, retail, and mixed-use developments.

Interior Design

Our teams draw inspiration from the potential of each project.

Stantec approaches to interior design is fundamentally collaborative and integrated with the other disciplines of their practice. Stantec's high-caliber team of designers draws inspiration from the potential of each project and works hard to exceed client goals. Stantec's award-winning portfolio features retail and commercial projects, workplace design, hospitality, health care and research, airports, and educational environments.

Stantec develops creative strategies for optimizing the use of real estate, brand, technology, furniture, and infrastructure to create dynamic, humancentered interior spaces. With an experienced and talented team, Stantec has the capacity and resources to complete projects of varying size and complexity with enthusiasm, vision, and success.

Planning

We provide comprehensive urban design and planning services to clients worldwide.

With an internationally- recognized portfolio, Stantec's experience includes large-scale urban design projects, master planning and design development of new towns, residential communities, urban mixed-use projects, major entertainment centers, and world-class destination resorts.

Stantec's range of client services includes facilitating visioning workshops as well as feasibility studies, design development, evaluation, and implementation strategies.

一个队伍——无限的解决方案

Stantec成立于1954年，提供城市规划、建筑设计、结构设备工程设计、工程评估和项目管理的专业和咨询服务。从概念设计和可行性研究到项目完工及后期服务的不同阶段的公共及私有业主的工程都是Stantec的业务范围。简单地说"Stantec的世界是我们喝的水，走的路，到访的建筑物，工作的厂房，和生活的社区。"

Stantec的服务由位于北美和加勒比地区的50个分支机构和超过4 000个雇员共同完成。Stantec在多伦多股票证券交易所上市交易，以STN为代号。Stantec意味着无限的解决方案。

在短短50年里，Stantec由位于阿尔伯达省埃德蒙顿市的一个人的工程咨询公司发展为4 000多名雇员、50个分支机构遍布北美的多方位的设计服务公司。

Stantec的目标是在2008年成为世界前十位的设计公司——10亿加币的年收入和1万个雇员。Stantec将通过出色的设计和杰出的后期服务建立实现其目标的坚实基础。Stantec以最佳的项目、最好的服务为目标。

建筑设计

我们创造永恒、智慧和可持续的建筑。

Stantec关注建筑物环境的设计，其设计经验涵盖建筑生命的全过程——从概念规划、建筑设计到室内设计。Stantec提供概念规划、建筑设计到室内设计服务，并享有创造高品质、经济高效的建筑和室内设计的声誉，他们的设计作品曾多次获奖。

Stantec提供复合型、有经验的专业服务——建筑师、项目负责人、室内设计师、规划师和其他技术人员，以保证方案的创造性和完整性。Stantec的经验集中在：飞机场、地标建筑、商业和居住建筑、医疗建筑、研究建筑、社区工业、交通建筑、教育建筑、旅馆、零售和多功能开发项目。

室内设计

我们从每一个项目的可能性中吸取灵感。

Stantec认为室内设计与建筑设计有本质上和不可分割的联系。他们的室内设计工程是由每一个项目的可能性中吸取灵感，以满足业主的要求。Stantec的获奖室内设计包括：零售和商业工程、工作室、旅馆、医疗建筑、研究建筑、机场和教育建筑。

Stantec提供创造性的策略，以优化建筑本身、品牌、科技、家具和基础设施的价值，以创造动感、以人为本的空间。充足的资源是Stantec成功、优质、有远见地完成不同设计项目的基础。

城市规划

我们为世界范围内的业主提供先进的城市规划设计。

在世界范围内，Stantec参与过的城市规划项目包括：大规模城市设计、新城区的概念规划和详细规划、居住区及城市多功能区规划、大型娱乐设施规划和世界标准度假休闲设施。

Stantec的规划服务包括：可行性研究、初步设计、项目评估和实施策略。

Ottawa Macdonald Cartier International Airport

渥太华 Macdonald Cartier 国际机场

业主：Ottawa Macdonald Cartier International Airport Authority
地点：渥太华市
规模：62 000m²

1

The new Ottawa Airport Terminal is located adjacent to the existing terminal, with a temporary connection to it in order to maintain operations and provide a smooth transition from the existing to the new. The new terminal has been designed to expand incrementally over the footprint of the existing as passenger volume increases.

The terminal has a two-level approach roadway, with departures on the upper level, and arrivals on the ground level. The approach roadway has been carefully designed and landscaped to give maximum prominence to the terminal while minimizing the visual impact of the parking structure. A generous landscaped space between the terminal and the parking structure (also an integral part of the design concept) provides ample natural air and light for the arriving passengers on the ground level.

The departures roadway and ticketing/check-in concourse are at elevation ten meters above ground level, which is double the height of conventional terminals. Apart from providing the benefits of visual transparency described below, this innovation enables weather- protected bridges with automated self-check-in facilities to be provided between the parkade and terminal for the benefit of day- passengers traveling the Ottawa-Montreal-Toronto corridor.

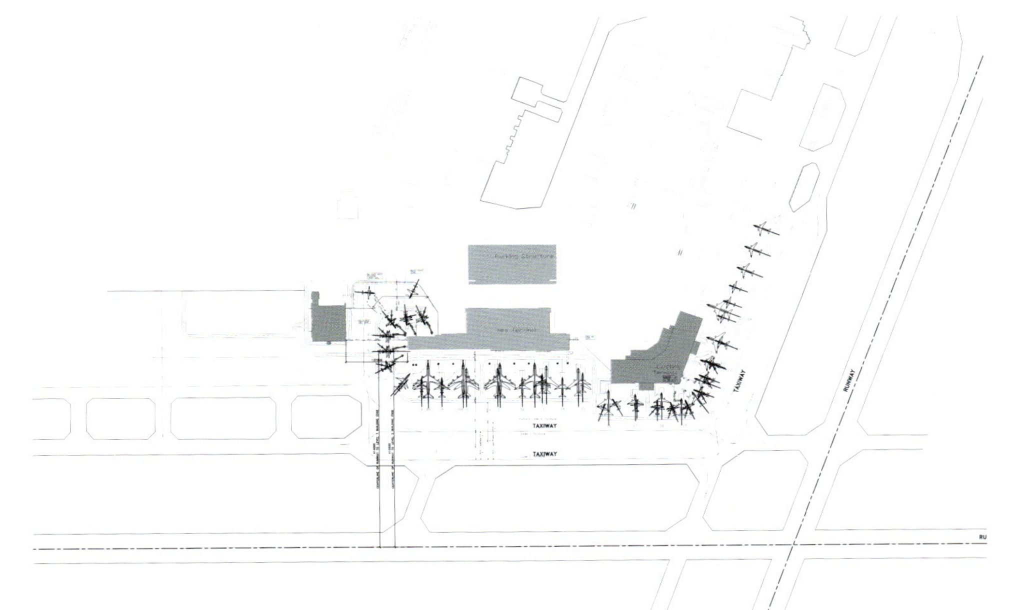

2

1　机场夜景
2　总平面图

3

4

渥太华新机场候机楼与已有的候机楼相邻，并有一个临时的连接部分方便新旧两个候机楼之间的交通联系。由于客流的快速增长，新机场的建筑超越了原有的基地范围。

候机楼有一个双层的通道，出发厅在上，到达厅在下。通道的设计非常精心：最大限度地加强了与候机厅的连接，并尽可能减小了对停车场建筑的冲击。作为设计理念的一部分，在停车场和候机厅之间有一个真正的绿化空间，给到达旅客提供了大量的新鲜空气和自然光线。

出发通道和出发厅位于地面上10m，是旧候机厅的两倍高。除了给下层的空间提供清晰的视野之外，这个创新化的防风雨的桥设有自动登机设施，极大地方便了使用渥太华、蒙特利尔到多伦多飞行通道的旅客从停车场到候机厅的交通。

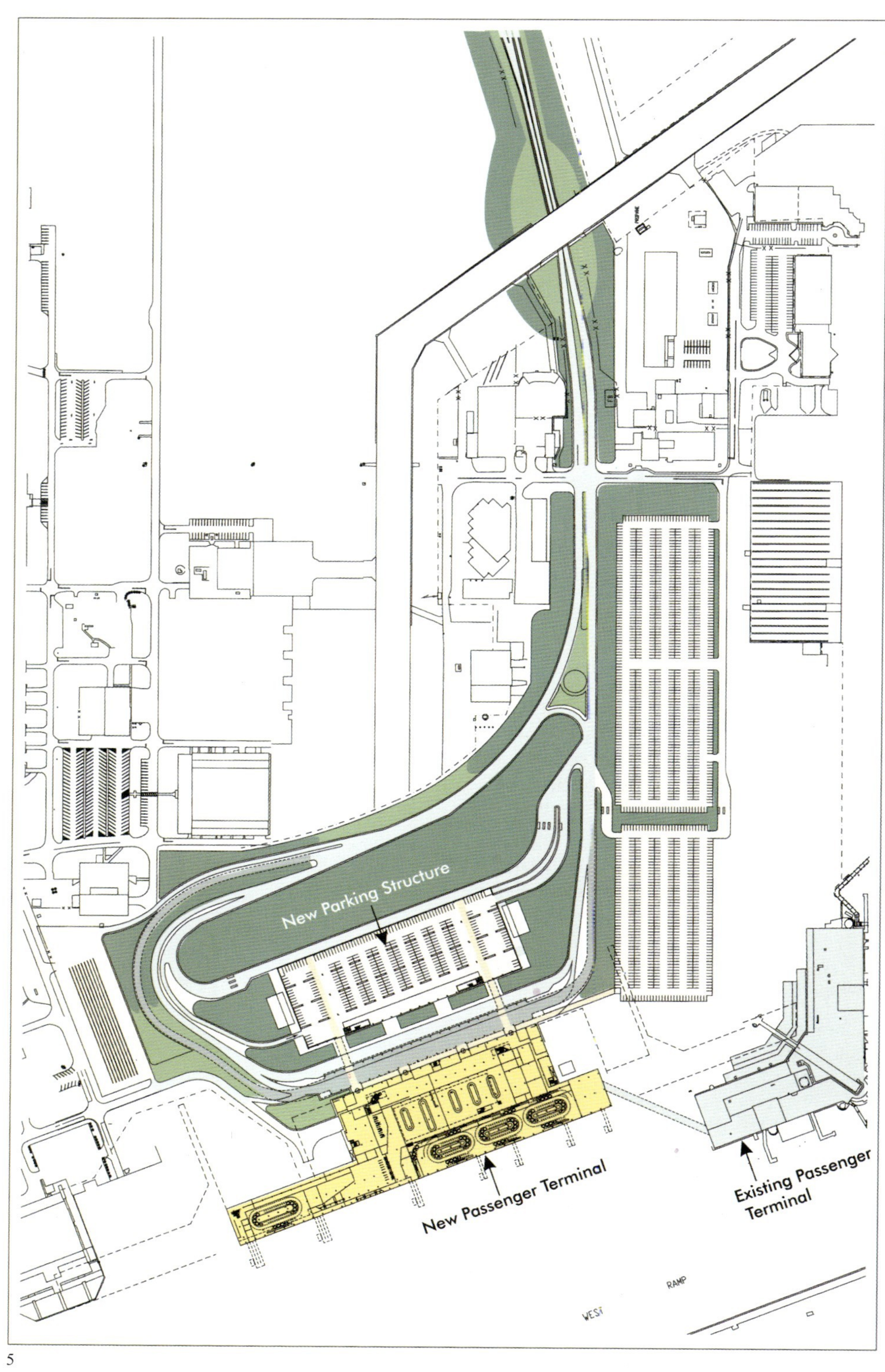

5

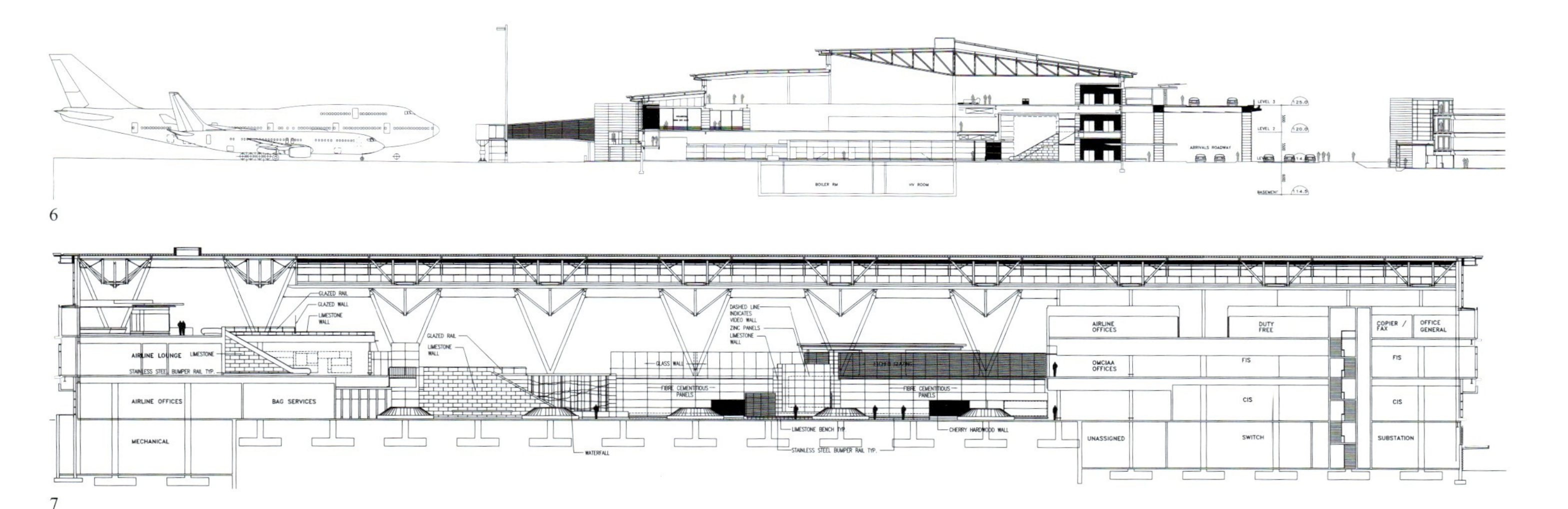

6

7

This unique terminal reconciles four major design objectives:

- Designing a building that is open and transparent,
- Creating a sequence of memorable spaces,
- Creating an appropriate gateway to the city,
- Resolving sophisticated security, airport and airline functional requirements in a manner that is unobtrusive to passengers.

Typical two-level terminals do not provide visibility from ticket counter to aircraft, nor do they provide architecturally memorable arrival or departure experiences. Through its highly innovative cross-section, the design of this terminal enables departing passengers to see and understand the relationship between the ticketing/check-in counters and the aircraft, virtually from the moment they are dropped off at the curbside. Conversely the visual openness and intuitive planning enable arriving passengers to understand and find their way through the entire building as soon as they leave their aircraft and enter the terminal.

Both departing and arriving passengers experience a unique sequence of spatial experiences. Prior to moving through the security checkpoint, departing passengers enter the dramatic atrium that interconnects all three levels in the terminal. Once past security, they enter pre-boarding hold rooms that provide an attractive, relaxed environment with generous height and natural light.

Arriving passengers are guided along their intended route by a multi-level water- feature, culminating in their descent and ceremonial entry into the baggage-claim area, which unlike most airport terminals, is the focus of the spectacular naturally- lit atrium.

References to the uniqueness of the city and region have been integrated into the design in a subtle, sophisticated manner that is intended to be evocative rather than literal. Inspiration has been drawn from unique aspects of Ottawa including the Byward Market, the Rideau Canal, the Chaudiere Falls, the parkways, and the seasonality of the city all of which find their expression in a contemporary architectural language. The palette of materials is refined and serene, with limestone and copper providing richness and connecting the terminal to the city's parliamentary and civic institutions.

8

这个独特的候机楼的设计有四个主要目标：

- 设计一个开放和透明的建筑；
- 创造一系列难忘的空间；
- 创造一个通向城市的良好通道；
- 在不影响乘客的基础上，提供精确的安全系统，以满足机场和航空公司的基本要求。

典型的两层候机厅在检票厅的位置上设有视角可以看见飞机和飞机的起落。高度创新的渥太华新机场候机楼的穿越设计，使出发者在交付行李的时刻起就可以感受到室外飞机场的活力。相反的开阔的视野使到达旅客一到达机场就很容易了解机场的空间结构和找到离开的通道。

到达和出发旅客在机场的经历是非常独特的。在到达安全检查站之前，出发旅客即已进入了可容纳所有三层的候机厅的动感空间。当通过安全检查之后，出发旅客进入巨大尺度的，拥有大量自然光线的令人心情舒畅轻松的候机空间。

到达旅客由占多层的水幕引导，向下进入行李提取处。与普通的行李提取处不同，渥太华机场的行李提取处位于自然光源下的大厅内。

渥太华的城市本身和机场四周区域的独特形制，对于机场建筑的形式的影响是以一种含蓄的，而不是直接的，以一种细致的形式表现出来。设计灵感来自渥太华城市本身，Byward市场、Rideau运河、Chaudiere瀑布、公园街和城市的鲜明四季。这一切都被演绎成建筑语言。材料的色调是精练和清晰的。石灰岩和铜的丰富质感与渥太华议会和政府建筑感觉一致。

3　机场到达厅外景
4　到达厅的水幕
5　总平面图
6　横剖面图
7　纵剖面图
8　到达厅行李提取处远景

Two major sustainability principles are integral to the design.

Firstly, clerestory windows separate the cascading roofs throughout the terminal, and in combination with the central atrium, enable natural light to penetrate into the deepest reaches of the interior, As a result very little artificial lighting is required during daylight hours.

Secondly, a low-velocity stratified approach has been taken to air conditioning. Most conventional terminals air condition-from the ceiling a wasteful approach that results in conditioning large volumes of air where people are not present. Our design provides low-velocity distribution through floor- level grilles that only serve the strata occupied by passengers, while warmer air rises and exhausted through the clerestory. Through this approach, only one-third of the entire volume of the terminal interior is air-conditioned, providing significant energy savings with no impact on passenger comfort.

9

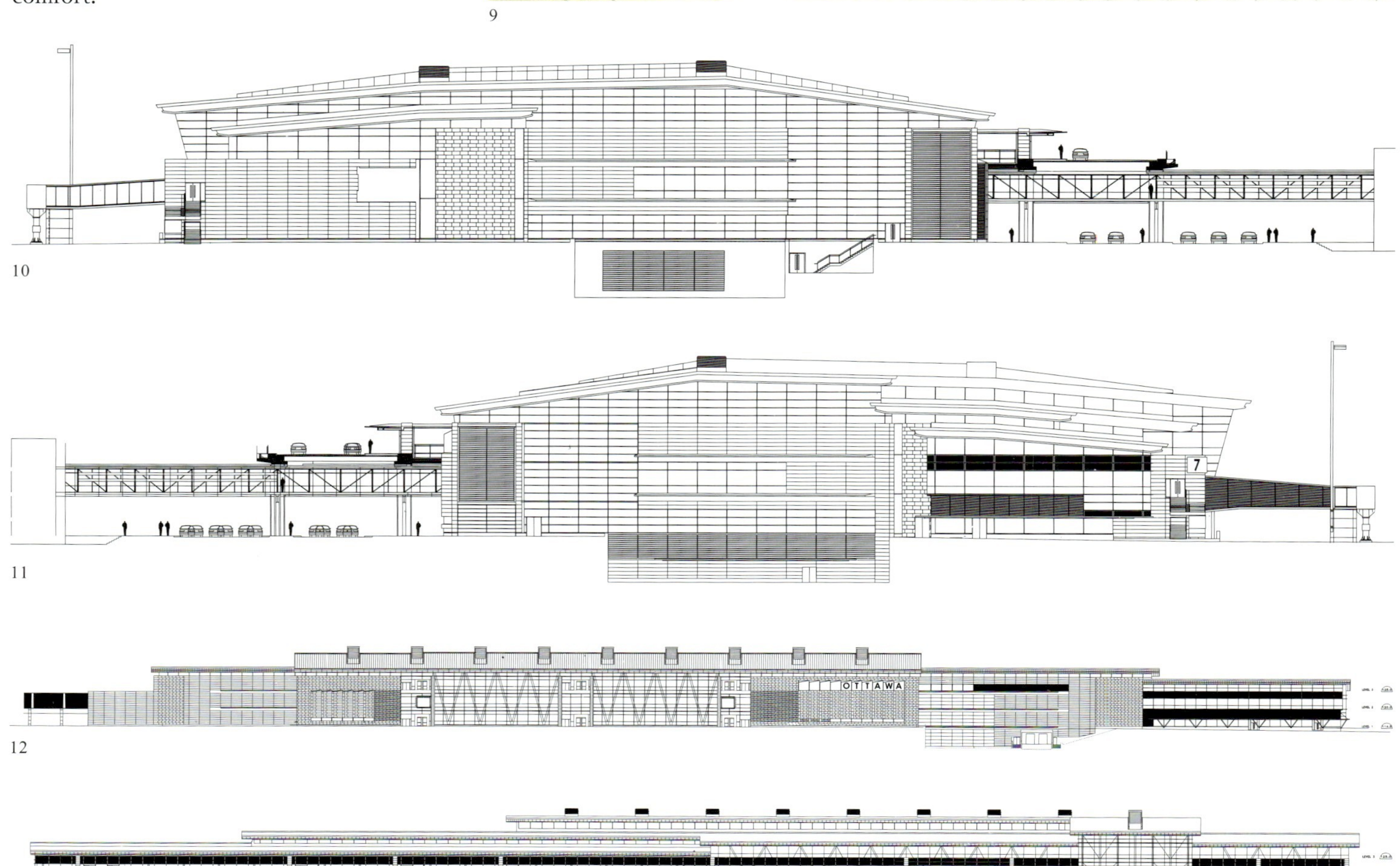

10

11

12

13

9　商业区
10　南立面图
11　北立面图
12　东立面图
13　西立面图
14　出发厅及等候区
15~17　剖面图

两个可持续化理念贯穿整个设计：

1.高窗设置于上下错落的屋顶之间，加之中厅的设计，自然光可以提供室内大部分的照明。整个机场在白天只需要很少的人工光源。

2.低速分层就近空调系统。传统的机场空调系统位于屋顶位置，当无人使用时，浪费了大量的能源。渥太华机场提供从地板部位的低速空调分配，而且只有在有使用者时才运转，热空气上升由高窗排出。利用这个设计，只有三分之一的空间有空调。在不影响舒适度的情况下，节省了大量的能源。

14

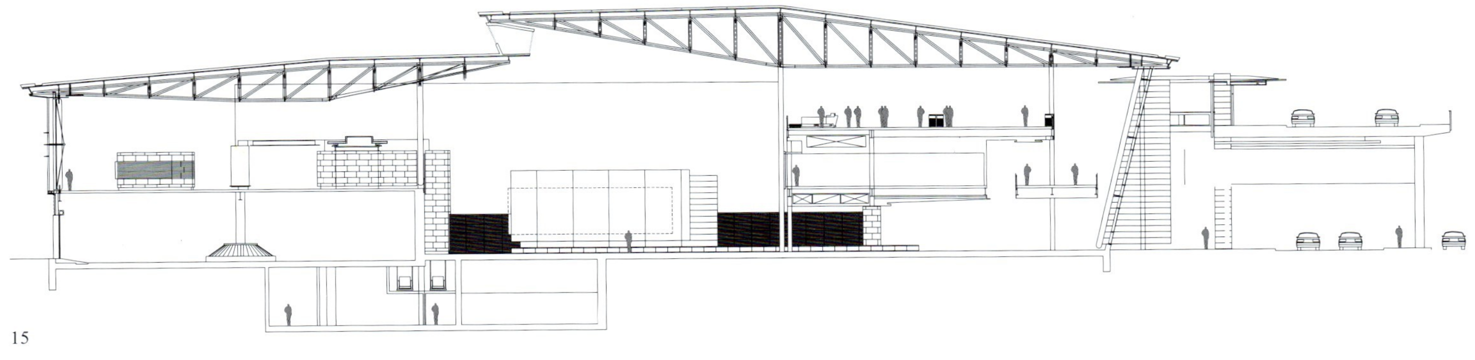

15

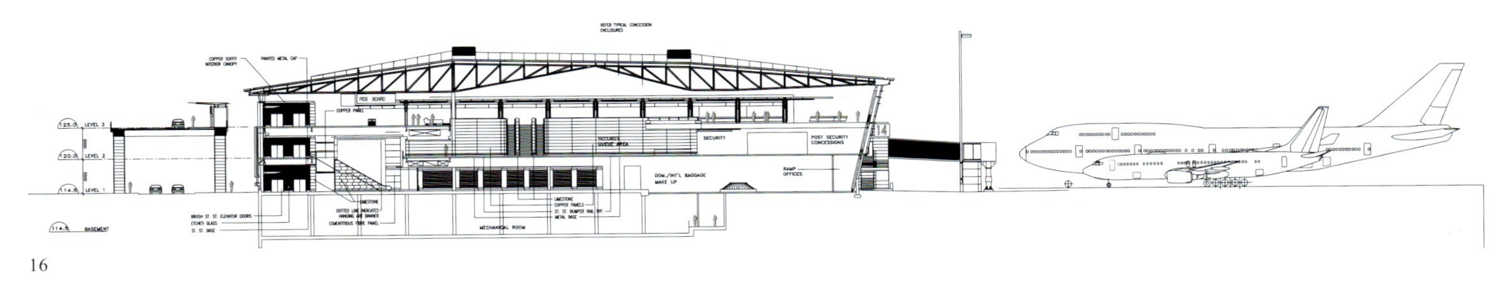

16

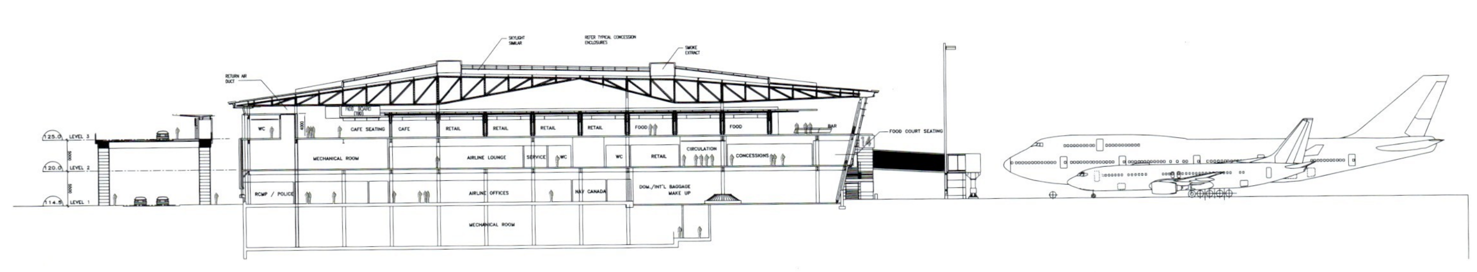

17

18

19

20

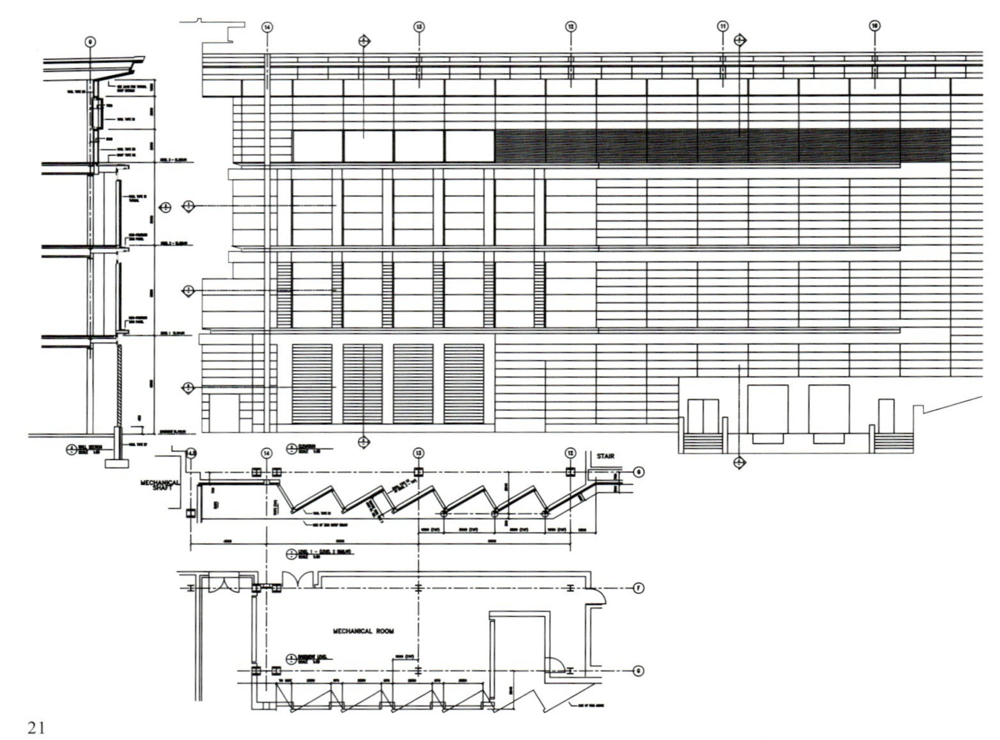

MECHANICAL SHAFT
STAIR
MECHANICAL ROOM

21

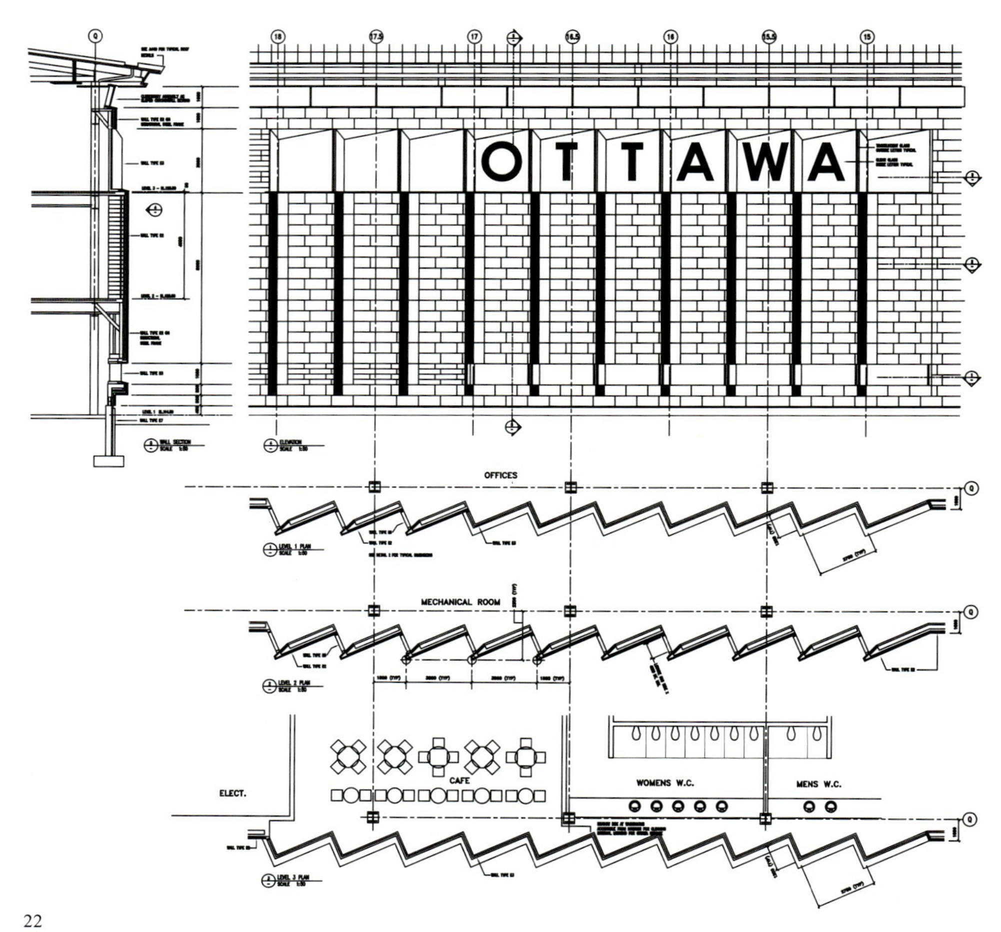

OTTAWA
OFFICES
MECHANICAL ROOM
ELECT.
CAFE
WOMENS W.C.
MENS W.C.

22

23

24

25

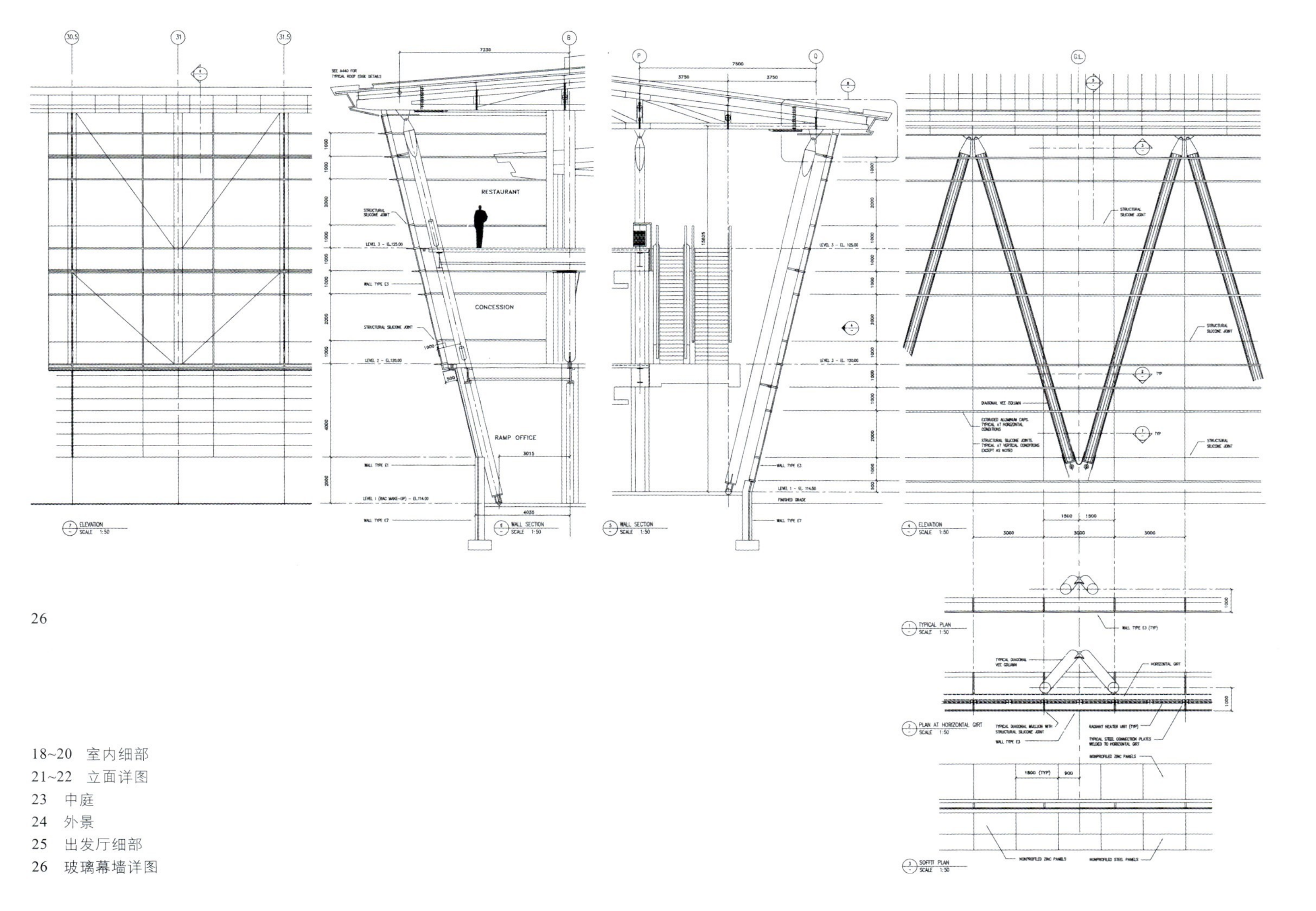

26

18~20 室内细部
21~22 立面详图
23 中庭
24 外景
25 出发厅细部
26 玻璃幕墙详图

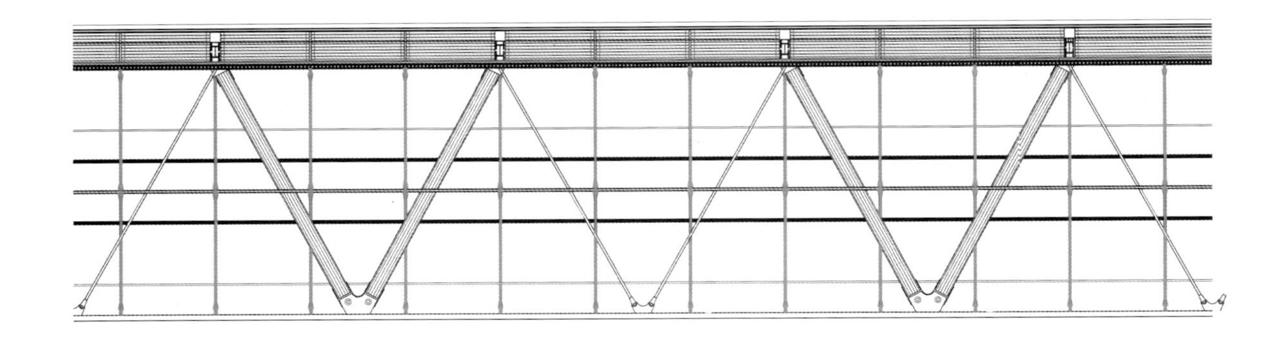

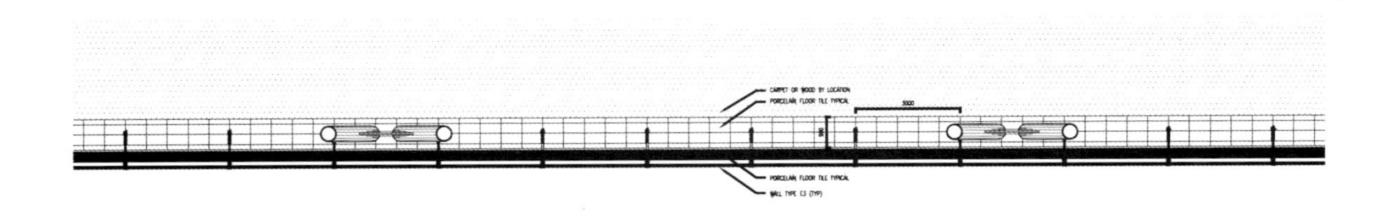
CARPET OR WOOD BY LOCATION
PORCELAIN FLOOR TILE TYPICAL
3000
PORCELAIN FLOOR TILE TYPICAL
WALL TYPE E3 (TYP)

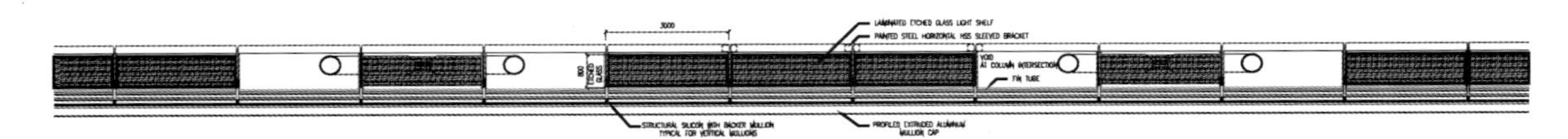
3000
LAMINATED ETCHED GLASS LIGHT SHELF
PAINTED STEEL HORIZONTAL HSS SLEEVED BRACKET
800
ETCHED GLASS
VOID
AT COLUMN INTERSECTION
FIN TUBE
STRUCTURAL SILICON WITH BACKER MULLION
TYPICAL FOR VERTICAL MULLIONS
PROFILED EXTRUDED ALUMINUM
MULLION CAP

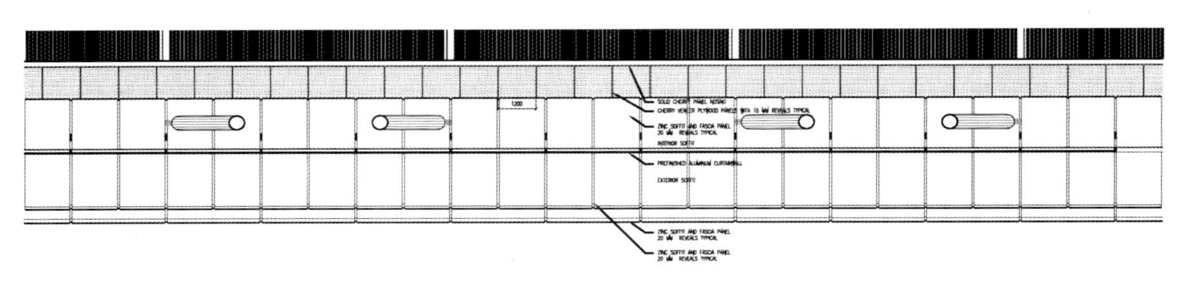
1200
SOLID CHERRY PANEL NOSING
CHERRY VENEER PLYWOOD PANELS WITH 16 MM REVEALS TYPICAL
ZINC SOFFIT AND FASCIA PANEL
20 MM REVEALS TYPICAL
INTERIOR SOFFIT
PREFINISHED ALUMINUM CURTAINWALL
EXTERIOR SOFFIT
ZINC SOFFIT AND FASCIA PANEL
20 MM REVEALS TYPICAL
ZINC SOFFIT AND FASCIA PANEL
20 MM REVEALS TYPICAL

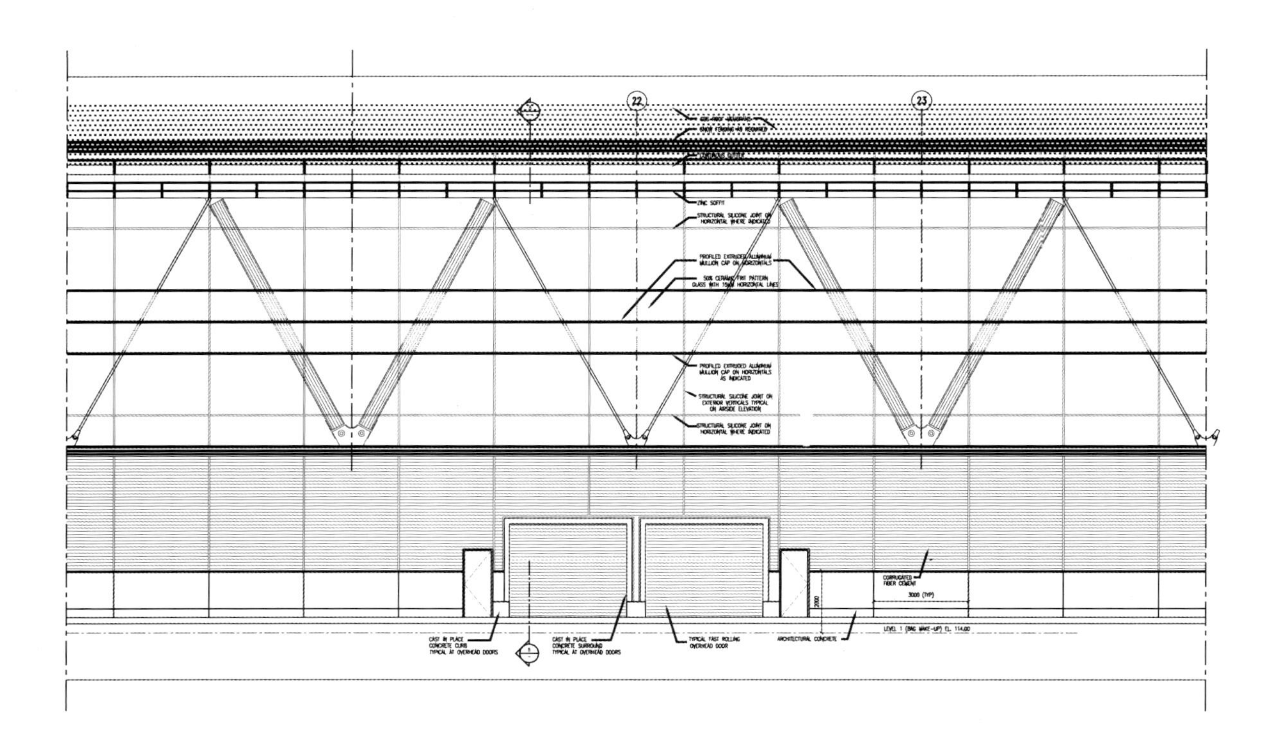
22
23
ZINC SOFFIT
STRUCTURAL SILICONE JOINT ON
HORIZONTAL WHERE INDICATED
PROFILED EXTRUDED ALUMINUM
MULLION CAP ON HORIZONTALS
50% CERAMIC FRIT PATTERN
GLASS WITH 15MM HORIZONTAL LINES
PROFILED EXTRUDED ALUMINUM
MULLION CAP ON HORIZONTALS
AS INDICATED
STRUCTURAL SILICONE JOINT ON
EXTERIOR VERTICALS TYPICAL
ON AIRSIDE ELEVATION
STRUCTURAL SILICONE JOINT ON
HORIZONTAL WHERE INDICATED
CORRUGATED
FIBER CEMENT
3000 (TYP)
2000
LEVEL 1 (BAG MAKE-UP) EL. 114.00
CAST IN PLACE
CONCRETE CURB
TYPICAL AT OVERHEAD DOORS
CAST IN PLACE
CONCRETE SURROUND
TYPICAL AT OVERHEAD DOORS
TYPICAL FAST ROLLING
OVERHEAD DOOR
ARCHITECTURAL CONCRETE

27

27,30 玻璃幕墙详图
28 室内走道
29 玻璃幕墙外的景观

28

29

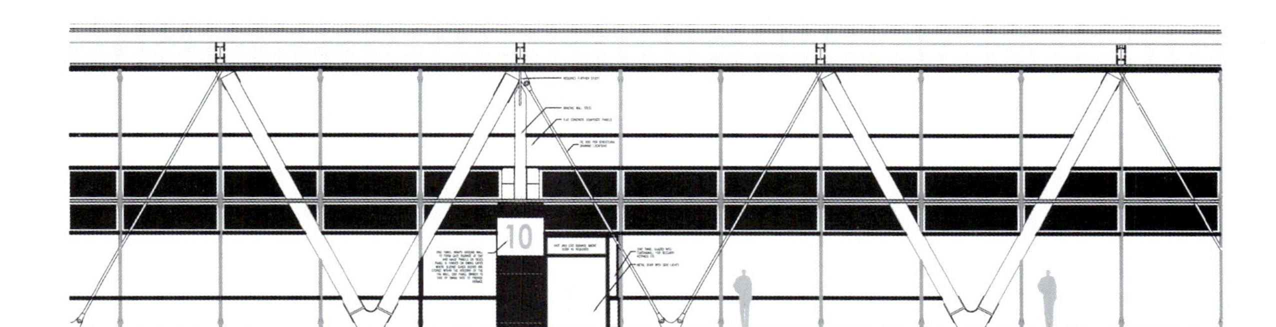

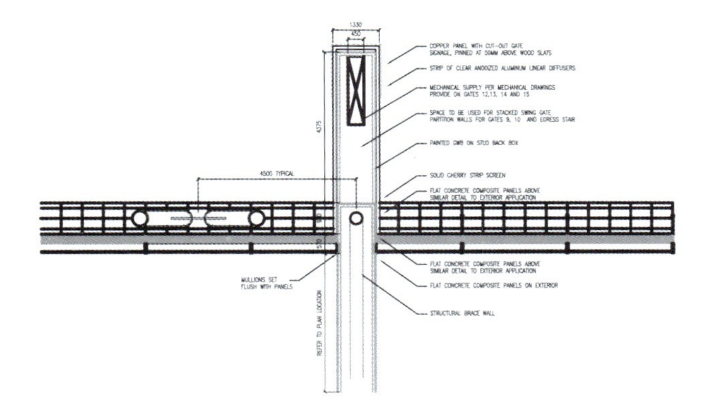

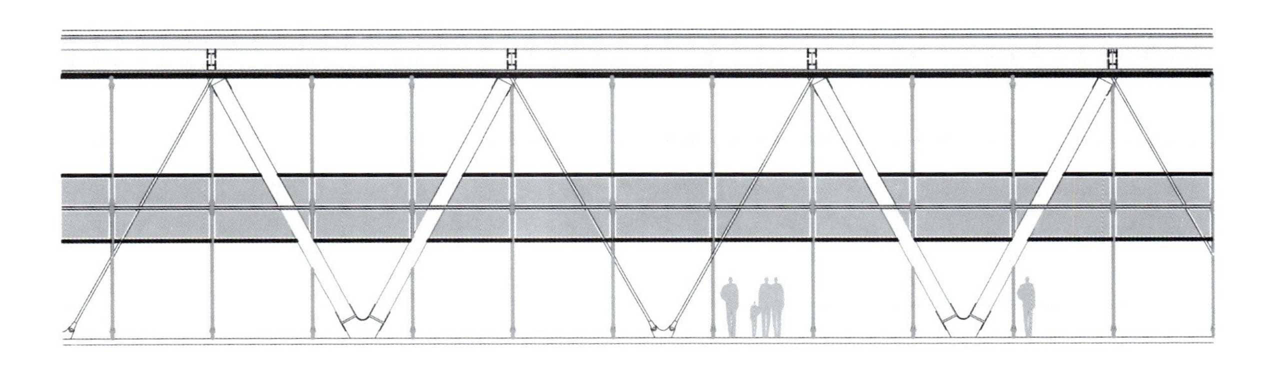
30

31

32

33

34

35

31 水幕细部
32 寓意鲜明的水幕与船造型
33 水幕占据两层空间
34 水幕与支撑构件
35 登机桥剖面图
36 屋顶构架详图
37 室内屋架细部
38 室内商业空间

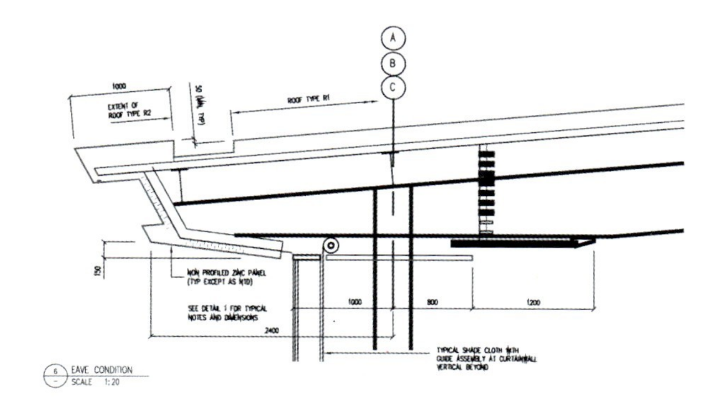
EAVE CONDITION
SCALE 1:20

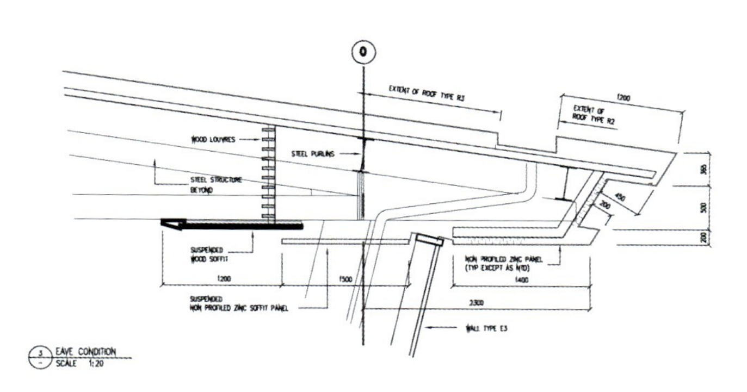
EAVE CONDITION
SCALE 1:20

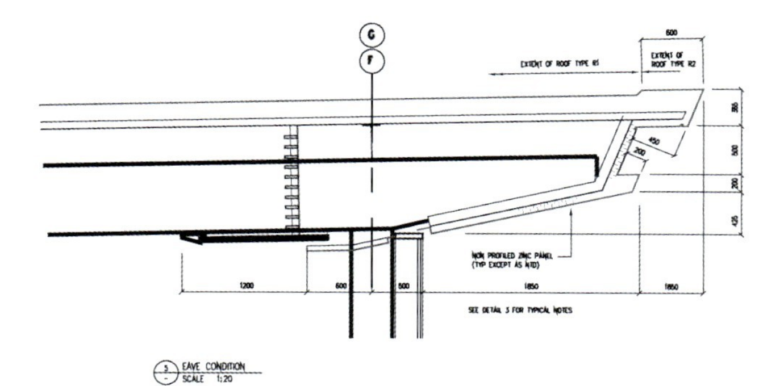
EAVE CONDITION
SCALE 1:20

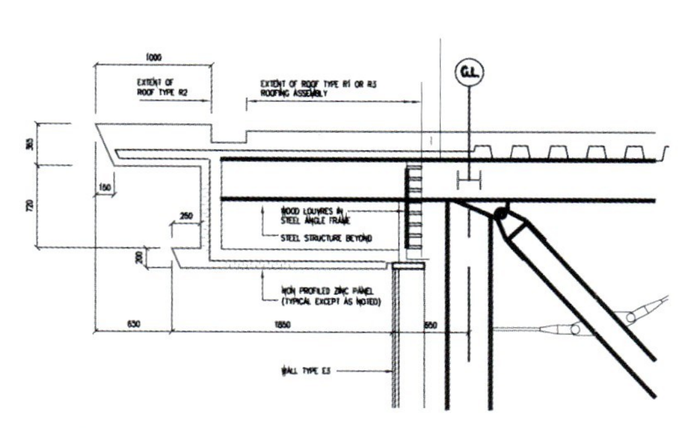
TYPICAL GABLE CONDITION
SCALE 1:20

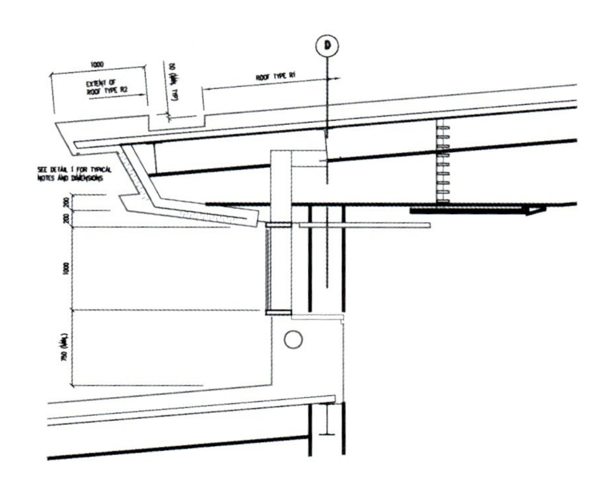
CLERESTOREY CONDITION
SCALE 1:20

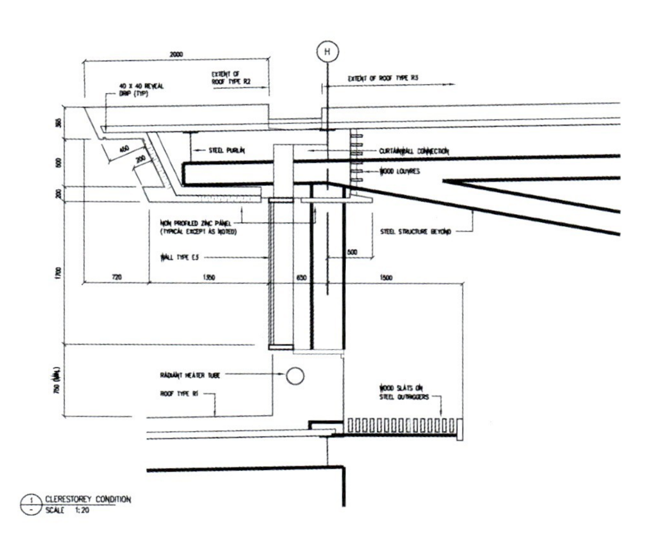
CLERESTOREY CONDITION
SCALE 1:20

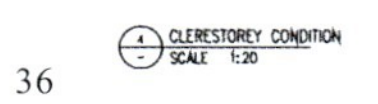

36

37

38

Edmonton International Airport Terminal
埃德蒙顿国际机场候机楼

地　　点：埃德蒙顿市
业　　主：Edmonton Regional Airport Authority
规　　模：58 500m²
合作建筑师：Barr Ryder Architects, Wood O'Neill O'Neill Architects.

1

As a member of EIA3, Stantec Architecture was selected to design a major expansion to the existing airport that will meet passenger projections of seven million for the year 2015. The addition and renovations were designed to provide flexibility within the building so that as many gates as possible are used interchangeably for domestic, international and trans-border flights. The airport's design celebrates the region's uniqueness by incorporating design elements inspired by the Prairie landscape, grasslands, and cultural heritage.

The addition and renovations were designed to be implemented incrementally without disruption to operations.

作为EIA3的成员，Stantec被选中设计埃德蒙顿国际机场的扩建工程，以满足到2015年年客流量达到700万的需求。要求空间设计灵活，能提供尽量多的通道以满足本土航班、国际航班和过境航班旅客转乘的需要。机场建筑设计的灵感来自大草原的自然景观、草地和对文化传统的承袭。

埃德蒙顿国际机场在扩建和改建的设计和施工过程中不影响机场的运作。

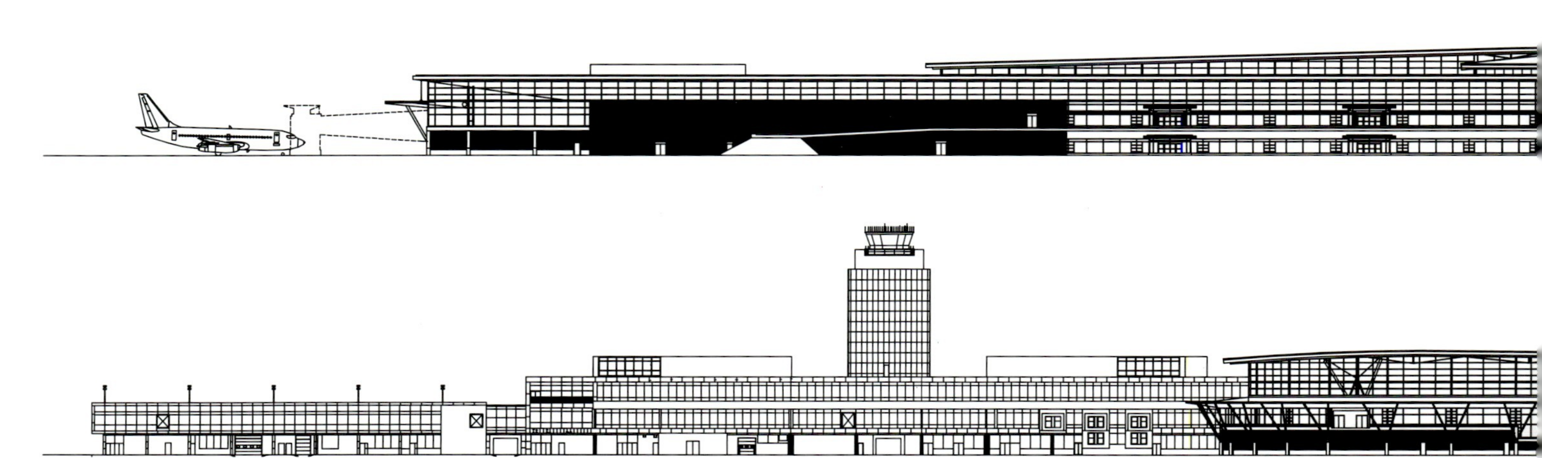

2

1 机场外景
2 立面图
3 机场夜景

3

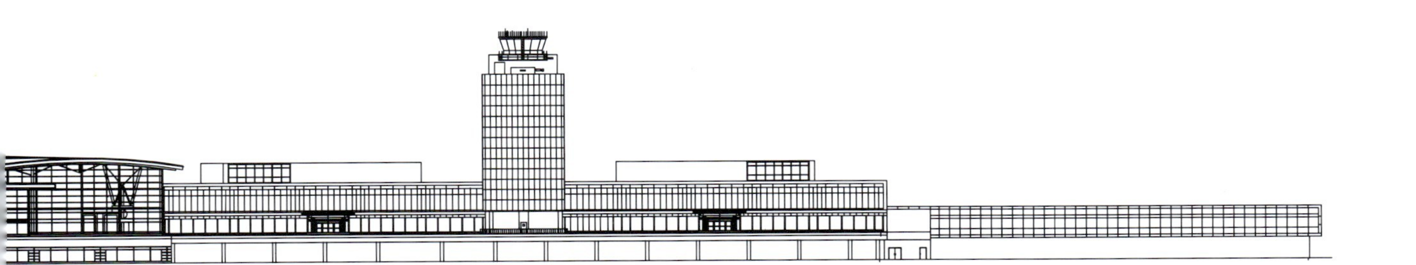

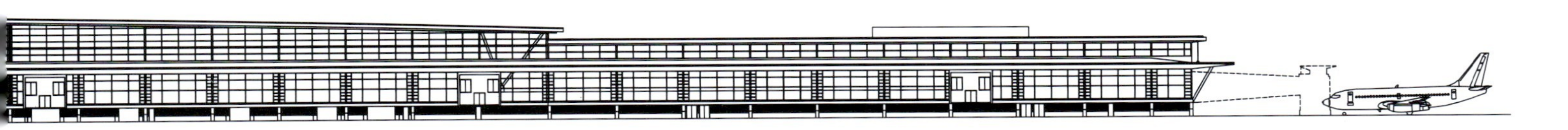

4 剖面图
5 反射屋顶顶棚平面
6 从反射屋顶下部仰视顶棚
7~8 剖面图
9 室内过道
10 室内通道细部

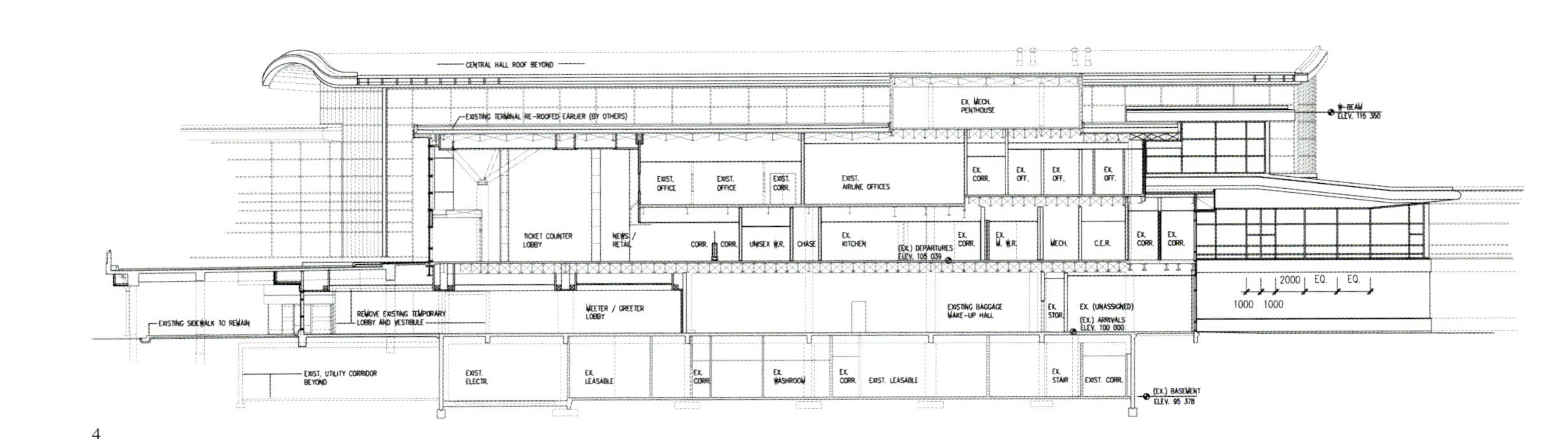

4

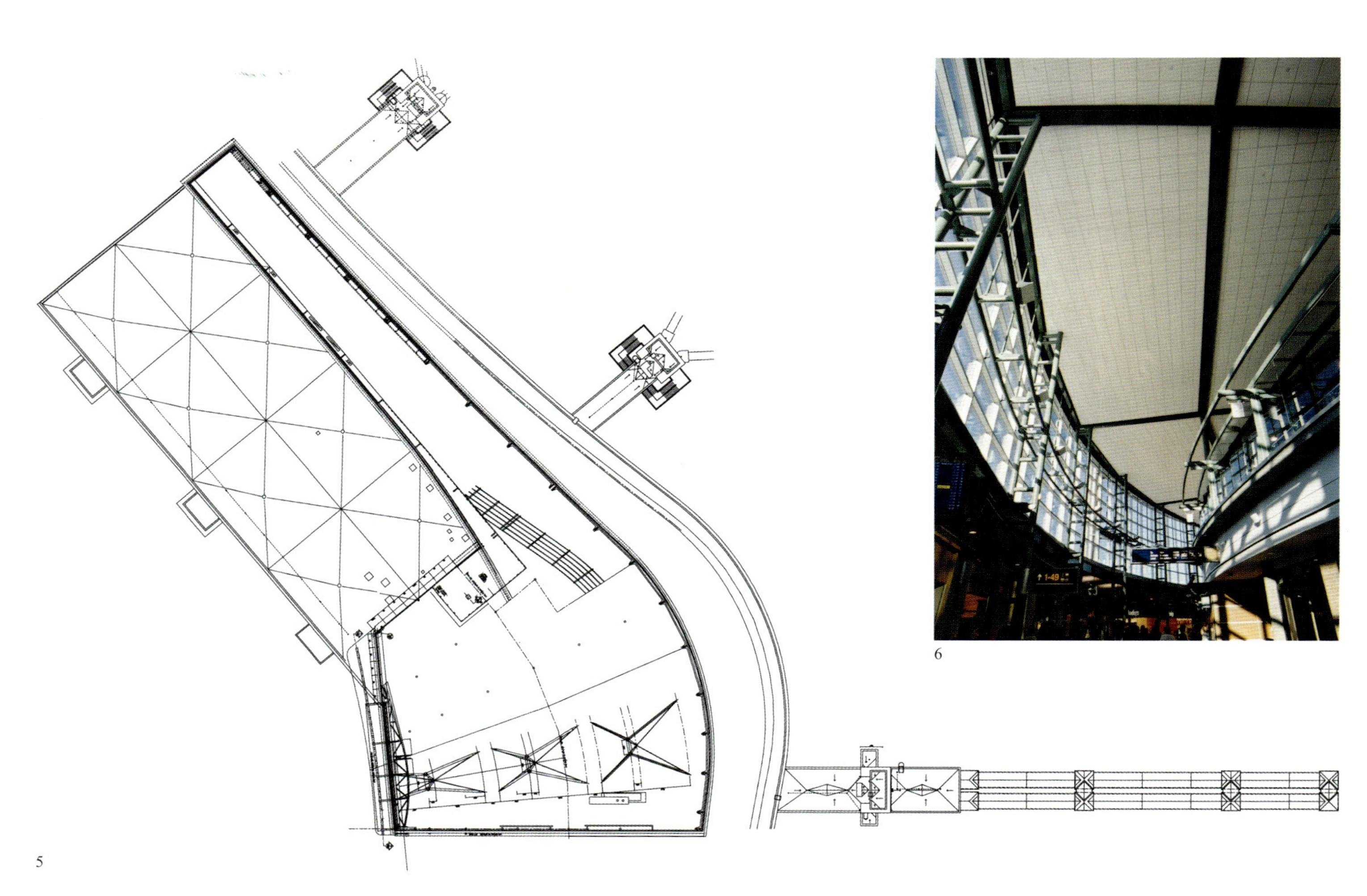
6

5

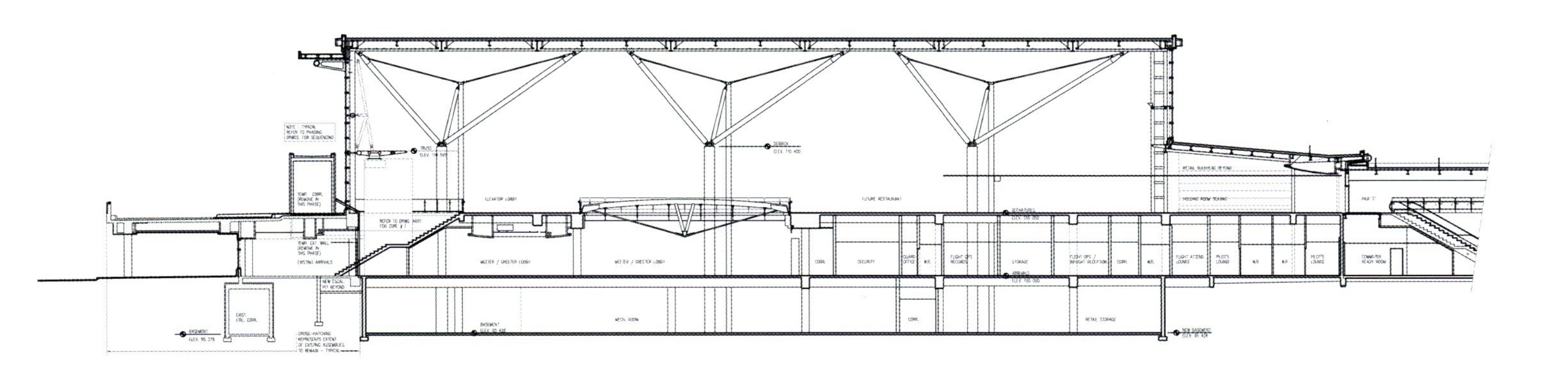

7

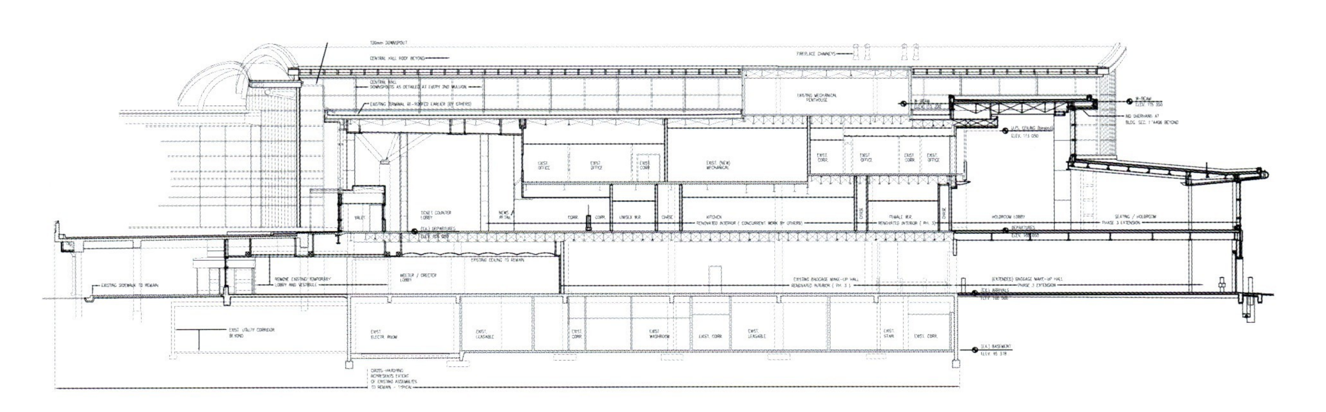

8

9

10

11

12

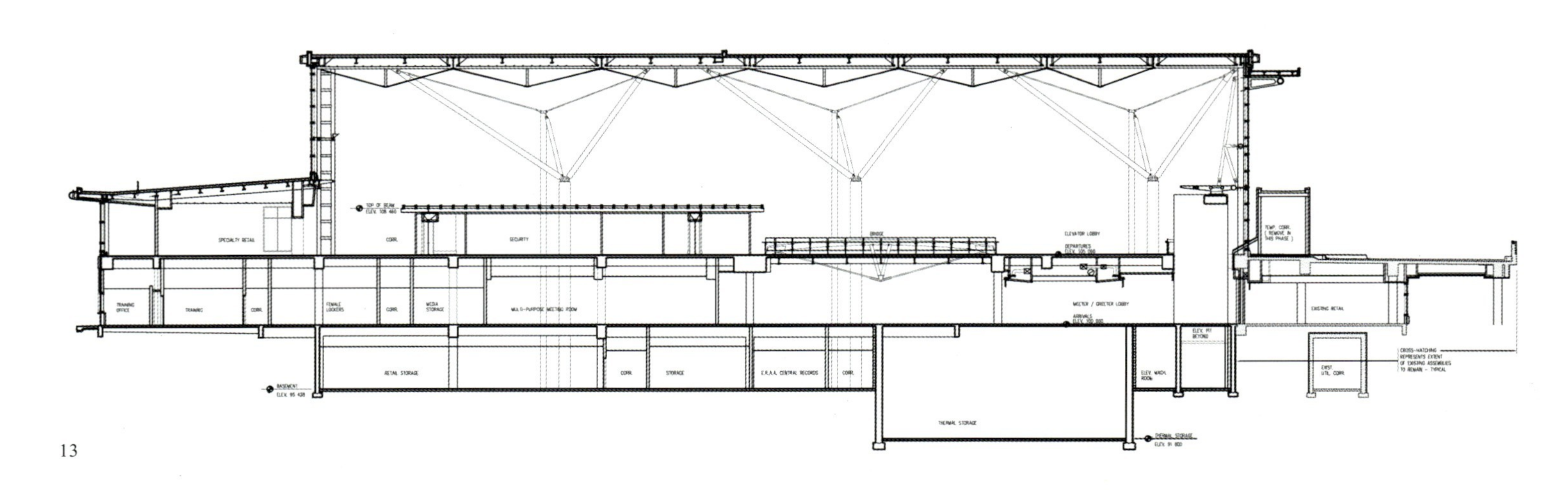
13

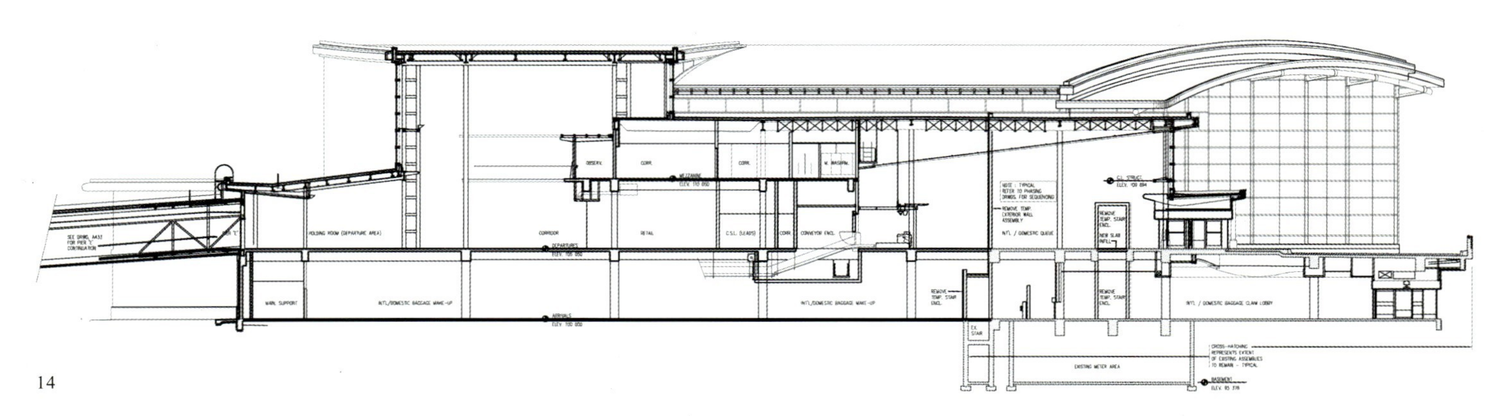
14

11　出发厅室内
12　出发厅细部
13~14　剖面图
15　候机厅细部
16　通道
17~18　剖面图

15

16

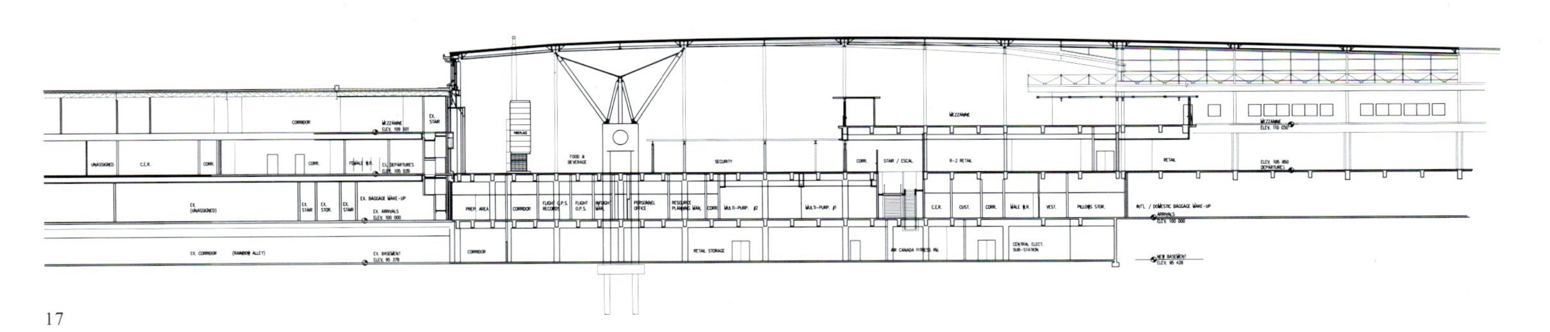

17

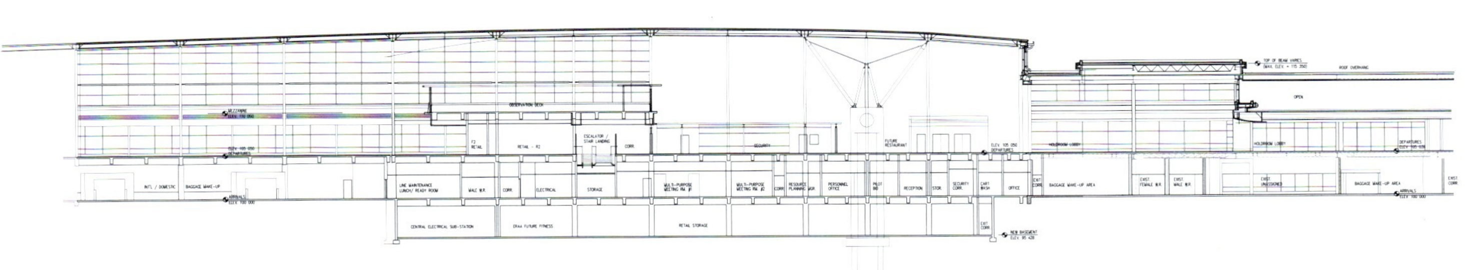

18

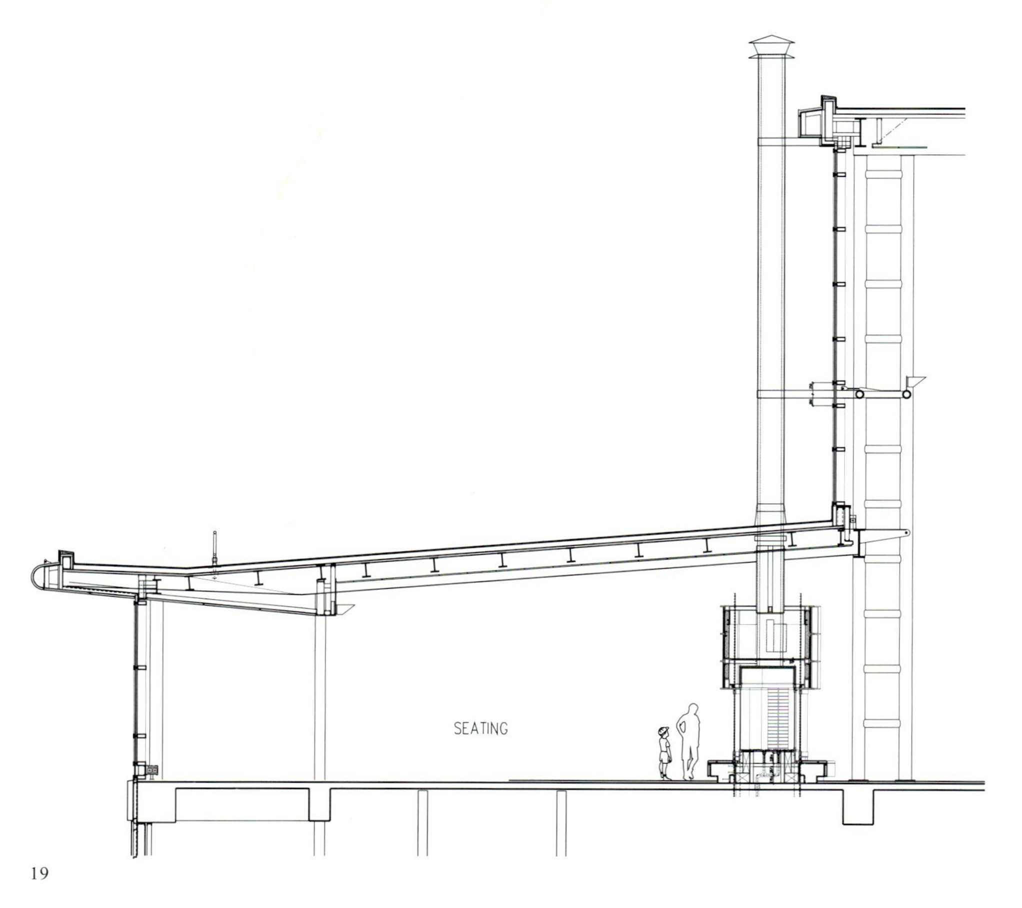

19

21

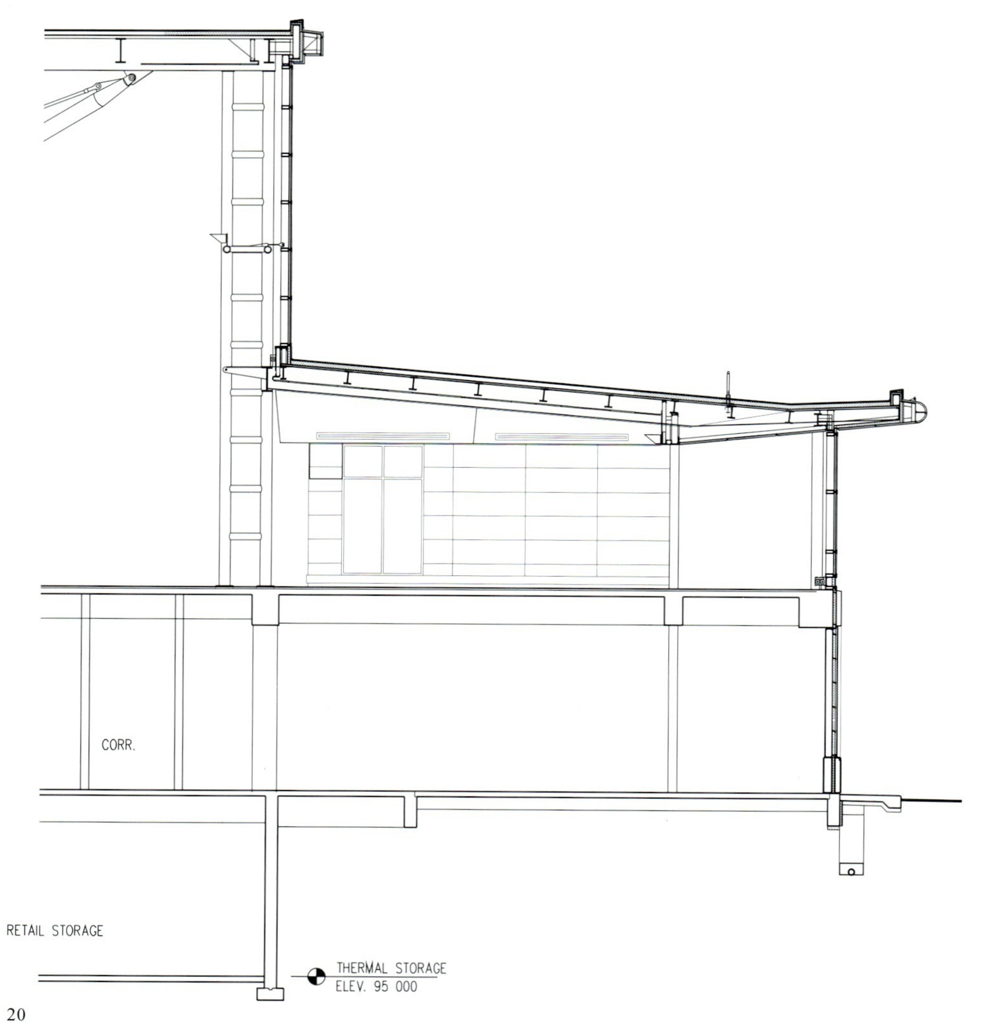

20

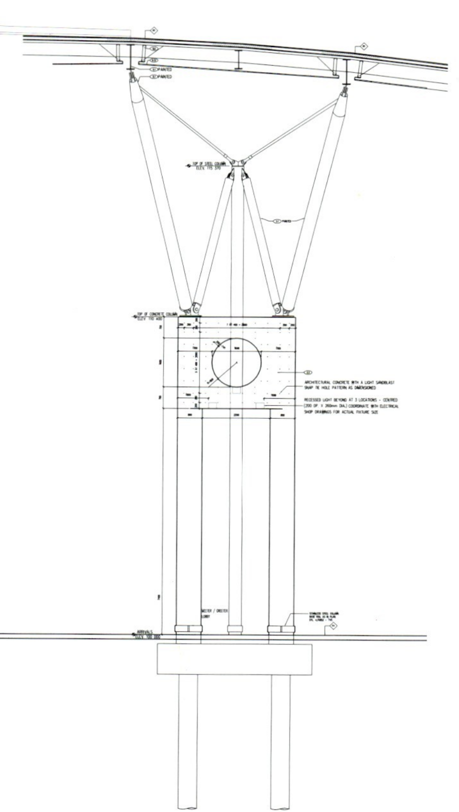

22

23

24

25

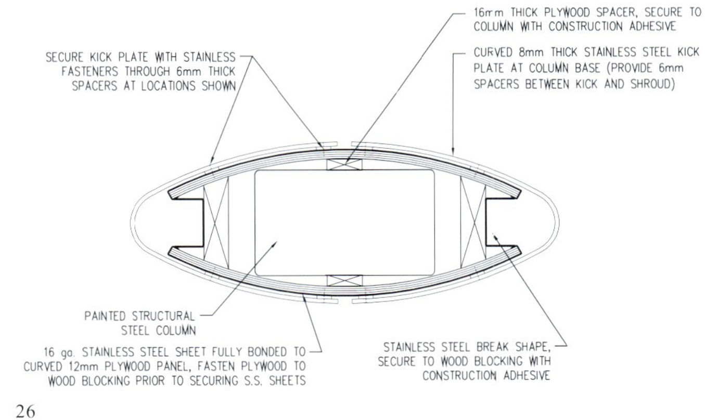

26

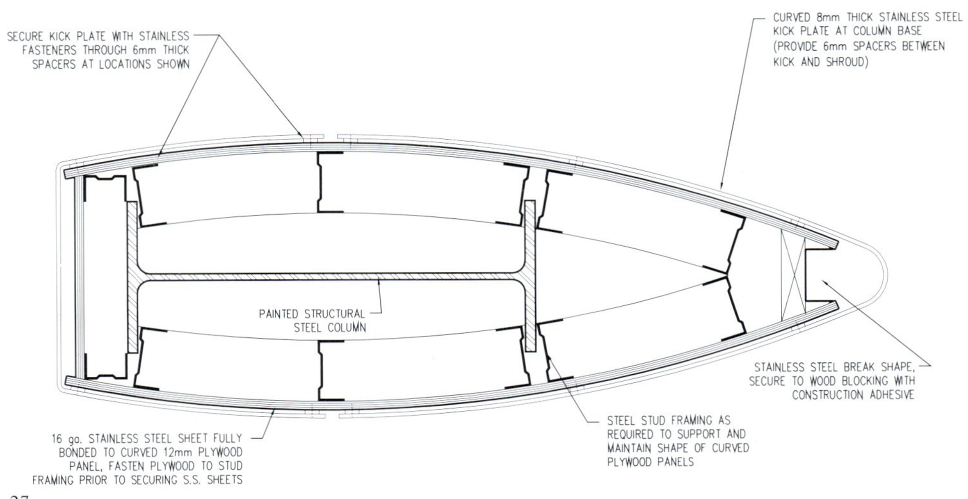

27

19~20 剖面详图
21 建筑细部
22 支架详图
23 檐口细部
24 支架细部
25 独特的柱装饰
26~27 柱饰面构造详图

Regina International Airport
里贾纳市国际机场

业　　主：Regina Airport Authority
地　　点：萨斯喀彻温省里贾纳市
规　　模：4 500m²
完成时间：2005年

The Canadian prairies of Southern Saskatchewan strike a stark, elemental landscape: the horizon, flat and expansive; sky and sun in abundance; and wind sweeping the space between. Seasonal change adds further drama to this landscape, and completes the context that the airport building attempts to represent.

The Regina Airport Terminal Building -like most airport terminals- has had a succession of additions to its early beginnings, the latest of which occurred in 1984. The handsome and robust red brick building required expansion of its inbound passenger arrivals area, inbound baggage handling system international arrivals and Canadian Inspection Services (CIS), and a variety of renovations to the existing building spaces to accommodate increased passenger traffic and security requirements and equipment. The airport terminal expansion seeks to continue the architectural vocabulary of the 1984 addition while adding a striking new form where old meets new.

Stantec provided a full range of integrated services, including prime consultant, architectural and engineering disciplines. The challenges inherent in this project were to design a series of cost-effective expansions and improvements to the existing terminal, while at the same time phasing the redevelopment so that construction could occur with minimal impact to day-to-day operations. The design solution adds extra capacity, improves customer service and passengerflow, and makes the terminal compatible with the latest explosive-detection and security requirements. A spectacular roof structure in the form of a giant sundial draws its inspiration from the quality of light and sky in the region, and creates a memorable welcome experience for arriving passengers as they descend into the baggage claim and meet-and-greet area.

1

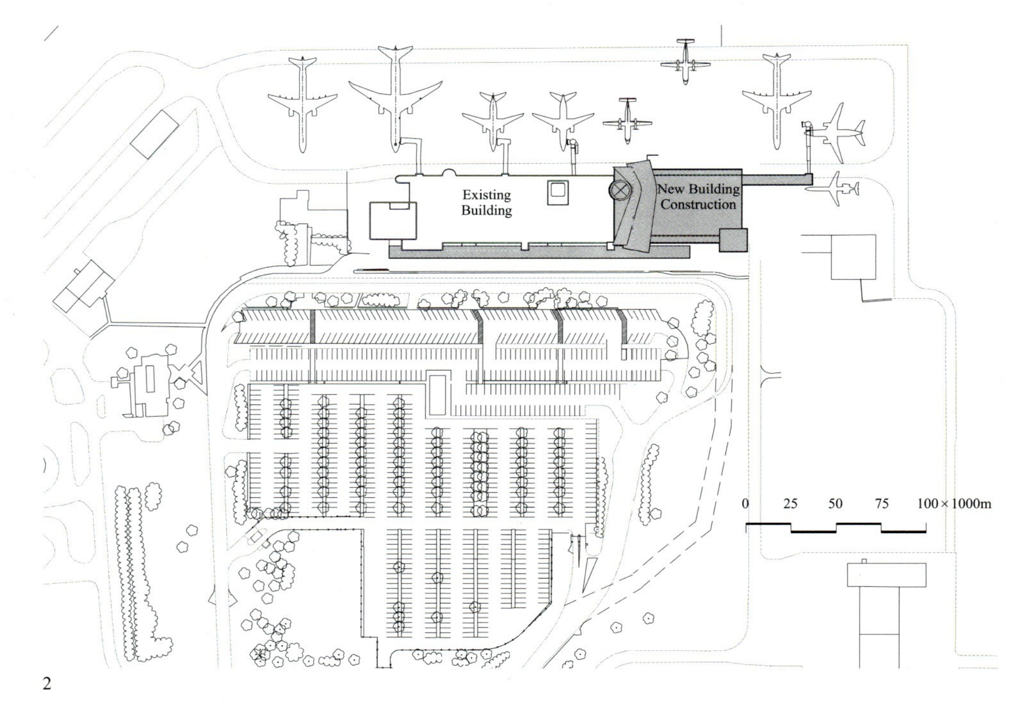

2

1　机场鸟瞰
2　总平面图
3　机场模型
4　平面图

加拿大萨斯喀彻温省南部的大草原的景色十分独特：平坦的草原、广阔的天空、充足的阳光，和吹过草原的风，还有戏剧化的四季变化。这一切都是里贾纳市国际机场建筑设计所要展示的。

像大多数机场一样，里贾纳市国际机场也是由旧机场扩建而来的，最近的一次改建是在1984年。旧机场是清秀坚固的红砖建筑，需要更大的到达区、交付行李系统、国际到达区、加拿大检查服务区（CIS），和一系列对已有的空间的改造，以满足旅客增长和安全检查的需要。

里贾纳市国际机场1984年的扩建在原有建筑的基础上增加了富有视觉冲击力的新形式，此次改扩建力图延续这种建筑语汇。

Stantec提供里贾纳市国际机场设计的全过程服务，包括策划咨询、建筑设计及其他配套专业设计。

设计的挑战性在于：在设计一整套经济高效的扩建和改造工程的同时，对原有部分的正常运作要尽量不产生影响。新的设计方案增大了机场的运营能力，改善了顾客的服务，优化了乘客的交通流线，满足最新的安全检查的需要。一项壮观的屋顶改造是加建了日冕，其灵感来自当地充足的阳光，当旅客在领取行李区和进入迎候区时，都可以感受到当地的热情。

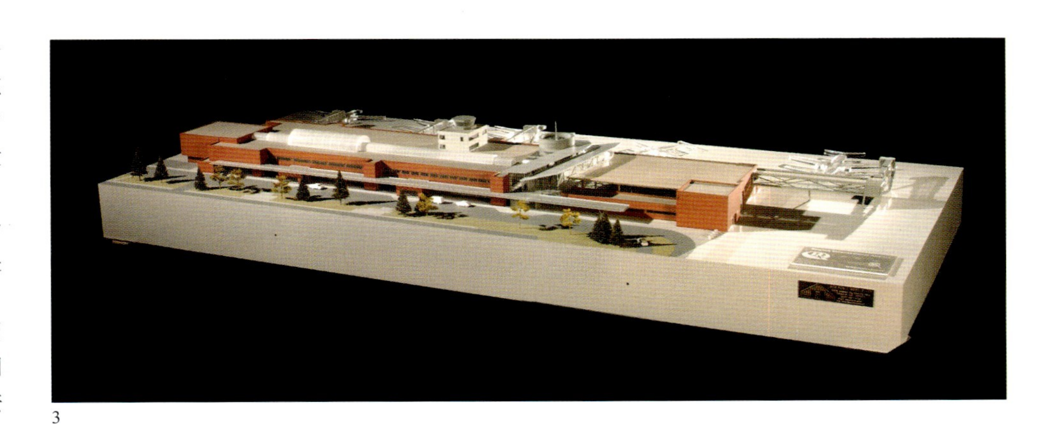

3

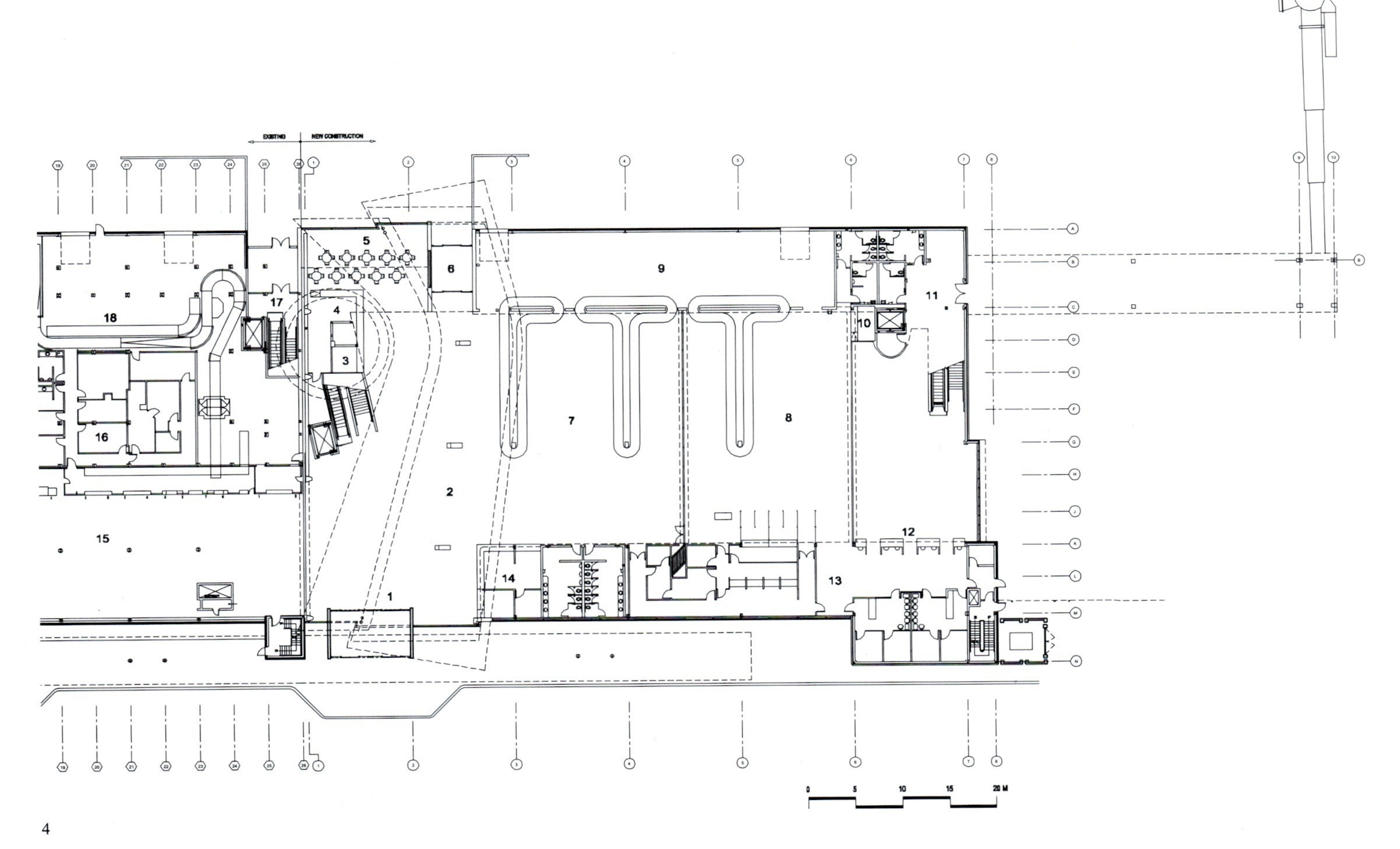

4

5

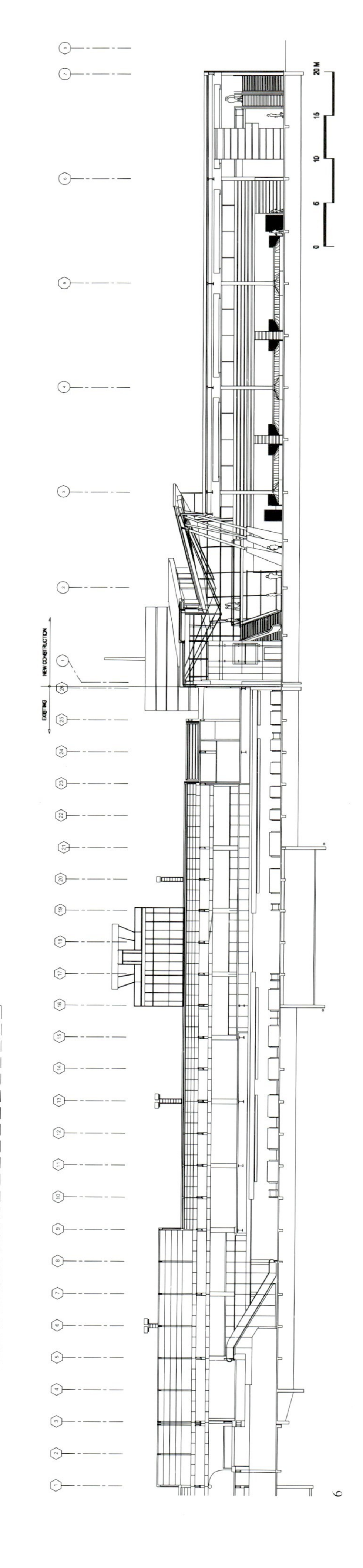

6

5 机场模型
6 剖面图
7 机场模型细部
8 北立面图
9 西立面图（空侧）
10 东立面图（陆侧）

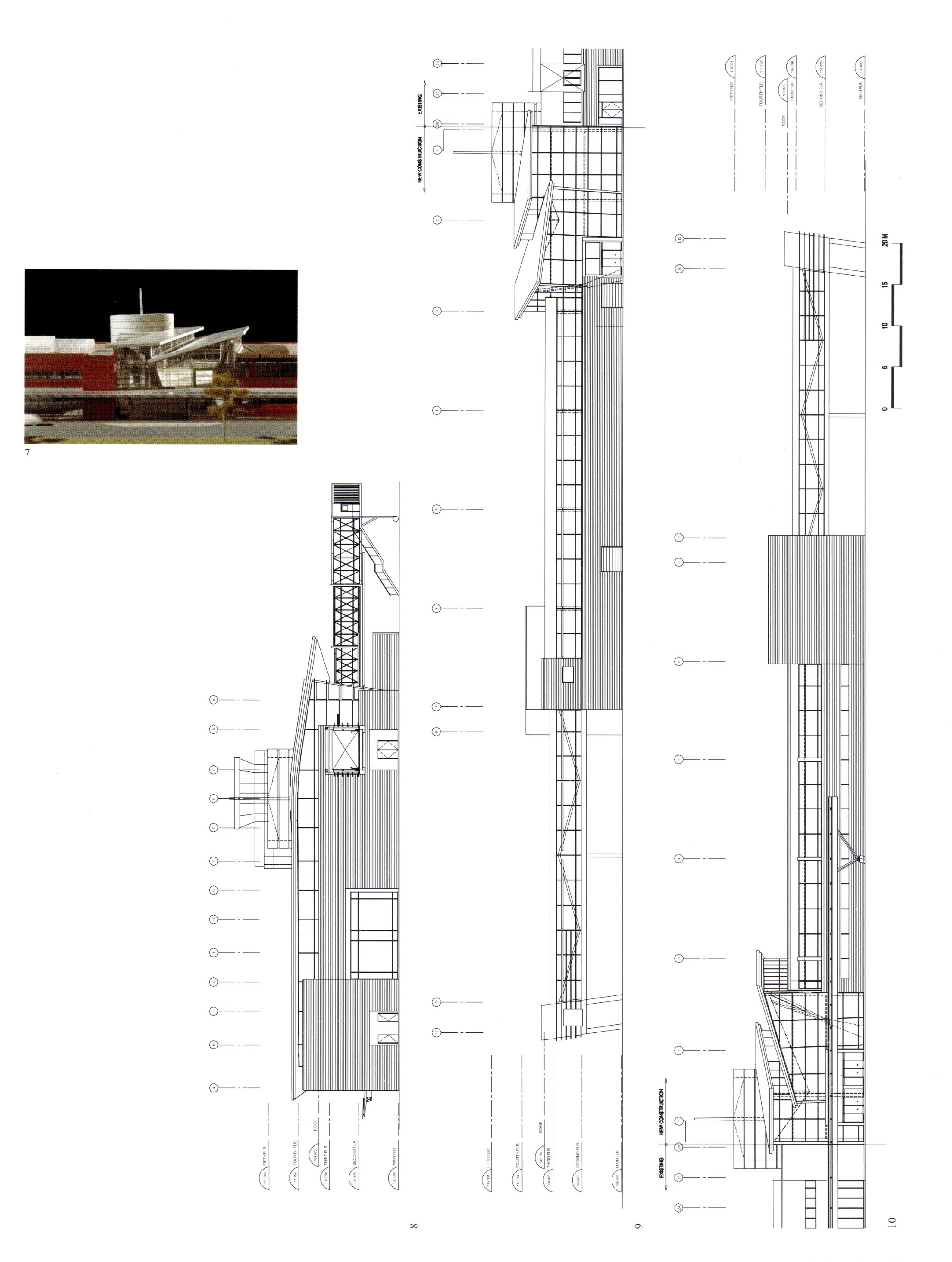

7

8

9

10

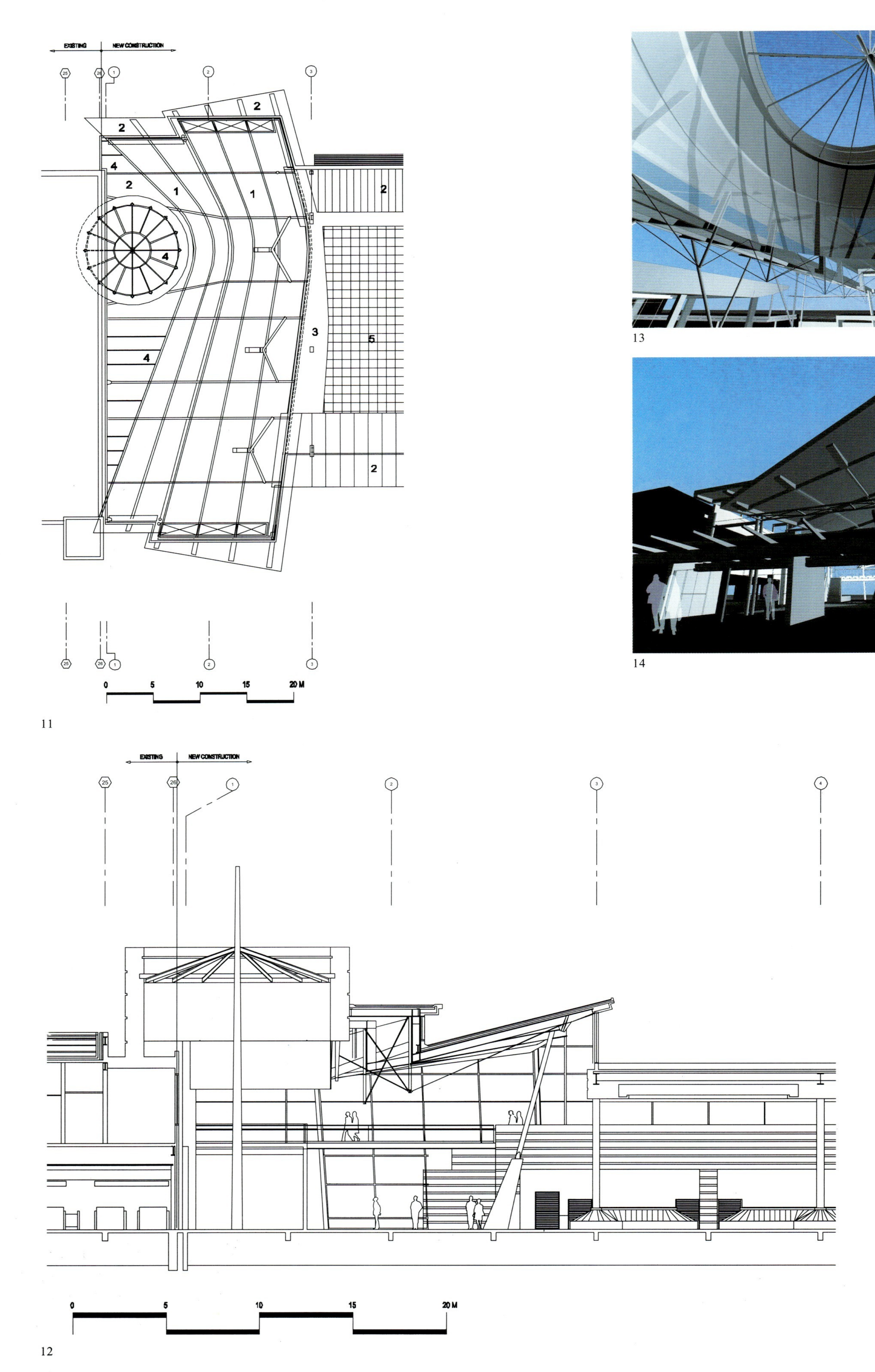

11

12

13

14

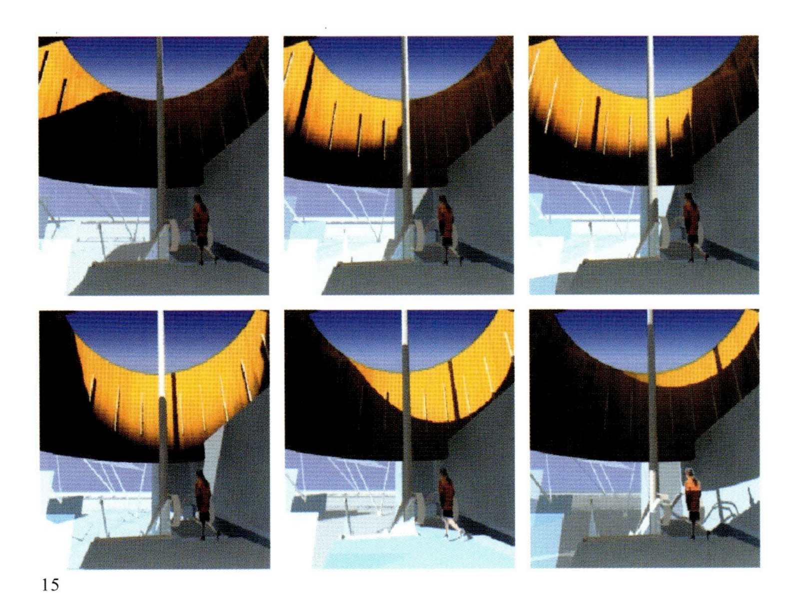

15

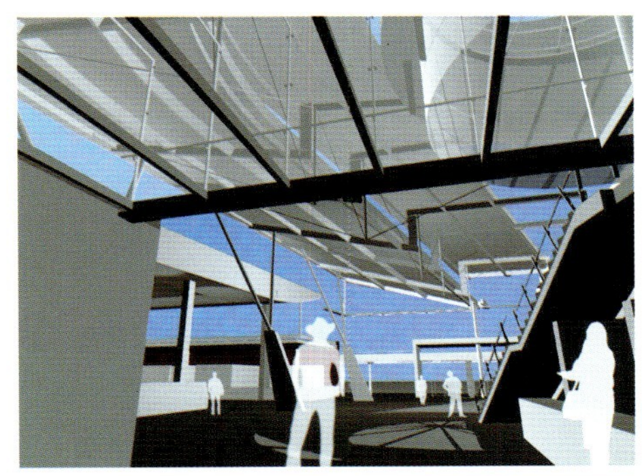

17

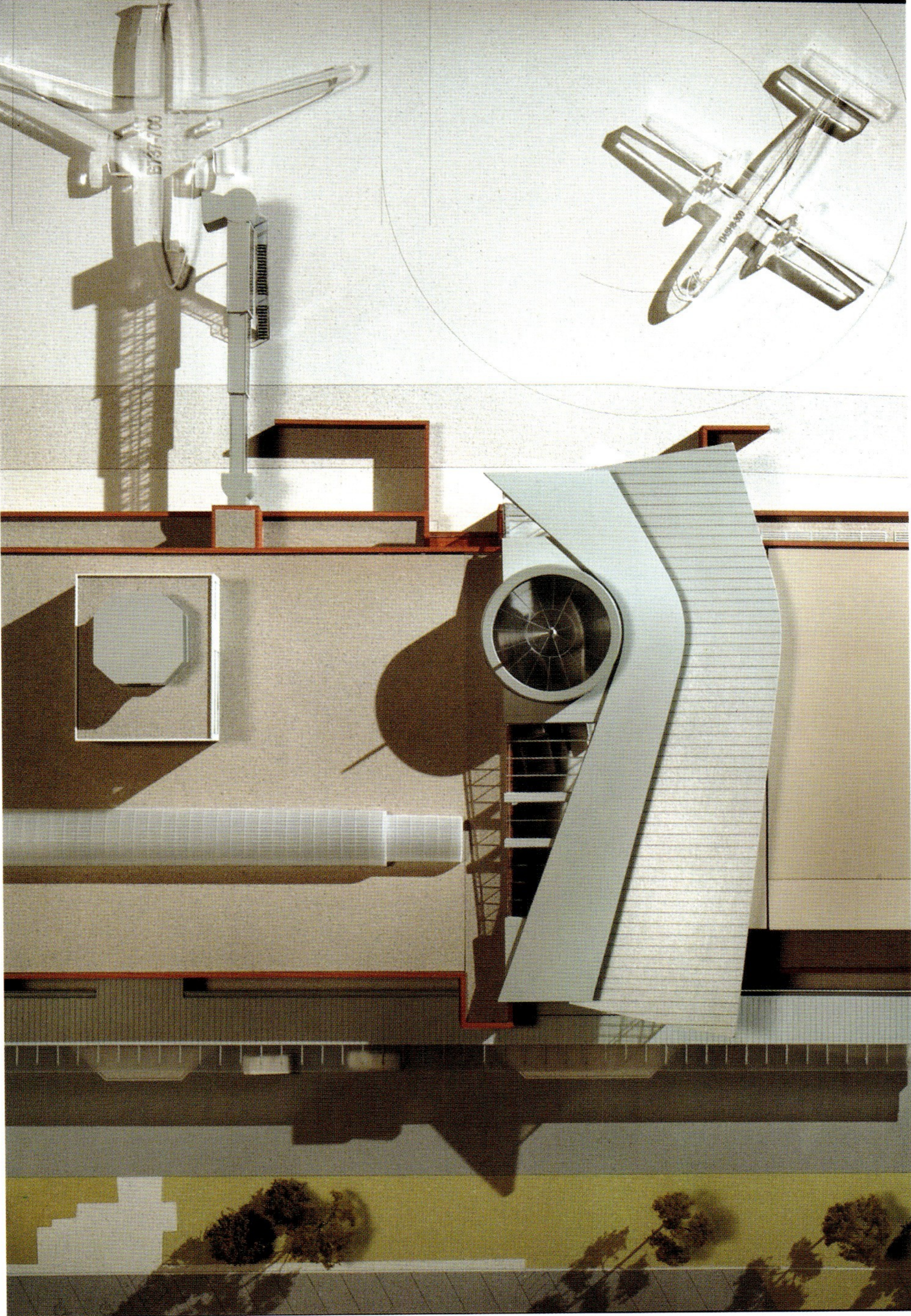

16

11 反射屋顶平面
12 反射屋顶剖面图
13 中庭构筑物室内效果图
14 中庭构筑物室外效果图
15 日冕电脑模拟效果图
16 模型细部
17 室内设计创意效果图

Lake City Station
Lake City 轨道交通车站

业　　主：Light Rapid Transit Project Office
地　　点：伯纳比市
完成时间：2003年
合作建筑师：Walter Francle Architects

1

Lake City Station is located on the freshly opened "Millennium Line" at the intersection of Lake City Way and Lougheed Highway in Burnaby, BC. Lougheed is a busy arterial route that crests in elevation to the east of the station and falls away west of the station. The station seen from an eastern approach along Lougheed appears to climb from the ground; approached from the west, it appears as dramatically hovering over the ground. Lake City Way is the entrance to a "High tech" business park currently under construction. The immediate site under and adjacent to the station is a simply sloping lawn that climbs to the north, away from the highway. South of the station, immediately across Lougheed is a continuous stand of mixed, mature deciduous and coniferous trees with an established residential neighborhood beyond.

2

The client's ambitions for the design of the Millennium Line stations were threefold:

- to build a significant, regionally identifiable portal to the neighborhoods through which the train passes,
- to provide a safe, transparent refuge for the traveler,
- to construct stations that were durable and long lasting.

1　车站全景
2　车站仰视
3　车站夜景
4　车站总平面图

3

Lake City轨道交通车站位于新投入使用的“新千年干线”上，伯纳比市Lake City大道和Lougheed高速公路的交叉口。Lougheed高速公路是一条繁忙的公路，在车站的东部，并向西倾斜而下。车站从东部看上去像是由地面沿Lougheed高速公路向上攀爬；从西部看上去，车站则以一种动感的造型盘旋于地面之上。Lake City大道是一条进入正在建设中的高科技办公中心的出入道路。车站下方，紧贴着一片向北倾斜而上的草坪。南向，在Lougheed高速公路的一侧是一片成熟的落叶松树林，更远些是已建成的居住社区。

业主对“新千年干线”车站的要求有三点：

• 建造一个在轨道沿线上，形象突出，易于识别的建筑；

• 提供一个安全和可以庇护旅客的车站；

• 一个结实和使用寿命长的车站。

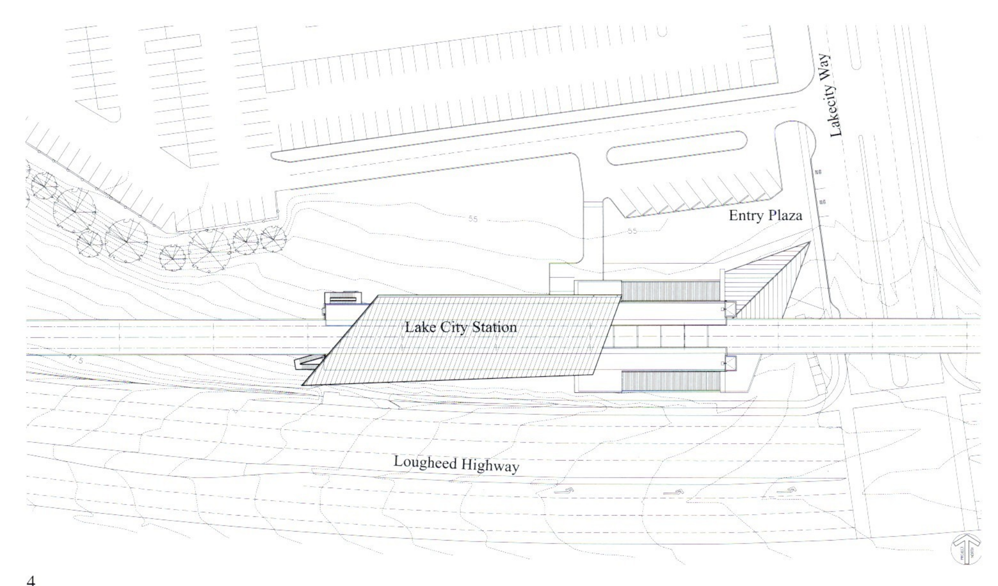

4

5

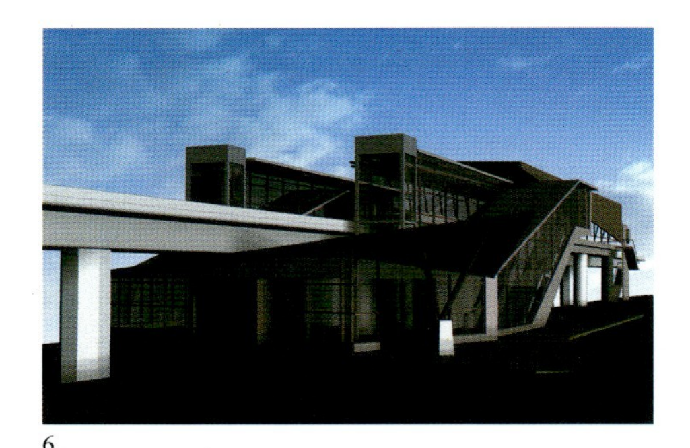

6

7

The Lake City Station is conceived as a wind-swept cloud, a dynamic Japanese origami-like shroud that is anchored to the ground at its eastern extent and leaping cantilevered above the ground to the west. It aspires to frozen dynamism and technological ambition. As a gateway structure, it affirms the character of the high tech park it fronts.

Steel -painted, gavalume finished and stainless combined with glass, aluminum, and concrete comprise the spare, long-lasting materials of the station's construction. Oval, darkly pigmented "Agilia" concrete platformsupport columns form the base of the superstructure. These are canted seven degrees to the west as are the sandblasted concrete fin walls supporting the stairs and escalators. The concrete base of the entry canopy and structural steel columns, springing from the top of the base to support the canopy cant dramatically. The platform roof canopies' steel trusses taper in depth and are framed in a diagonal bias to the guide way and platforms. The stair/escalator roof structure is framed high above the traveler's head at the concourse level, and pinches to a minimum at this platform level. A tall, angled "V" column strut surmounts a canted concrete base stretching skyward to support what seems to be an impossibly long cantilever of the upper roof's west end. In places, horizontal window mullions are seemingly stretched beyond their fixing points by speed itself.

All of these compositional elements are intended to reinforce the spirit of a dynamic search for technological invention and lend this spirit to skytrain travel.

8

车站的造型像一片被风吹动的云，一个动感的日本折纸式的覆盖物。东部根植于地面，西部则像悬臂一样伸出于地面之上。灵感来自冰冻力学和高科技园的象征，作为一个入口构筑物，车站的形态进一步确认了高科技园区的特点。

油漆的钢材、仿木的钢材、不锈钢、玻璃、铝合金和混凝土等长寿命的建筑材料一起围合成车站的宽敞空间。卵形的深色混凝土月台支撑物坐落于巨型支撑的底部。向西倾斜7°的喷砂混凝土装饰墙面是楼梯和自动扶梯的支撑物。混凝土基座的入口雨篷和基座上弯曲支撑雨篷的钢柱的倾角非常强烈。安装在通道和月台对角框架内的月台屋顶的钢架随着纵深渐细。楼梯和自动扶梯的屋面结构在大厅层高高地位于旅客之上，以减少对月台层的压迫感。一个高大的V字形支架位于倾斜的混凝土基座上，支撑一个看上去在西端上层几乎不可能的长悬臂。水平的窗棂则因列车的速度而拉长。

所有的这些措施都是为了加强建筑的高科技和富于动感的效果，并传递给旅客。

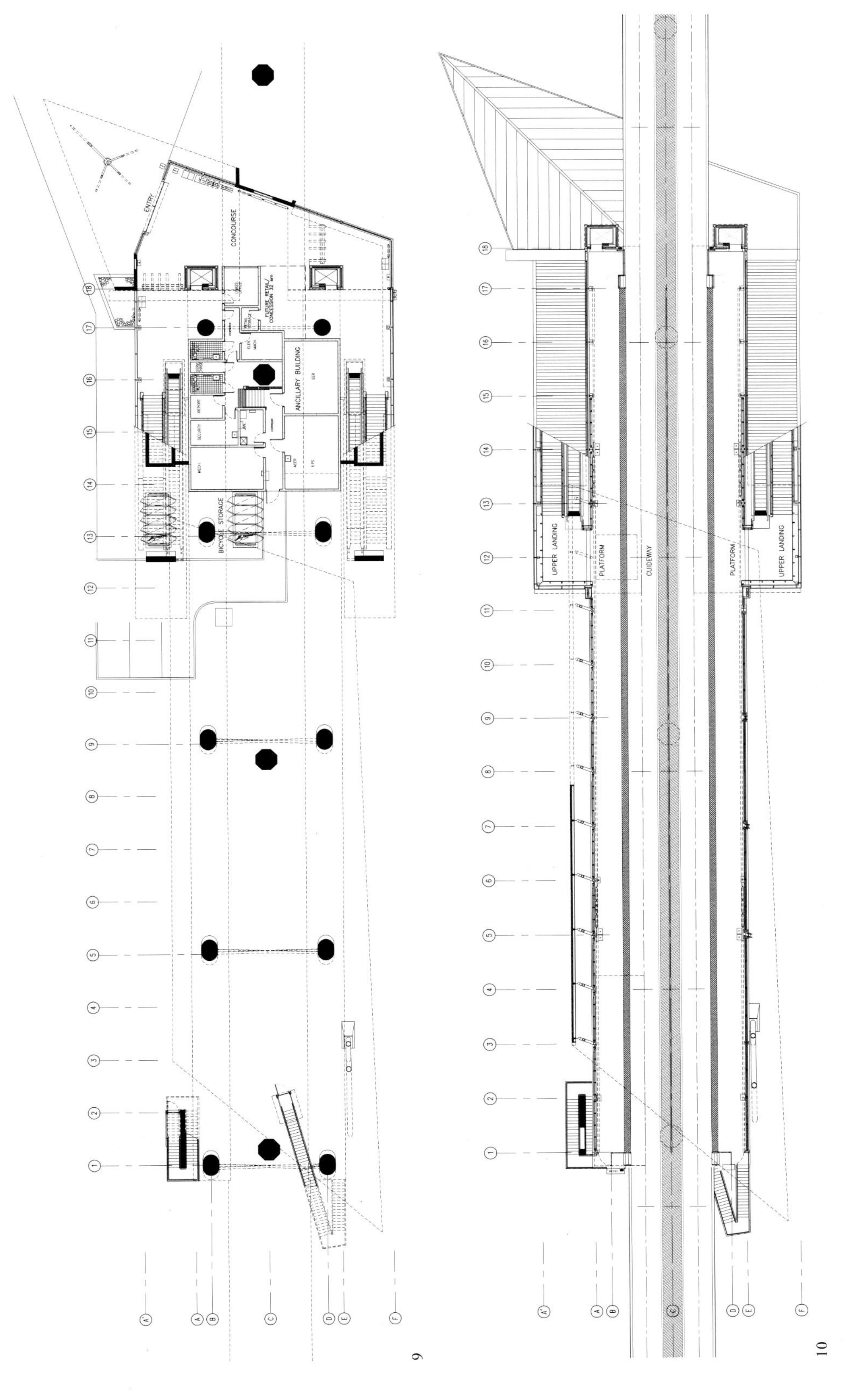

5~7 车站电脑效果图
8 车站模型
9 地面层平面图
10 月台层平面图

11

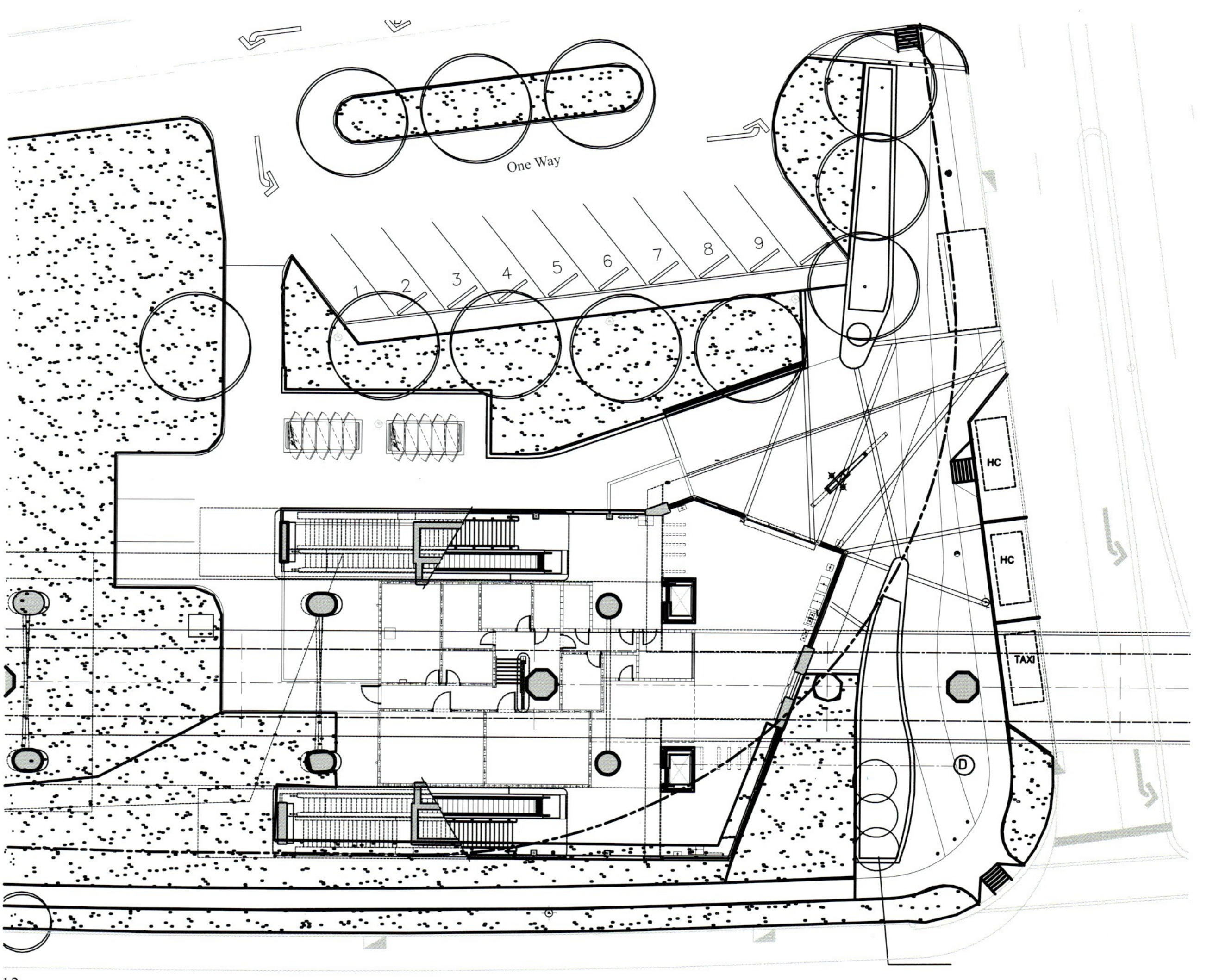

12

11　车站入口
12　平面图
13　横过街道的轨道桥梁
14　北向室内立面图
15　南立面图

At the completion of construction, the sloped lawn ground plane will be restored to its original state to emphasize that this is not a structure that rises from the earth but hovers over it. At the eastern extent of the site, travelers enter or exit the station at a relatively small grade level concourse where the paving pattern takes clues from the geometry of the roof canopies. Here the rainwater is channeled from the roof canopies and splashed onto a rock circle mediating the traffic noise from busy Lougheed Highway.

The station design embraces the need for passenger safety, with visual accessibility provided day and night. The station is designed to minimize areas for potential entrapment or undesired loitering. Intuitive way finding is also emphasized with a well-defined entry sequence and easily navigated circulation inside. Tile patterning on the platform level helps guide visually impaired commuters between the elevators and designated waiting areas.

当施工完成之后，斜坡上的草坪被植回原处，以显示建筑是架空于草地之上，而不是由地面生出。在东部的伸出体内，旅行者可以从相对较小的地面层的大厅出入车站。地面的铺地图案借鉴屋面雨篷的几何造型。雨水由雨篷落下，汇聚于一个石制的环状物内，以中和高速公路的噪声。

车站设有昼夜出入口标志，以保证乘客的安全。车站内部交通流线通畅，空间序列清晰，尽量减小了乘客在内部迷路和被围困的可能。月台层的地面图案设计同样给予乘客良好的、从自动扶梯到设计等候区的指引。

13

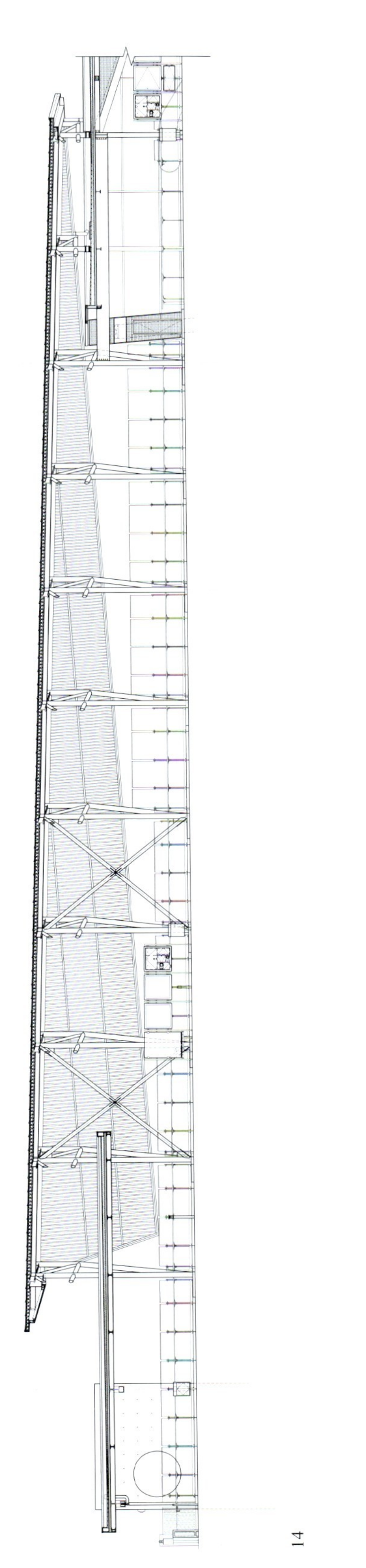
14

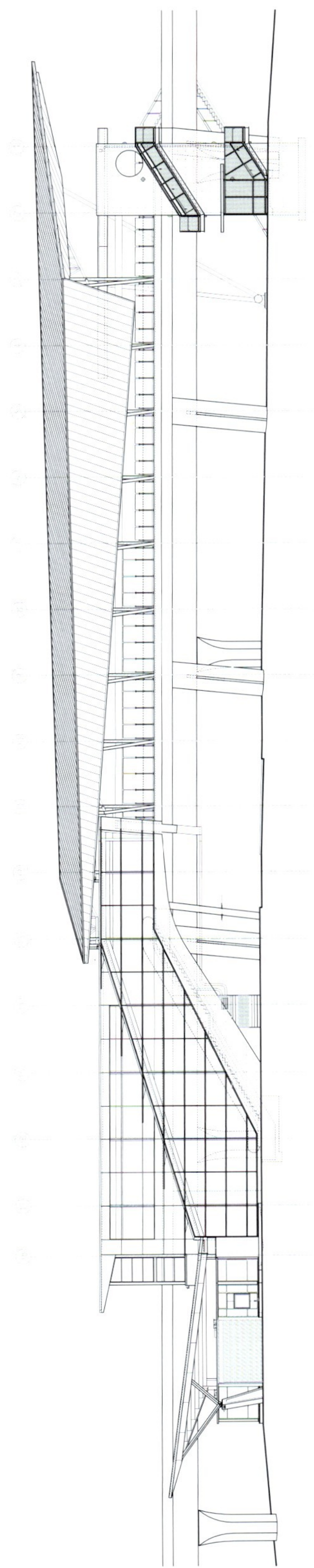
15

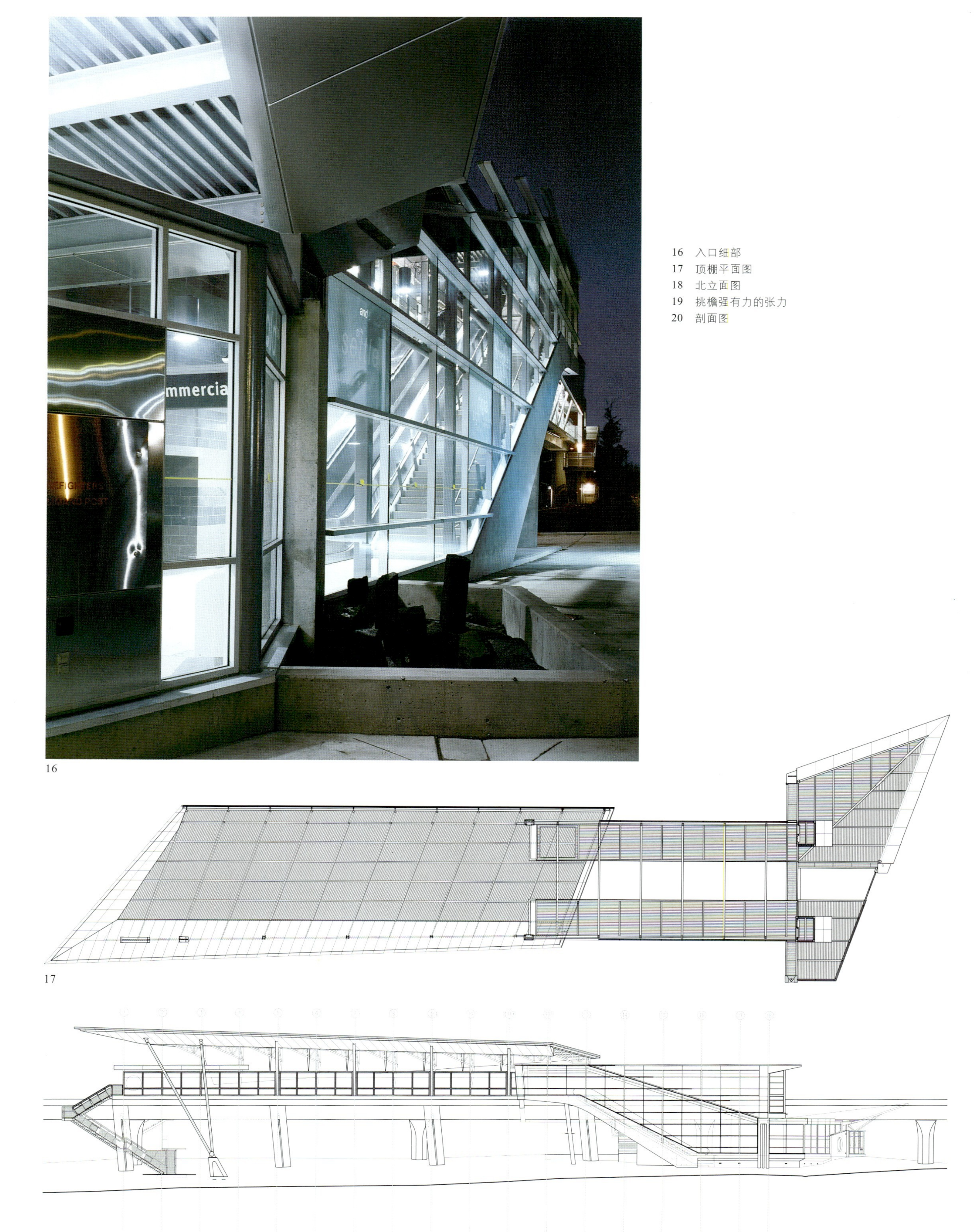

16 入口细部
17 顶棚平面图
18 北立面图
19 挑檐强有力的张力
20 剖面图

16

17

18

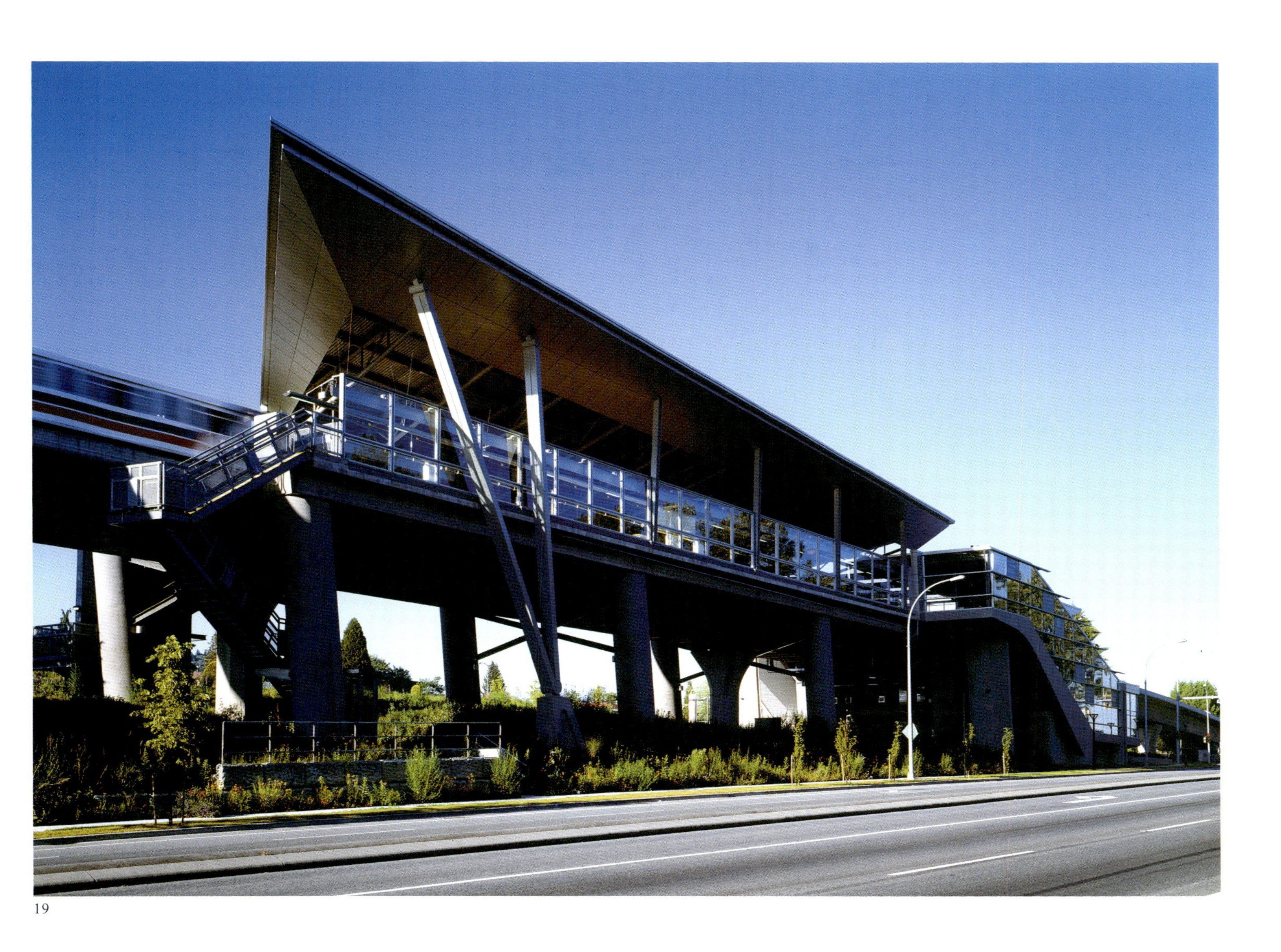

19

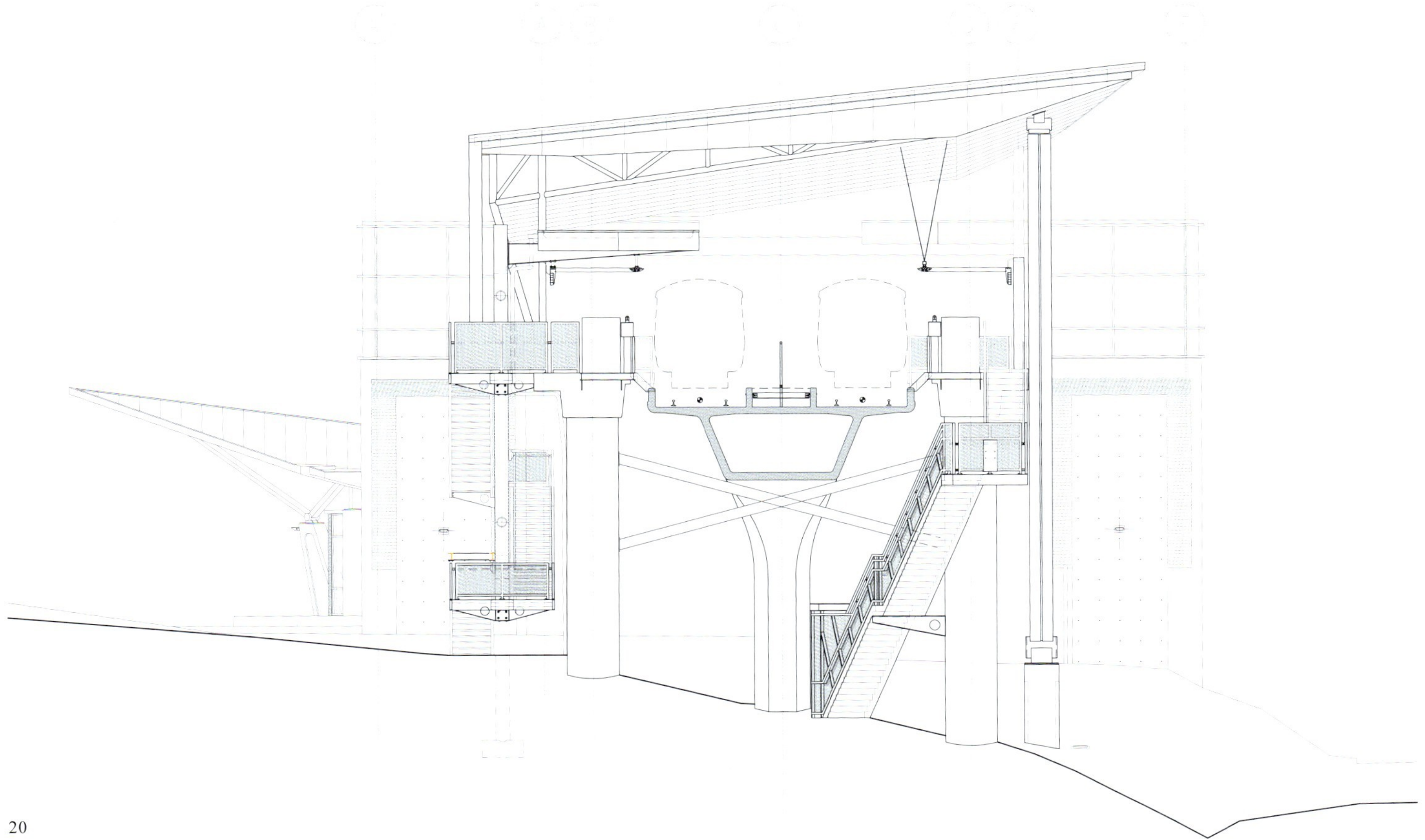

20

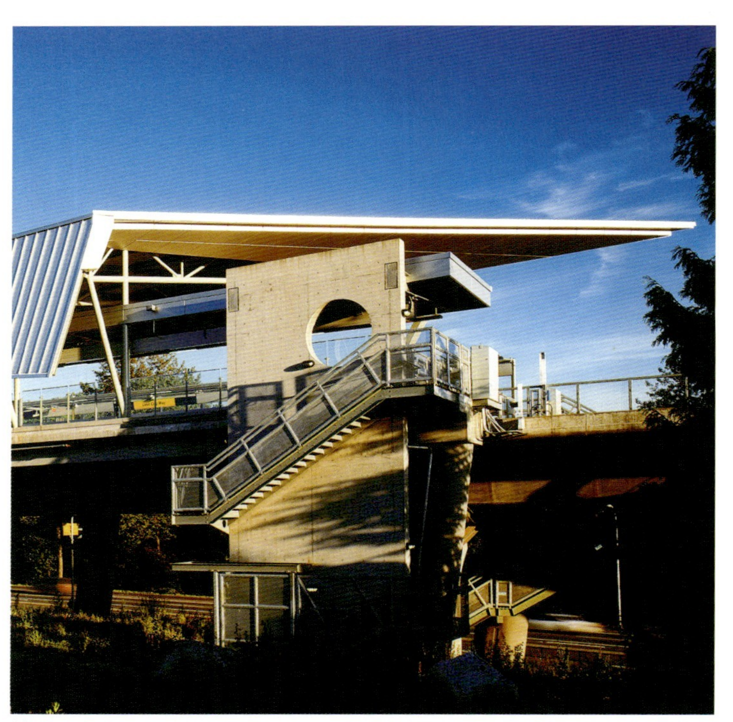

21

22

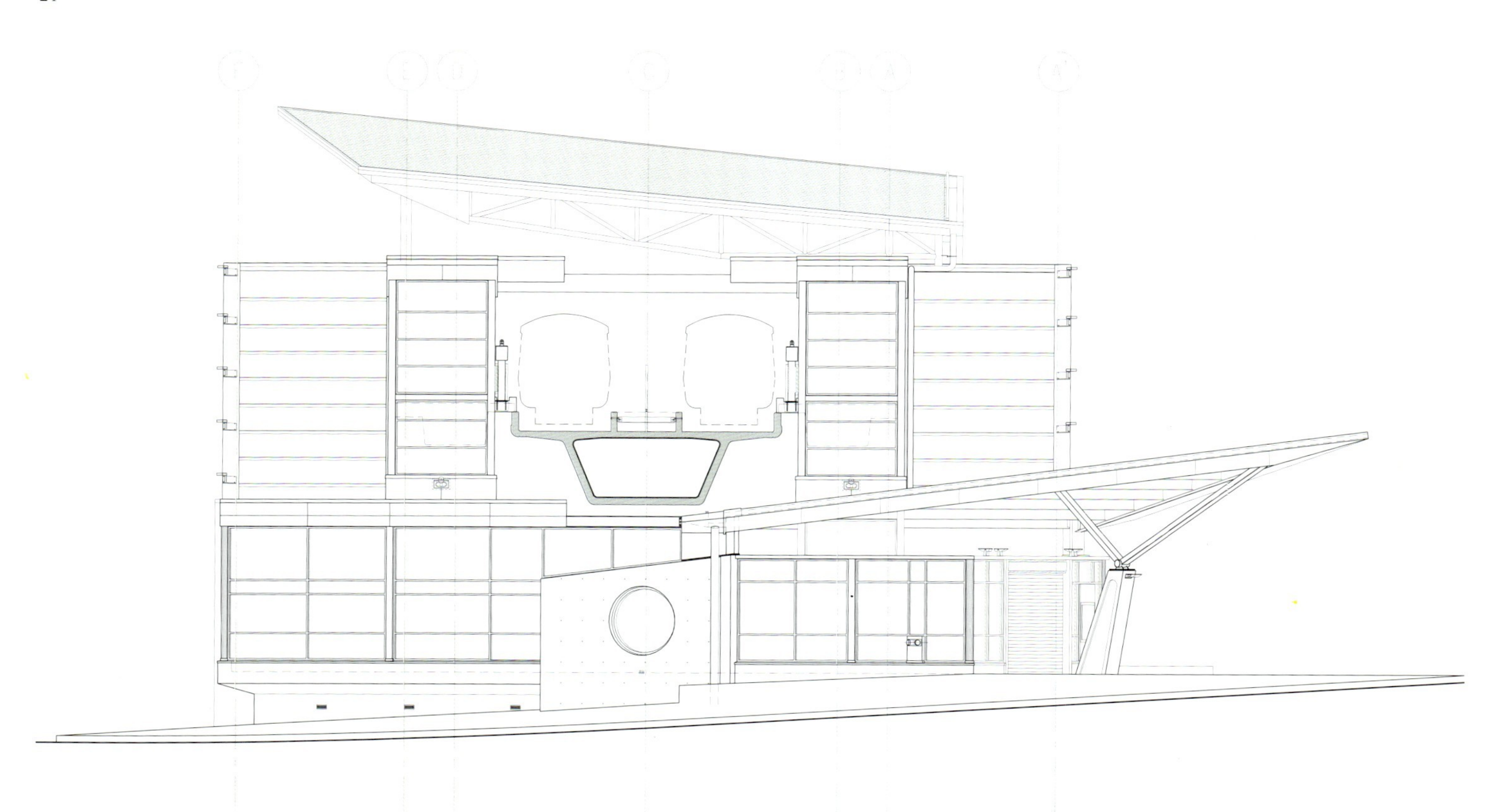

23

24

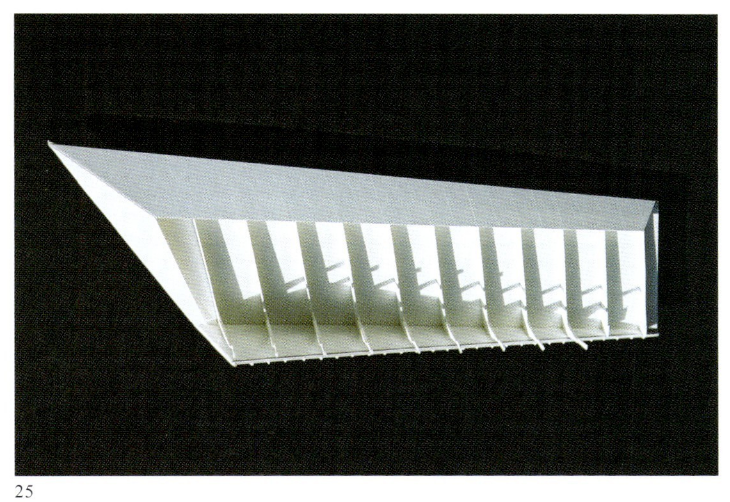

25

26

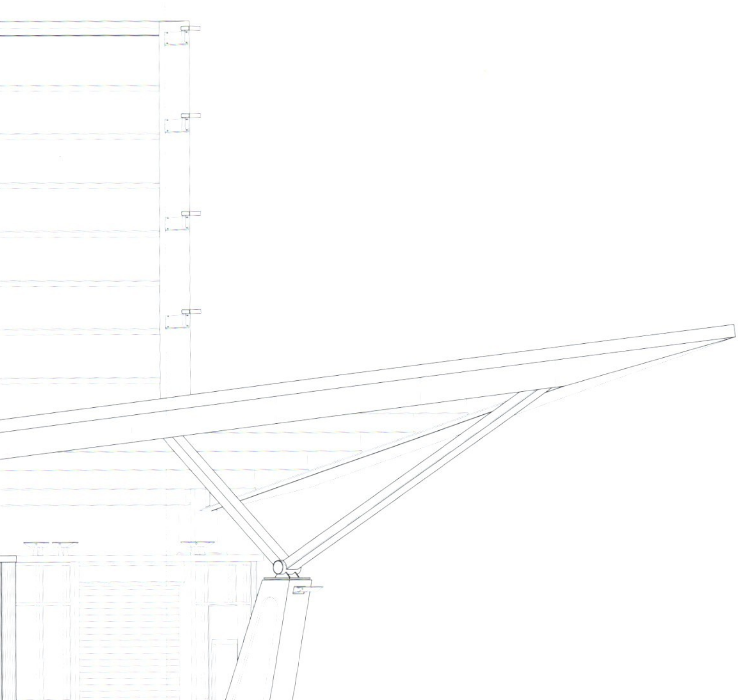

27

21 车站细部
22 顶棚细部
23 车站立面图
24 车站细部
25 屋顶模型
26 挑檐细部
27 挑檐细部详图

Braid Street Skytrain Station
Braid 街轨道交通车站

业　　主：Light Rapid Transit Project Office
地　　点：伯纳比市
完成时间：2001 年
合作建筑师：Walter Francle Architects

Braid Station is prominently located at the crossroads of major transportation arteries for the Fraser Valley. Immediately to the east are rail lines and the Trans-Canada Highway. To the south is a major highway which leads to the historical city of New Westminster and the Fraser River. To the west is Braid Street, which leads into a mature residential neighborhood. The station is designed as a portal to, and anchor for, a new civic space, which opens to the south and serves as the hub for future growth in the neighborhood.

The immediate context is predominately large-scale industry, including trucking operations and lumber production, which has been integral to the fabric of the community since the beginning of last century.

Braid 街轨道交通车站位于菲沙河谷的主要交通枢纽的交会点上，紧贴着其东部的是火车道和横穿加拿大的高速公路。南部是一条连接历史悠久的新敏斯特市和菲沙河谷的主要高速公路。西部的Braid街通向一个成熟居住社区。轨道车站设计为一个通向成长中的社区中心的大门。

最直接的设计构思来自大型的工业建筑，包括卡车运输和木材生产。他们从20世纪初就是当地社区的一部分。

1

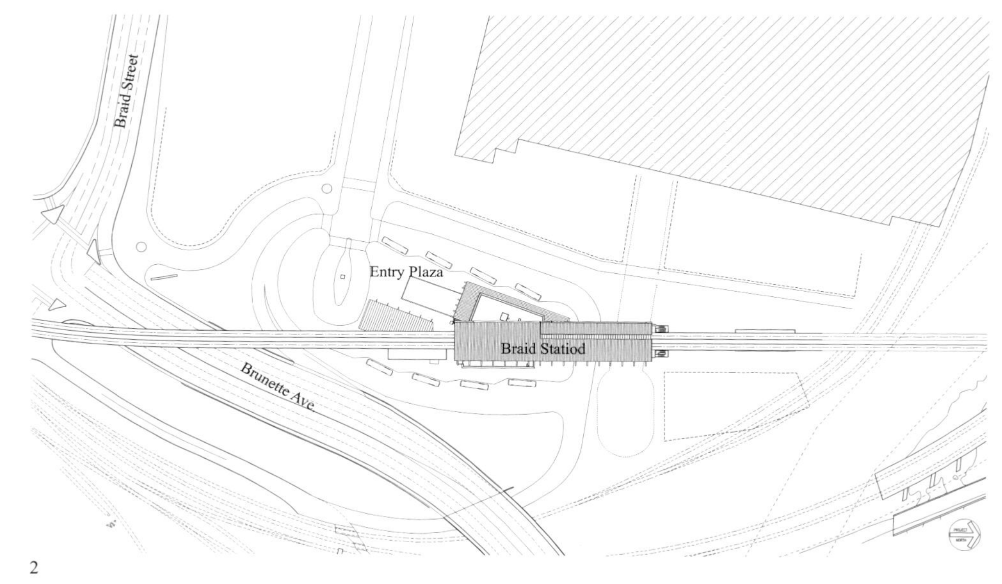

2

1　车站模型
2　总平面图
3　车站全景
4~5　车站夜景

3

4

5

Braid Station is a composition of simple, large planar elements held apart by carefully crafted, painted steel connectors. These large planes and steel supports refer not only to the super scaled industrial buildings around the station, but also to bridges spanning the Fraser River and to the folded plates of the mountains beyond. Roof and wall planes slip by and overlap each other, allowing clear station orientation and visibility while contributing to a variety of spatial experiences.

While glulam wood was employed to recall the regional timber heritage, steel details refer to the neighboring bridges and provide a compelling human scale where they are immediately adjacent to travelers.

The client's ambition was to provide a safe haven for travel. To that end, a significant amount of glazing was employed, which aids visual accessibility to both the concourse and platform levels.

The Concourse form is askew of the guide way orientation in order to more frontally address the new urban plaza to the south. The resulting form is a large welcoming canopy and a significant Concourse level room that are both easily identifiable from the neighboring streets and reminiscent of grand, historical transportation buildings.

Inside, the high volume of the Concourse is open and airy, the space compressing only as passengers ascend to the Platform level to board the Skytrain. At the top landing, the robust details of wood, steel and concrete can be experienced at close hand. Navigating the station is purposefully simplified to aid passengers on their journeys. On the platform, passengers are lifted above the frenetic station context and provided with a view of the valley beyond.

6

7

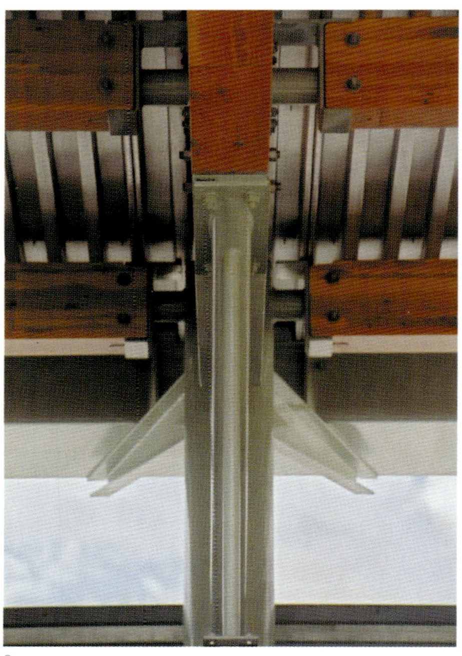

8

6 月台层内景
7 檐口细部
8 屋架细部
9 月台层内景
10 入口室外景观
11 月台层平面图

Braid 街轨道交通车站一系列简单平行的构件由精心地构造和粉刷的连接体连接在一起。这些大型的平板和钢的支撑物不仅仅是对四周巨型的工业化建筑的回应，也是对菲沙河大桥和山脉运动产生的倾斜平原的呼应。屋顶和墙面相对交叉重叠使车站的方位感明确，在视觉空间中十分突出。

木材的运用反映了当地木构件使用的历史沿革，钢构件的细部则呼应周围的桥梁和提供一个近人的空间尺度，可以与旅行者直接沟通。

业主希望车站成为一个旅行者的安全天堂。设计中大量运用玻璃，使得车站大厅层和月台层的视线无阻隔。

为了充分使用南部的新城市广场，大厅设计为一个向南倾斜而下的坡地。巨大的友好的雨篷构造和明显的月台层与周围社区的街道相比尤为突出。即使与体量巨大的历史性交通建筑相比，Braid街轨道交通车站也与众不同。

大体量的车站大厅是开放和通风良好的，只有在乘客由大厅进入月台层上车时才有压迫感。顶层壮硕的木材、钢和混凝土的细部像合起的手掌。简单化的交通流线设计有效地帮助乘客识别行动路线。在月台层，乘客可以离开车站喧闹的一面，有一个安静的区域可以眺望远处的山谷。

9

10

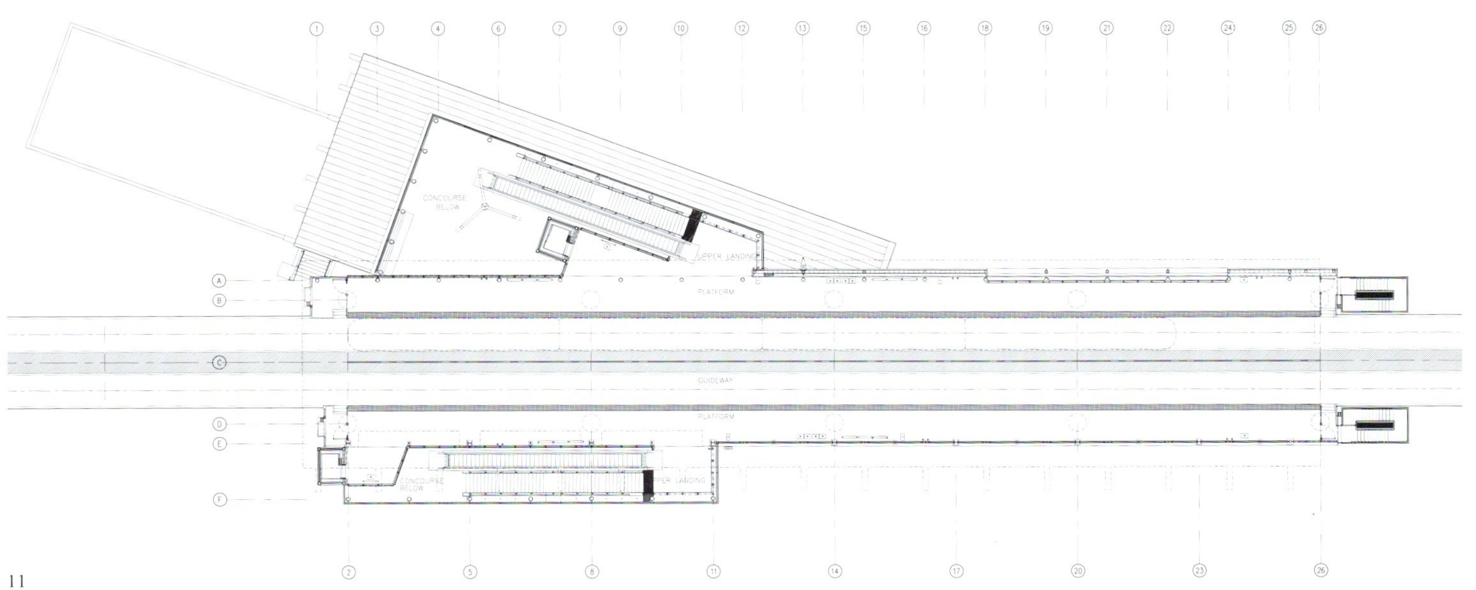

11

12　入口广场
13　地面层平面图
14　车站夜景
15~16　立面图
17　屋顶平面图

12

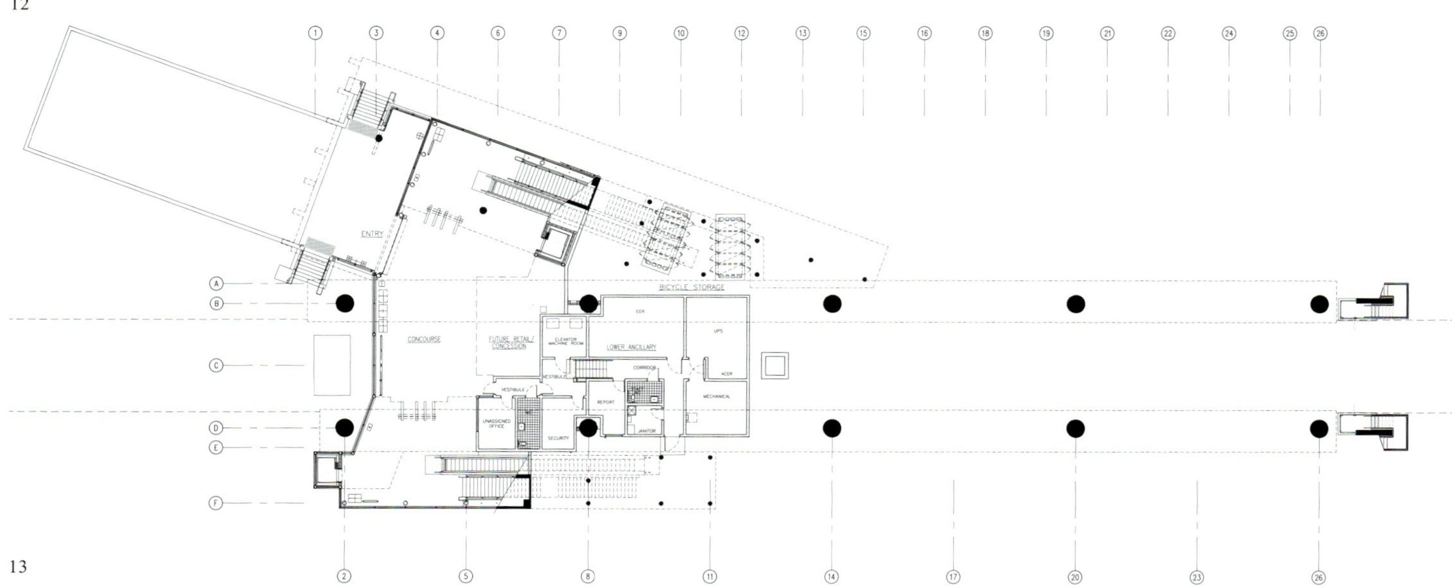

13

14

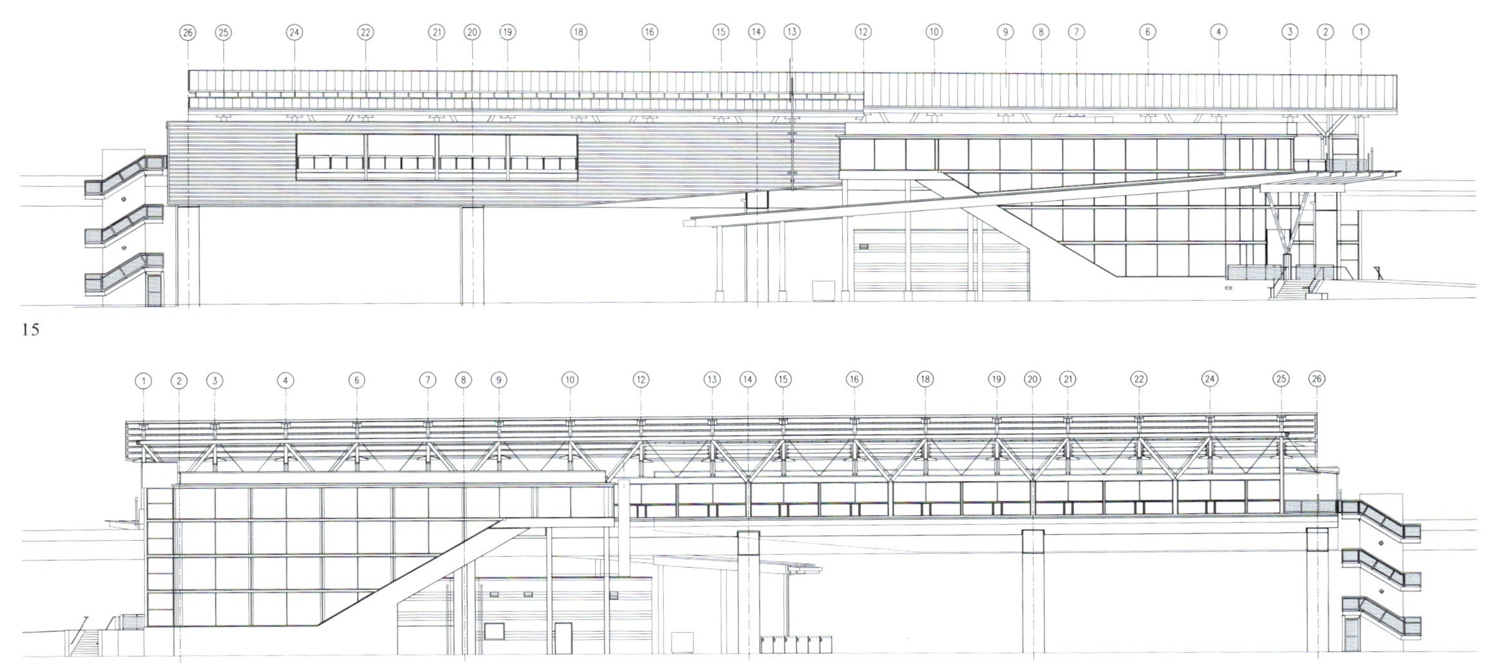

15

16

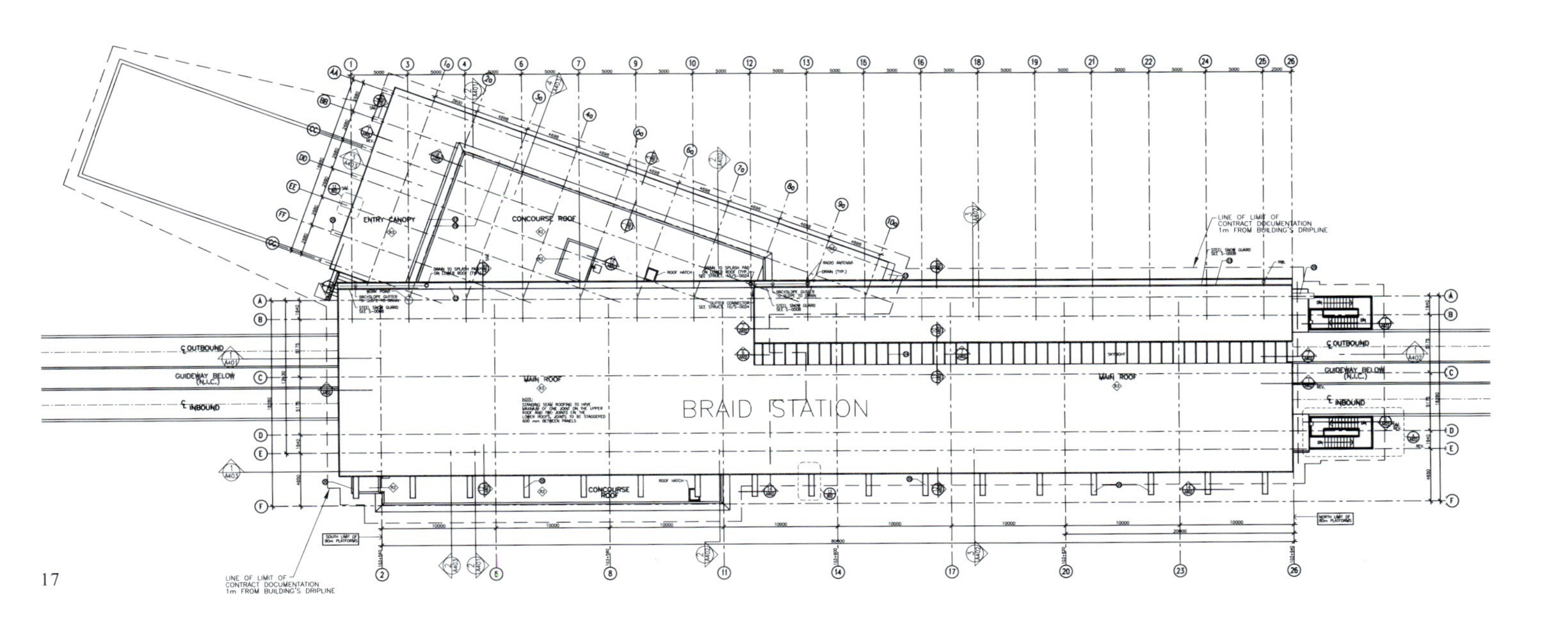
ENTRY CANOPY
CONCOURSE ROOF
MAIN ROOF
BRAID STATION
GUIDEWAY BELOW
CONCOURSE ROOF
LINE OF LIMIT OF CONTRACT DOCUMENTATION 1m FROM BUILDING'S DRIPLINE

17

18 入口细部
19 墙体细部
20 横剖面图
21~22 扶手细部
23 地面图案细部
24 入口处剖面图

Conspicuous, refined connection details express the virtues of the palette of steel, wood and concrete. The overall composition is an efficient balance between long spans and lightweight construction. Connection details add a layer of texture easily appreciated from close range, unfolding as passengers move through the station. Simple repeated structural bays minimize the size of structural members, while creating a pleasant rhythmic order.

Braid Station is designed as a 'post disaster' building. The materials and design are robust enough for that purpose, and will outlast the 50 year intended lifespan of the building. All the major components are easily maintained and, when the time comes, easily recycled.

The station design embraces the need for passenger safety, with visual accessibility provided day and night. The station is designed to minimize areas for potential entrapment or undesired loitering. Easy way finding is also emphasized with a well-defined entry sequence and easily navigated circulation inside. Tile patterning on the Platform level helps guide visually impaired commuters between the elevators and designated waiting areas.

The tender was awarded for 10% under budget. The station design was issued for tender 6 months after the initial concept sketches were started. Station construction took 12 months.

18

19

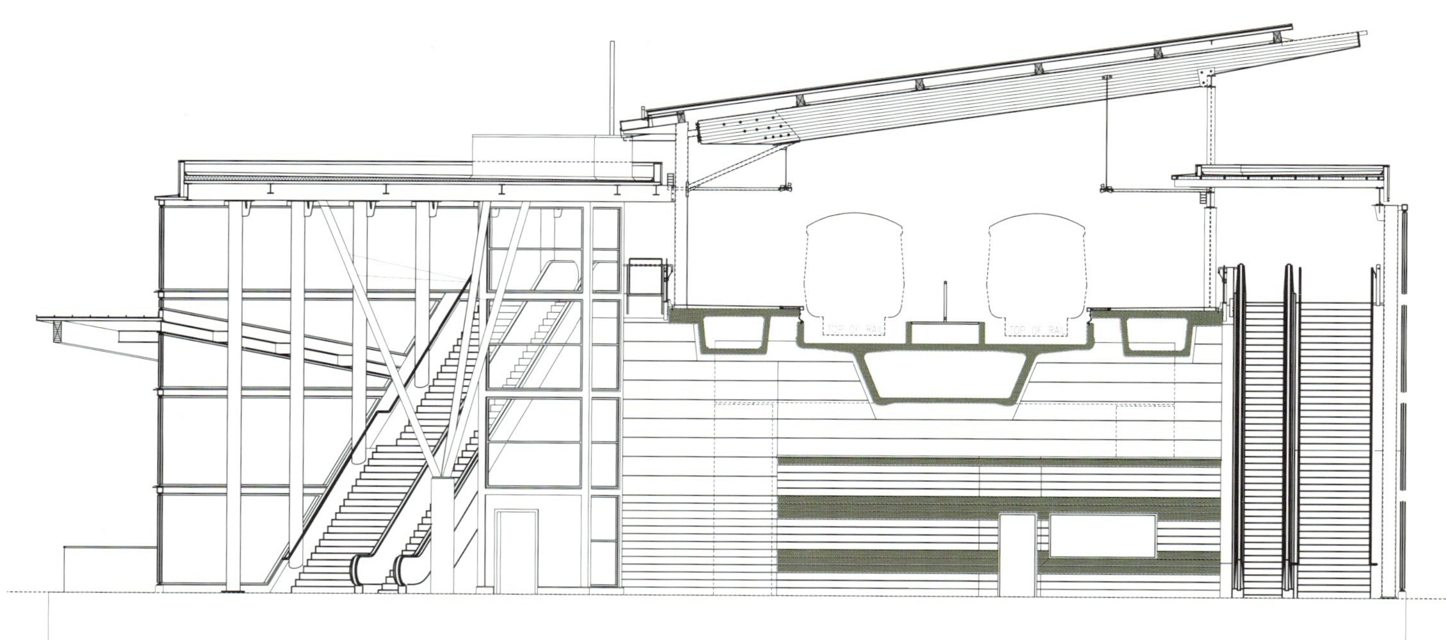

20

出众、精致的细部设计突出了木材、钢和混凝土各自特有的质感。大跨度与轻质的构造的均衡来自整体的良好组合。

连接点的细部设计使平坦的空间平添了精致的质感。简单与重复简化了结构构件，创造出了音乐般的韵律。

Braid街轨道交通车站有抗灾设计。材料和设计有足够的强度，可以满足50年的使用寿命。所有的构件都易于维护和循环使用。

Braid 街轨道交通车站设计强调使用者的安全，在白天和夜晚都有明显的视觉标识。车站设计了小型的逗留区域、一系列显而易见的出入口标志和简单的内部流线。从自动扶梯到等候区的流线也由月台地砖标识出来。

承包商节省了10%的工程造价。从概念设计到施工图完成一共耗时 6 个月，施工耗时 12 个月。

21

22

23

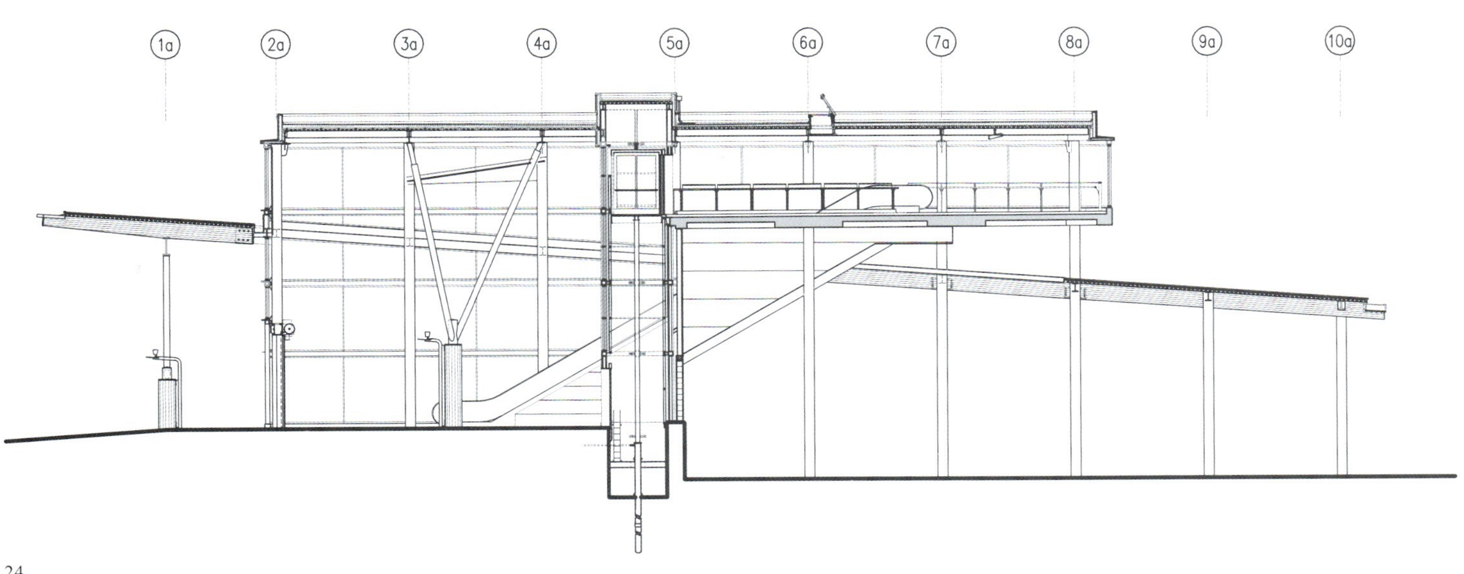

24

Emergency Operations and Communications Training Center (E-COMM)

紧急救援通信培训中心（E-COMM）

地　　点：温哥华
业　　主：温哥华市政府
规　　模：5 600m²
完成时间：1998年
合作建筑师：Ross Drulis Architects and Planners

E-COMM is one of the largest and most advanced facilities of its type in Canada. Stantec Architecture designed the Emergency Operations and Communications Center (E-COMM) to withstand a post-disaster situation and allow for the coordination of a large number of emergency response agencies throughout the Lower Mainland. The Center is a highly program intensive facility with a strong emphasis on internal communication flow and state-of-the-art integrated computer, media, and communications support. The result is a two-part building expression that clearly represents its components while at the same time creating an interesting play of forms.

紧急救援通信培训中心（E-COMM）是加拿大同类型设施中最先进和最大的。Stantec设计的中心是灾难应急建筑，并有应付温哥华低陆平原地区灾后大规模应急通信的能力。紧急救援通信培训中心（E-COMM）的设计强调内部通信和满足对最先进的电脑设施、传媒和通信的支持。紧急救援通信培训中心（E-COMM）的建筑造型有一个鲜明的形体的同时，被清晰地分为两个部分。

1

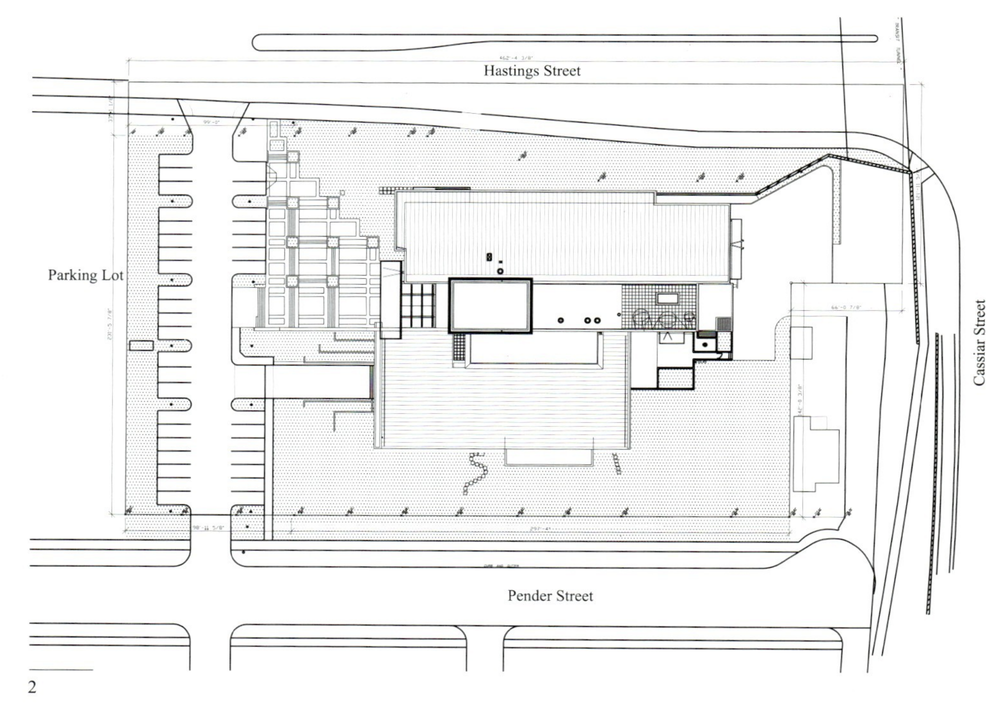

2

3

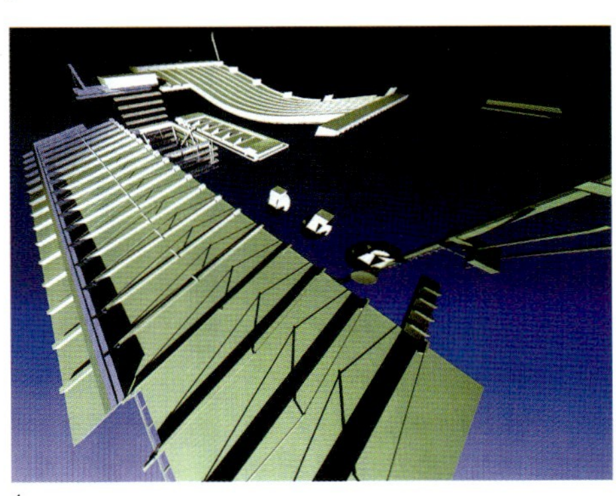

4

5

1 模型
2 总平面图
3 中心夜景
4 屋面构造电脑模型
5 中心全景

6~9　建筑模型
10　南立面图
11　平面图
12　中心夜景

6

7

8

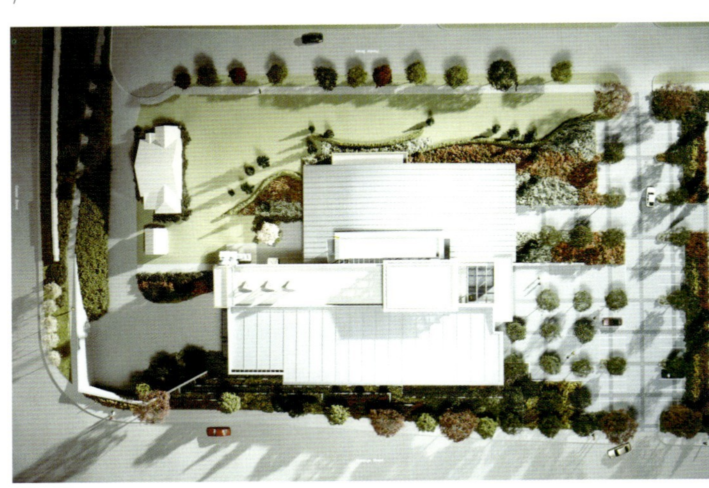
9

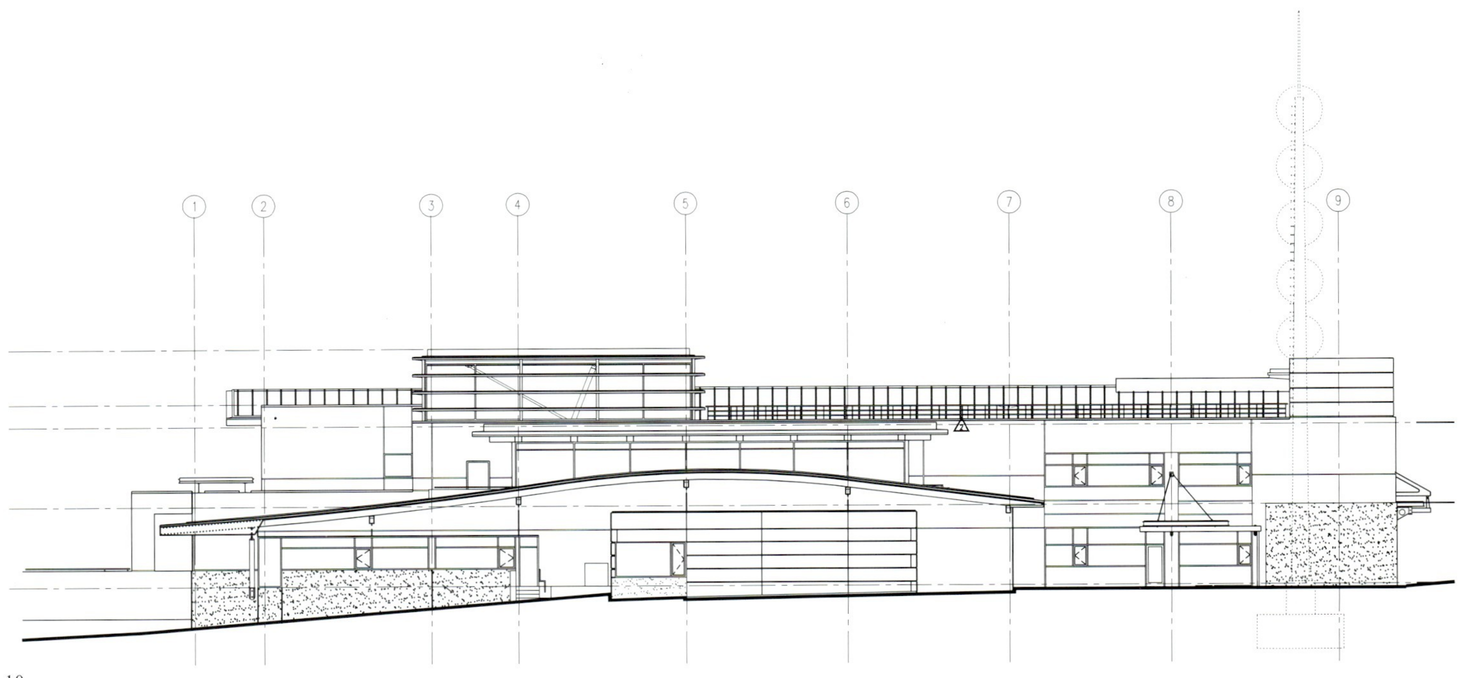

10

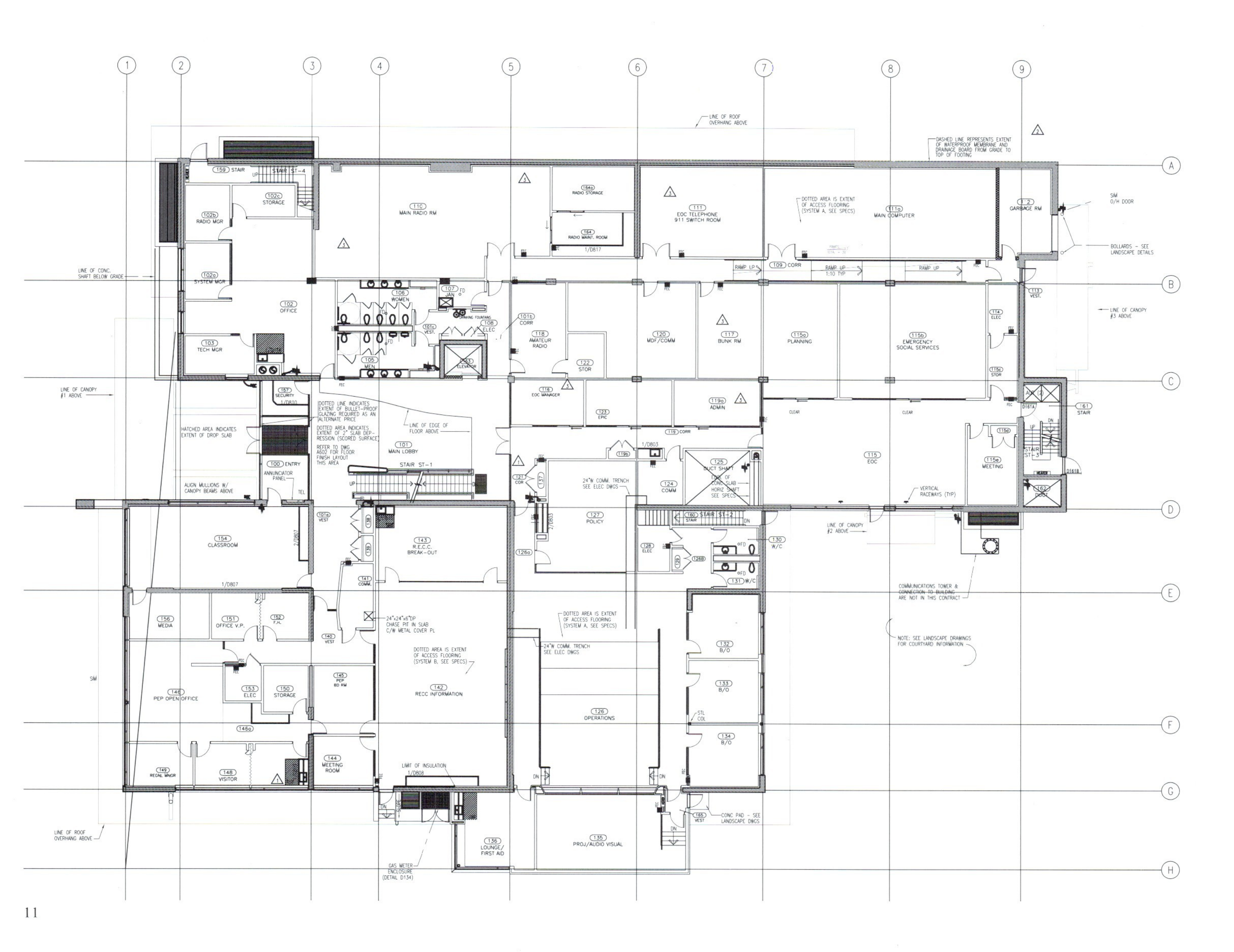

LINE OF ROOF OVERHANG ABOVE
DASHED LINE REPRESENTS EXTENT OF WATERPROOF MEMBRANE AND DRAINAGE BOARD FROM GRADE TO TOP OF FOOTING
STAIR ST-4
STAIR
STORAGE
RADIO MGR
SYSTEM MGR
OFFICE
TECH MGR
MAIN RADIO RM
RADIO STORAGE
RADIO MAINT. ROOM
EOC TELEPHONE 911 SWITCH ROOM
DOTTED AREA IS EXTENT OF ACCESS FLOORING (SYSTEM A, SEE SPECS)
MAIN COMPUTER
GARBAGE RM
O/H DOOR
BOLLARDS - SEE LANDSCAPE DETAILS
LINE OF CONC. SHAFT BELOW GRADE
RAMP UP
CORR
RAMP UP 1:10 TYP
WOMEN
JAN
VEST.
ELEC
MEN
ELEVATOR
CORR
AMATEUR RADIO
STOR
MDF/COMM
BUNK RM
PLANNING
EMERGENCY SOCIAL SERVICES
LINE OF CANOPY #3 ABOVE
LINE OF CANOPY #1 ABOVE
SECURITY
EOC MANAGER
EPIC
ADMIN
CLEAR
EOC
STAIR
HATCHED AREA INDICATES EXTENT OF DROP SLAB
LINE OF EDGE OF FLOOR ABOVE
MAIN LOBBY
STAIR ST-1
ENTRY
ANNUNCIATOR PANEL
TEL
ALIGN MULLIONS W/ CANOPY BEAMS ABOVE
24"W COMM. TRENCH SEE ELEC DWGS
DUCT SHAFT
COMM
POLICY
MEETING
VERTICAL RACEWAYS (TYP)
STAIR ST-2
LINE OF CANOPY #2 ABOVE
W/C
CLASSROOM
R.E.C.C. BREAK-OUT
COMMUNICATIONS TOWER & CONNECTION TO BUILDING ARE NOT IN THIS CONTRACT
MEDIA
OFFICE V.P.
24"x24"x6"DP CHASE PIT IN SLAB C/W METAL COVER PL
DOTTED AREA IS EXTENT OF ACCESS FLOORING (SYSTEM A, SEE SPECS)
NOTE: SEE LANDSCAPE DRAWINGS FOR COURTYARD INFORMATION
24"W COMM. TRENCH SEE ELEC DWGS
B/O
DOTTED AREA IS EXTENT OF ACCESS FLOORING (SYSTEM B, SEE SPECS)
PEP OPEN OFFICE
ELEC
STORAGE
RECC INFORMATION
OPERATIONS
MEETING ROOM
VISITOR
LIMIT OF INSULATION
VEST
CONC PAD - SEE LANDSCAPE DWGS
LINE OF ROOF OVERHANG ABOVE
LOUNGE/ FIRST AID
PROJ/AUDIO VISUAL
GAS METER ENCLOSURE (DETAIL D134)

11

12

E-Comm

13

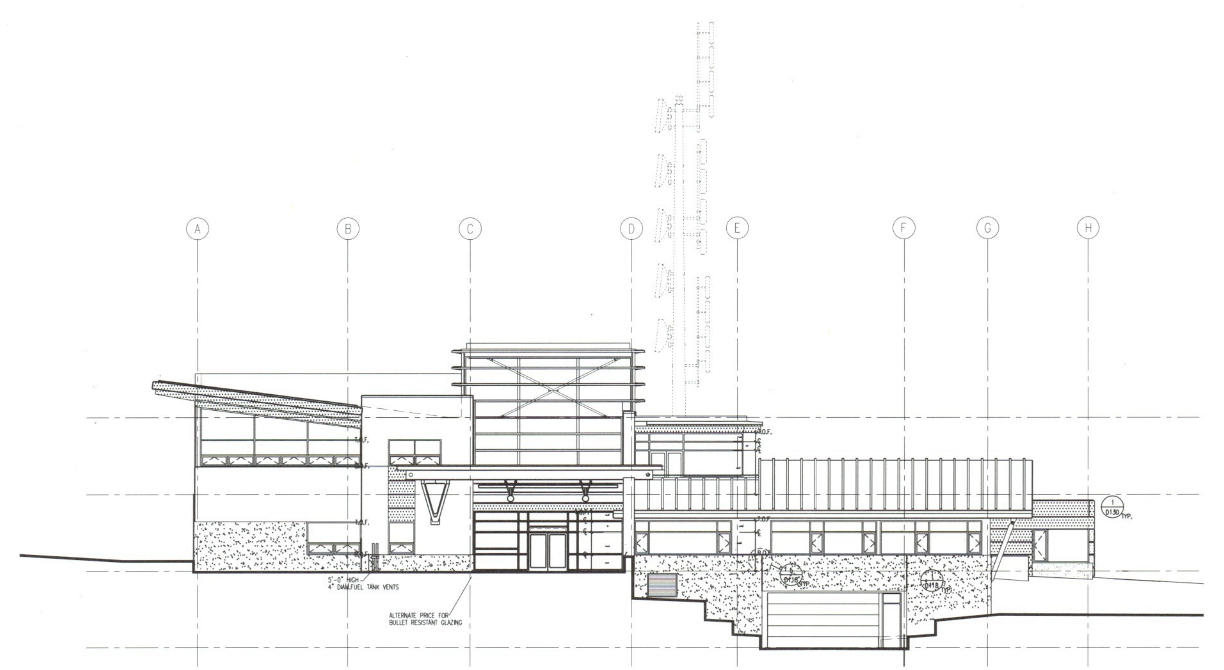
A
B
C
D
E
F
G
H
5'-0" HIGH
4" DIAM. FUEL TANK VENTS
ALTERNATE PRICE FOR
BULLET RESISTANT GLAZING

14

13　中心建筑全景
14　西立面图
15　建筑入口
16　计算机室
17~18　入口细部

15

16

17

18

19

20

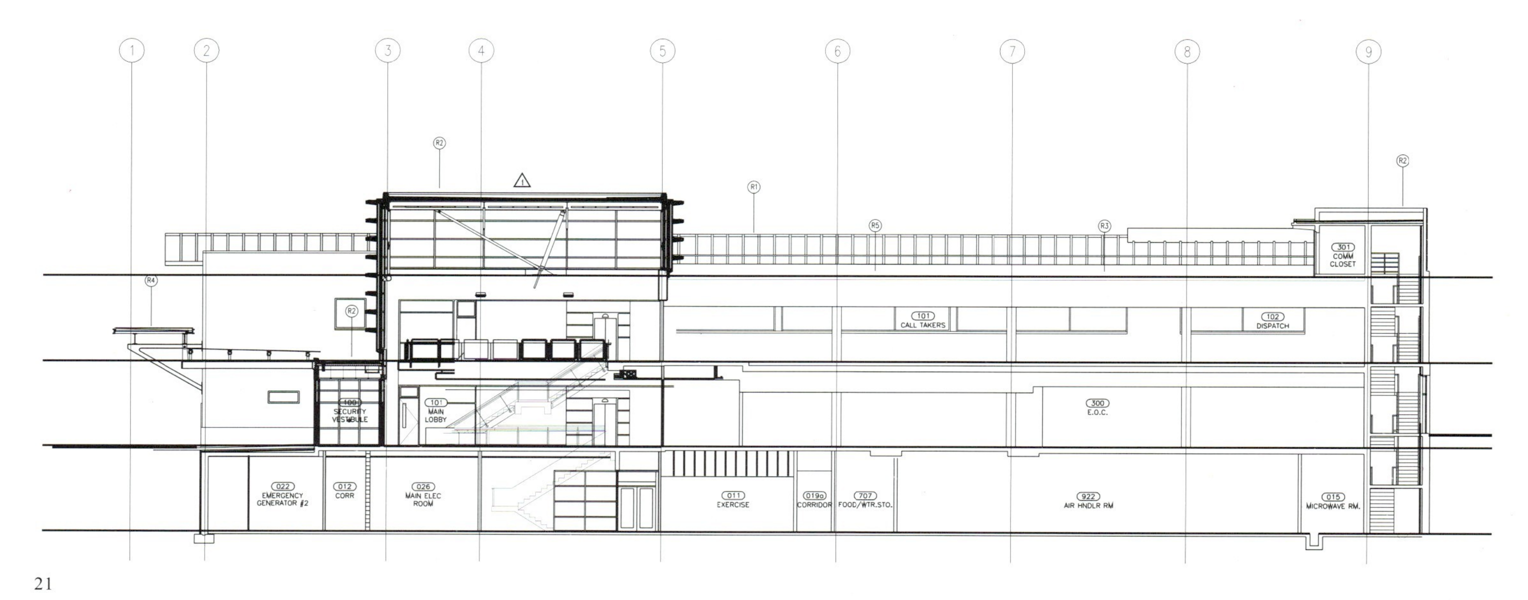

301 COMM CLOSET
101 CALL TAKERS
102 DISPATCH
100 SECURITY VESTIBULE
101 MAIN LOBBY
300 E.O.C.
022 EMERGENCY GENERATOR #2
012 CORR
026 MAIN ELEC ROOM
011 EXERCISE
019a CORRIDOR
707 FOOD/WTR.STO.
922 AIR HNDLR RM
015 MICROWAVE RM.

21

22

19 建筑外观
20 建筑室内细部
21~22 剖面图
23~24 剖面详图

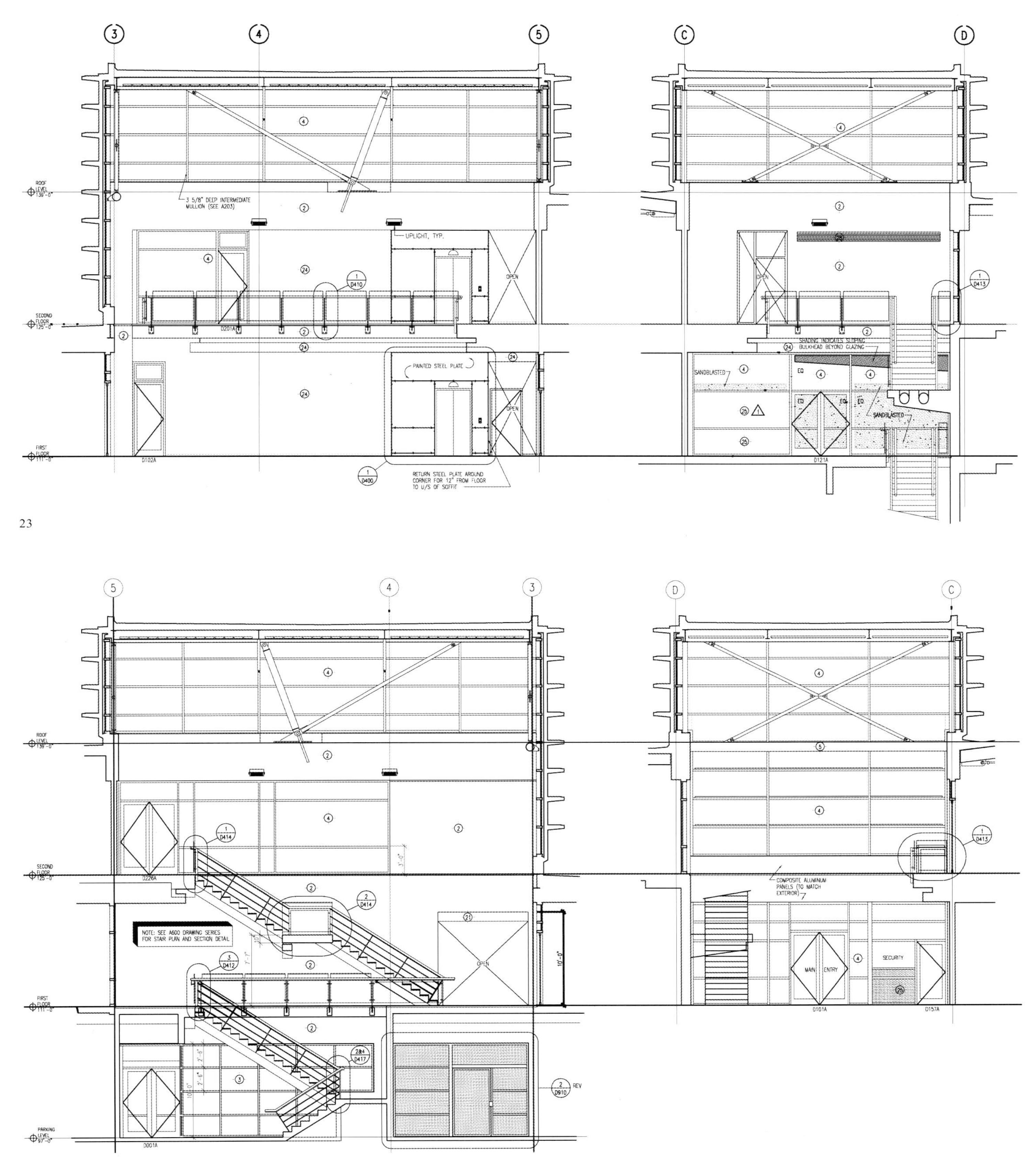

23

24

25

26

27

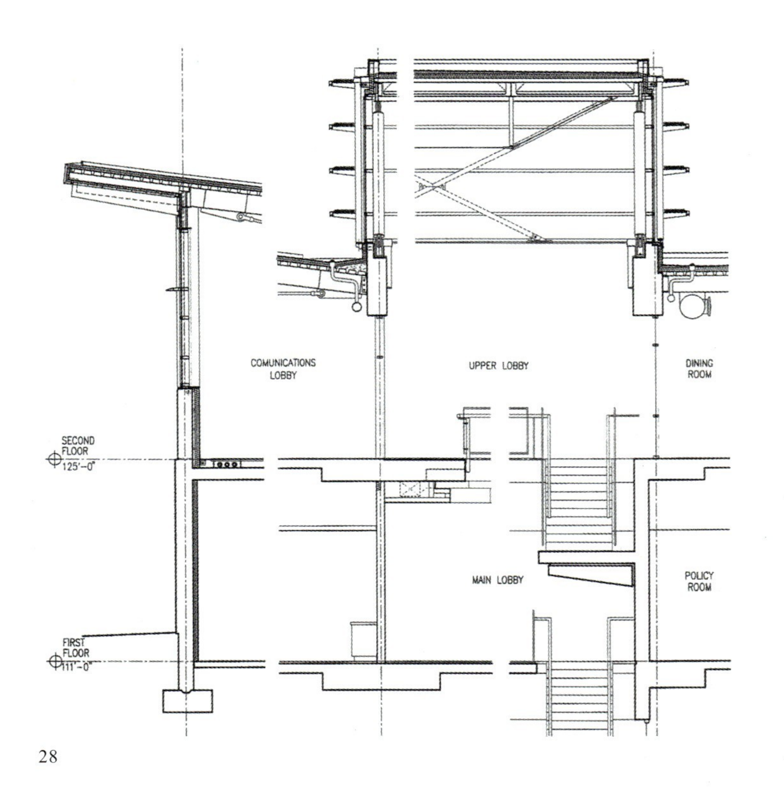

28

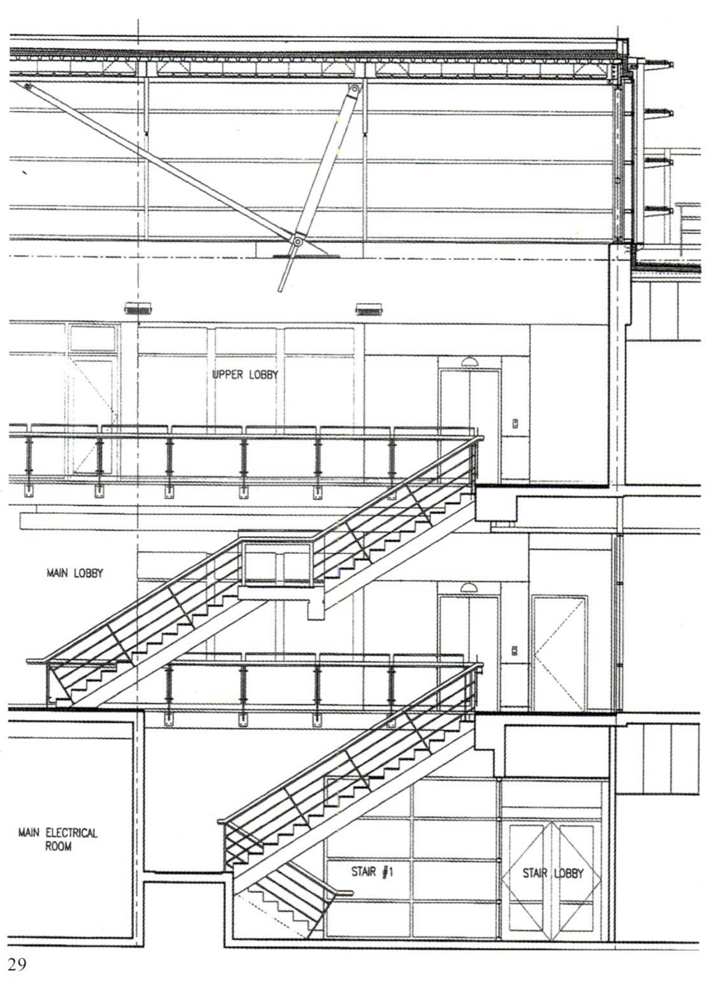

29

25~27　楼梯细部
28　剖面详图
29　构造详图
30~31　墙面细部
32~33　剖面详图

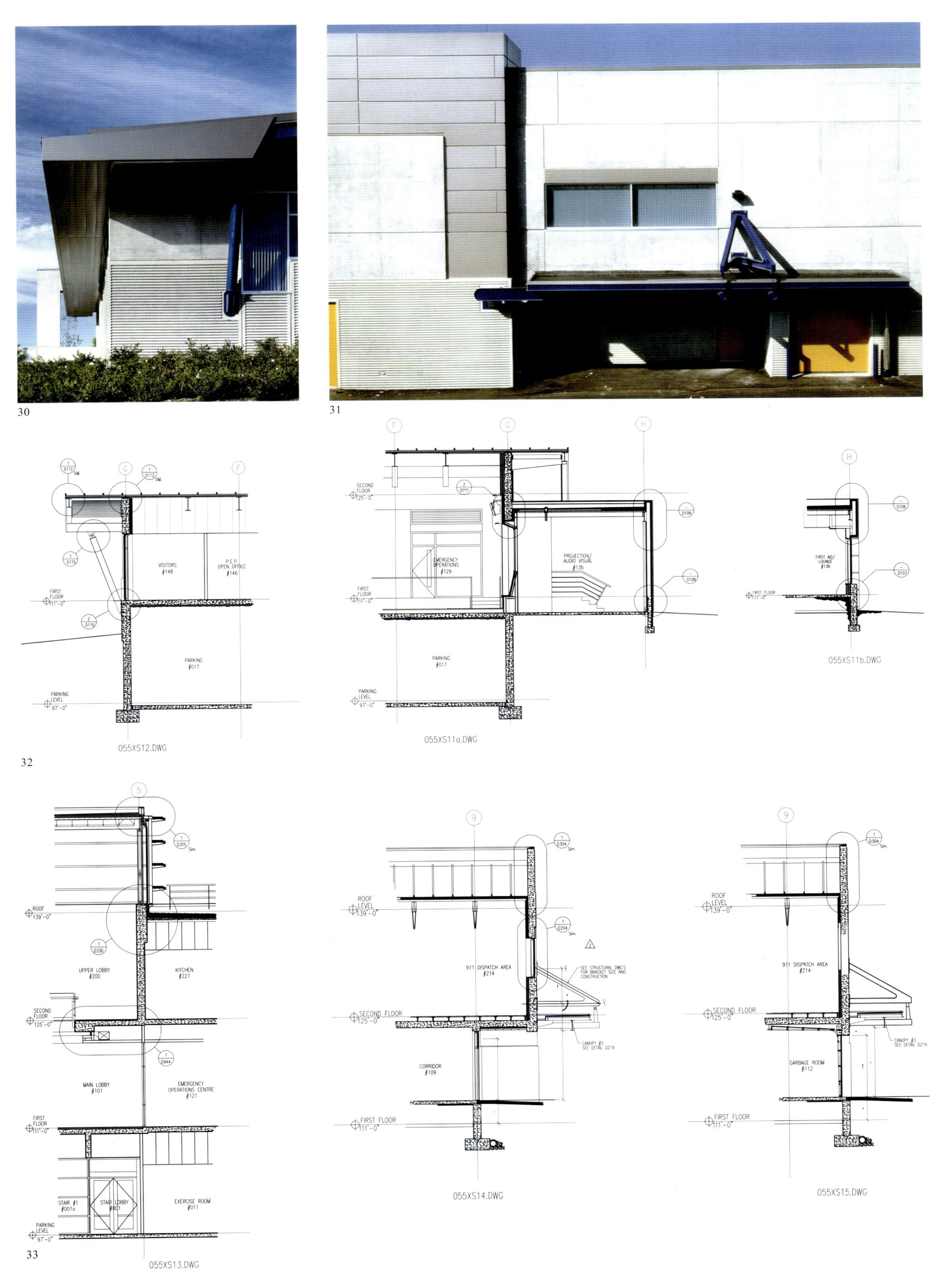

SECOND FLOOR
125'-0"
EMERGENCY OPERATIONS
#129
PROJECTION/ AUDIO VISUAL
#135
FIRST FLOOR
111'-0"
PARKING
#017
PARKING LEVEL
97'-0"
055XS11a.DWG
FIRST AID/ LOUNGE
#136
FIRST FLOOR
111'-0"
055XS11b.DWG
VISITORS
#148
P.E.P. OPEN OFFICE
#146
FIRST FLOOR
111'-0"
PARKING
#017
PARKING LEVEL
97'-0"
055XS12.DWG
ROOF
139'-0"
UPPER LOBBY
#200
KITCHEN
#227
SECOND FLOOR
125'-0"
MAIN LOBBY
#101
EMERGENCY OPERATIONS CENTRE
#121
FIRST FLOOR
111'-0"
STAIR #1
#001a
STAIR LOBBY
#001
EXERCISE ROOM
#011
PARKING LEVEL
97'-0"
055XS13.DWG
ROOF LEVEL
139'-0"
911 DISPATCH AREA
#214
SEE STRUCTURAL DWG'S FOR BRACKET SIZE AND CONSTRUCTION
SECOND FLOOR
125'-0"
CANOPY #3
SEE DETAIL D219
CORRIDOR
#109
FIRST FLOOR
111'-0"
055XS14.DWG
ROOF LEVEL
139'-0"
911 DISPATCH AREA
#214
SECOND FLOOR
125'-0"
CANOPY #3
SEE DETAIL D219
GARBAGE ROOM
#112
FIRST FLOOR
111'-0"
055XS15.DWG

30

31

32

33

401 Burrard Building
Burrard 街 401 号大厦

业　　　主：Canada Lands. Co Ltd.
地　　　点：温哥华
规　　　模：20 000m²
完成时间：2002 年

The 401 Burrard Building is a 19 story office building located in the core of downtown Vancouver's financial district. The site is afforded prominent visibility from all directions, enjoys dramatic views to the Burrard Inlet water front and the north Shore Mountains beyond, and is adjacent to several significant landmark buildings, most notably the art deco Marine Building immediately to the north.

The primary objective for the project was to provide a commercially viable, highly attractive triple "A" office development with 215,000 square feet of net rental area. Specific design objectives were: To create a distinct architectural expression and site development responsive to the neighboring commercial office tower precinct. Design references to the noteworthy Custom House Building, which previously occupied the site, are integral to the creation of a contextually responsive and unique architectural expression that neither competes with nor overpowers the adjacent landmark Marine Building;

To enhance the public pedestrian realm and streetscape by providing generous, visible and accessible spaces for people to gather sit, eat, and watch. Clearly identifiable approaches and entries to the, building from all directions are essential to providing maximum accessibility to the public;

To reinforce pedestrian linkages to neighboring sites and streetscapes on all frontages of the site with;

An open raised public plaza, broad stair and primary building entries on the Burrard Street frontage;

A weather protected pedestrian arcade structure along Pender Street;

A strong commercial street wall structure along Hastings Street;

A landscaped pedestrian plaza and access points to the building lobby fronting Oceanic Plaza to the west.

To optimize and enhance views to and from the building.

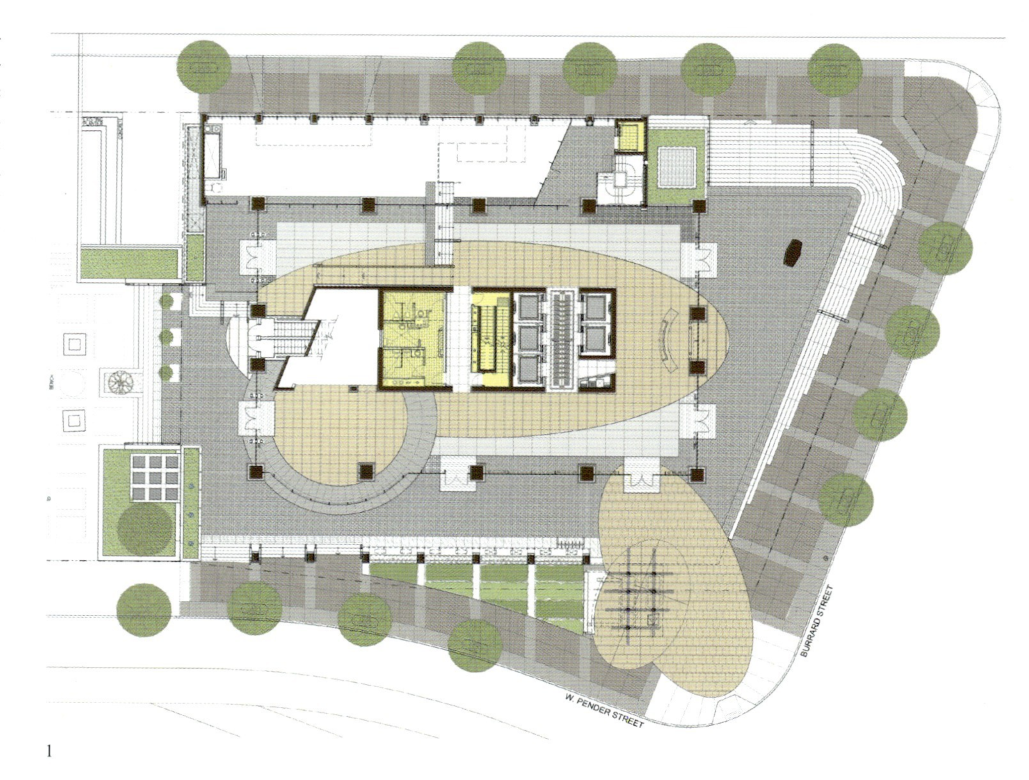

1

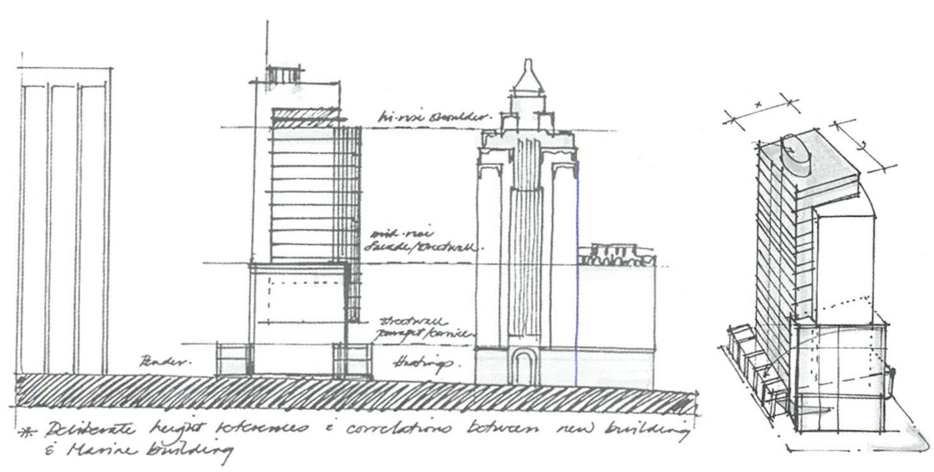

2

Burrard街401号大厦是一栋独特的19层办公建筑，位于温哥华市中心商业区。优越的地理位置，提供给使用者瞭望Burrard湾和北岸山脉的良好景观视角。Burrard街401号大厦紧贴着温哥华市中心的几个重要地标，最显著的是北部具有历史纪念意义的海运大厦。

Burrard街401号大厦的主要任务是提供20 000m²可出租的三A级办公空间。其主要设计理念有：

创造一个出众的可以与周围的商业办公区域共处的建筑物。建筑形态上与众不同，借鉴同一区域内的海运大厦，却不显示过大的体量以压迫海运大厦。

通过提供真正可见可进入的空间，增加公共步行区域、协调街道尺度。创造公共聚集、闲坐、餐饮的场所。各个角度都有清晰可见的标志使公众可以最大程度地亲近建筑。

加强与街区内同其他一切建筑因素的步行连接。

一个开放的、抬高的公共广场，宽大的阶梯和主要出入口位于Burrard街的正面。

一个有上盖的走廊位于Pender街的方向。

建筑意味强烈的玻璃墙位于建筑物沿Hasting街的一侧。

一个景观步行广场连接大厦的西部大堂入口和西面的海洋大楼。

优化建筑物的形态，提供更多有景观的使用空间。

3

1　总平面图
2　构思草图
3　大厦沿街景观

This project is responsive to current and future community planning goals and exemplifies a high quality of architectural design, sensitivity to its urban context, and a significant enhancement of the public pedestrian realm.

Employing a fully integrated design approach, 401 Burrard has also been designed to reduce its overall impact on the environment; use water and energy more efficiently, protect worker and occupant health, and improve employee productivity.

Burrard街401号大厦考虑到现在和将来的城市规划目标，以高质量的设计去适应城市敏感的结构，并扩展了公共步行的区域。

Burrard街401号大厦的设计同样也减少了建筑对环境的冲击，更有效地利用水、能源，关注使用者的健康，提高使用者的工作效率。

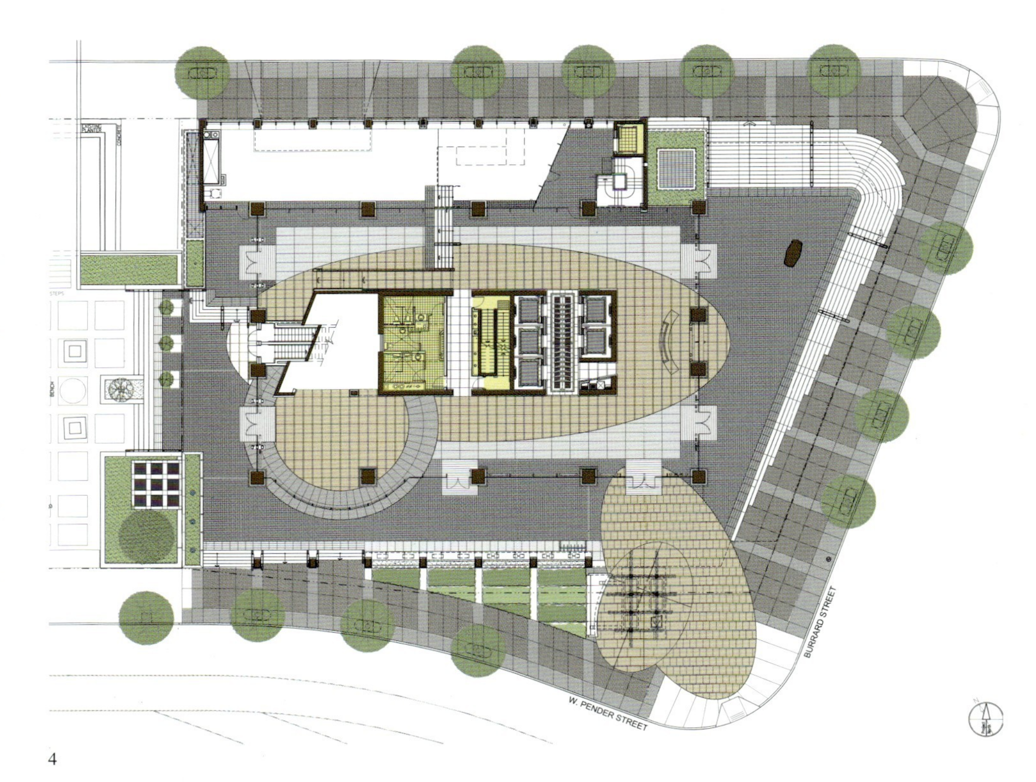

4

5

6

4　首层平面
5~6　大堂细部
7　精致的玻璃幕墙
8　标准层平面图
9　剖面图

7

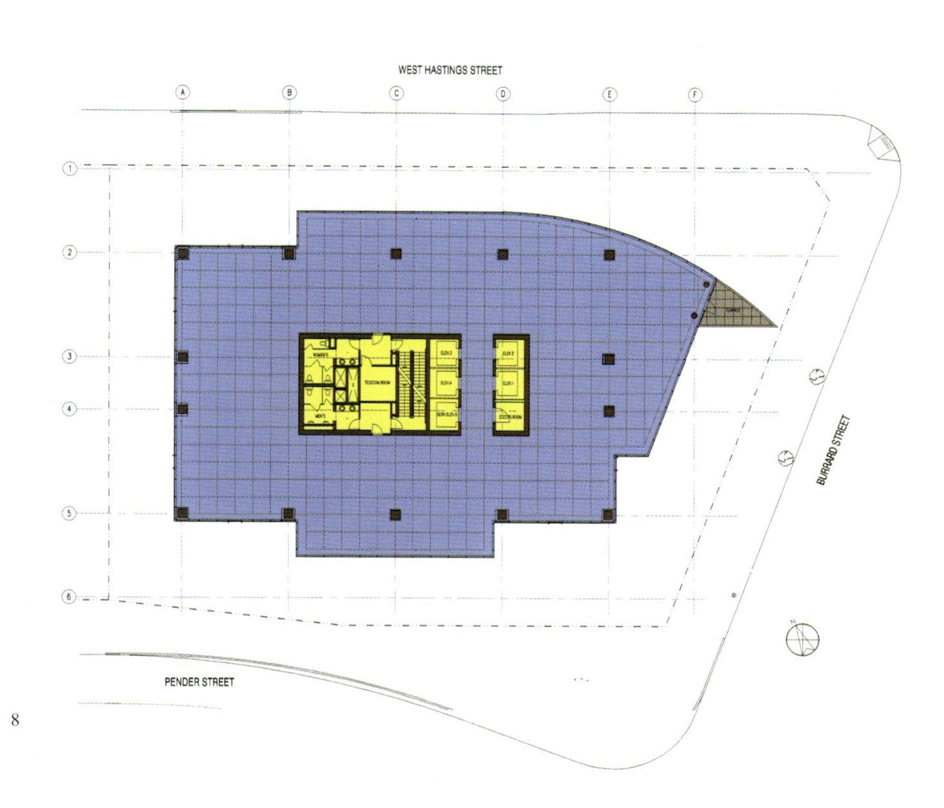

8

9

10　仰视大厦
11　大厦入口
12　立面图
13~16,18~20　雨篷详图
17　玻璃雨篷细部

10

11

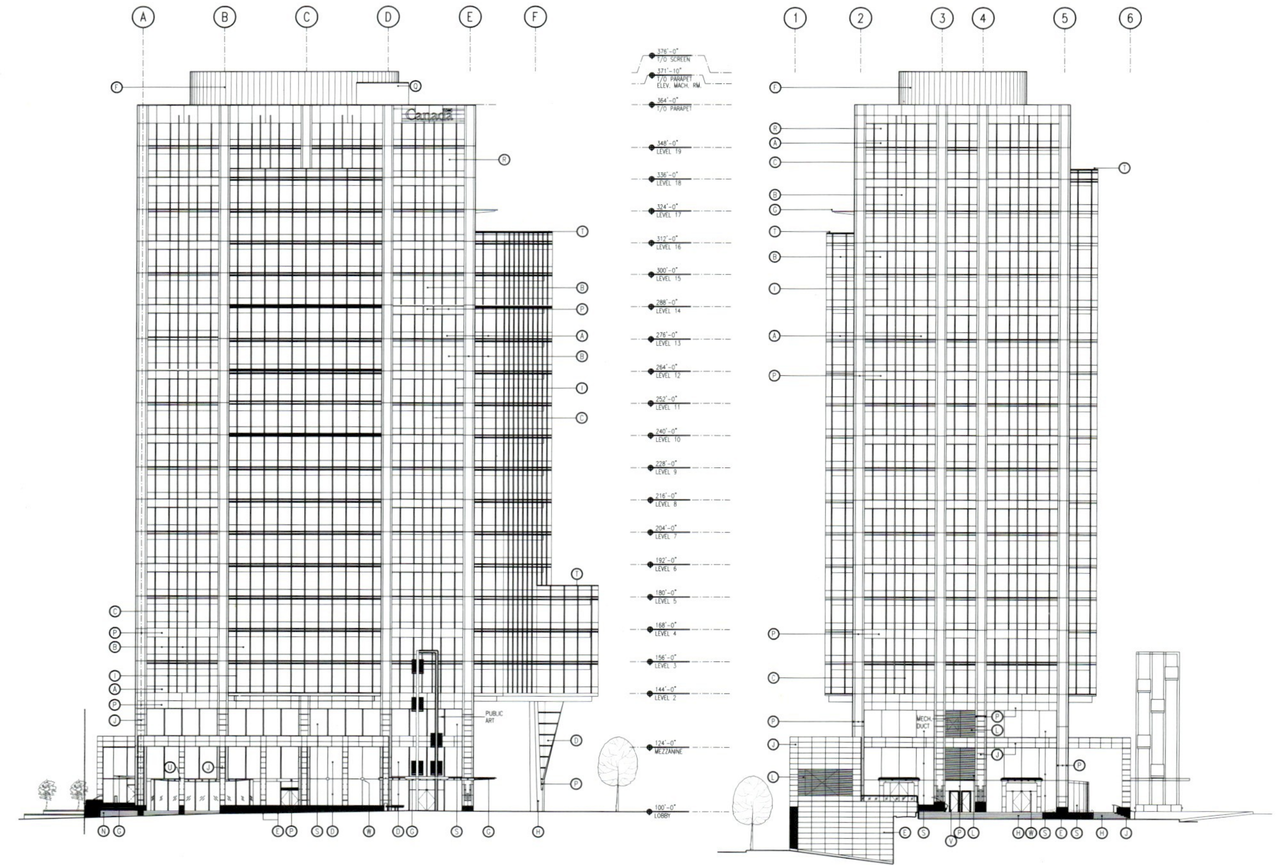

12

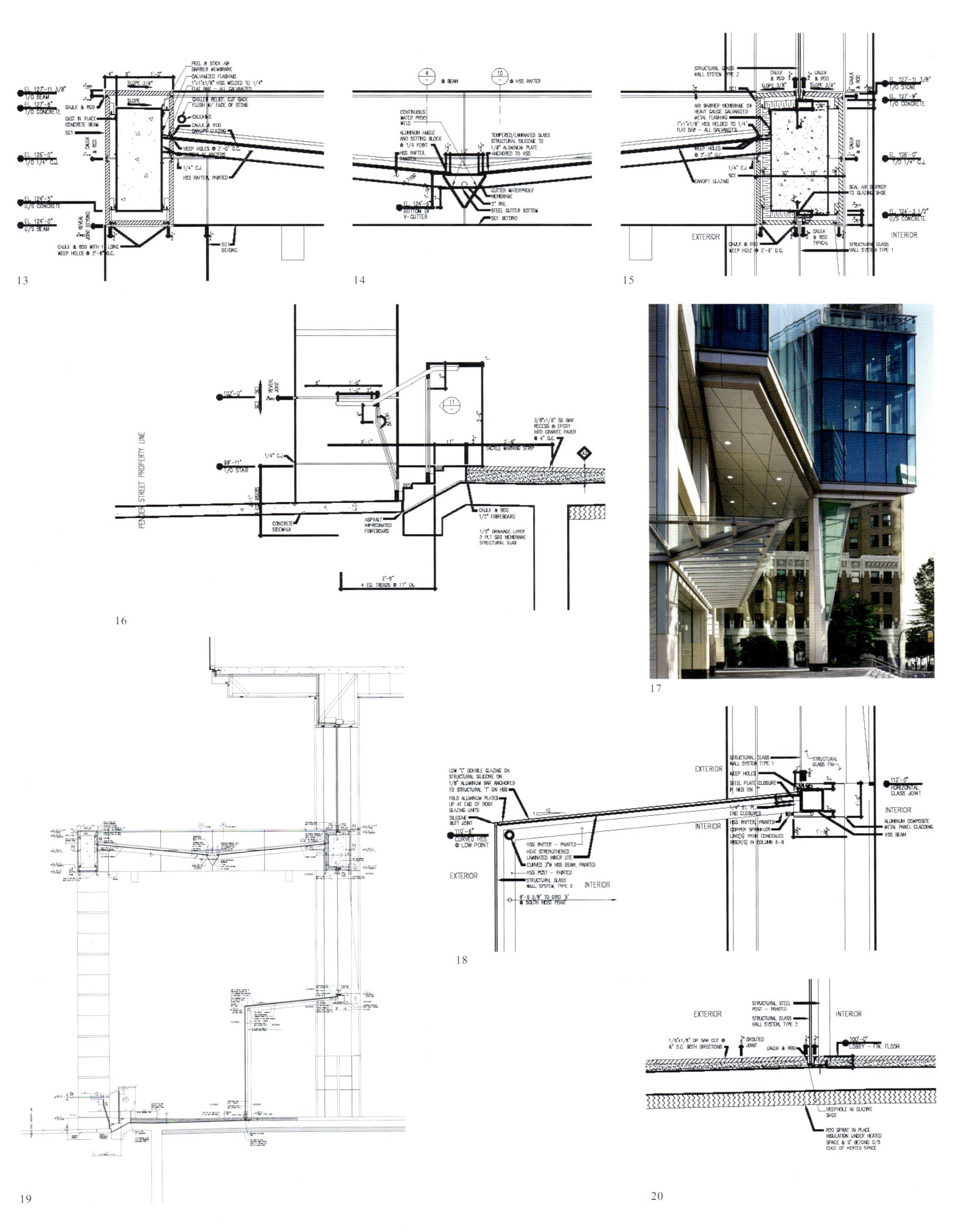

13

14

15

16

17

18

19

20

21

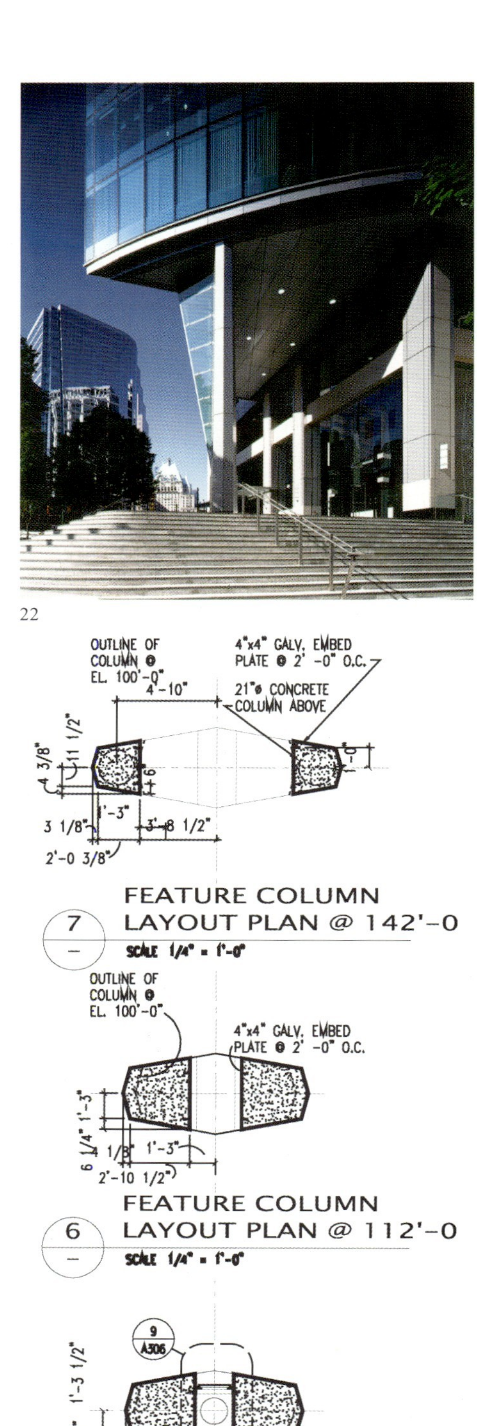

OUTLINE OF COLUMN @ EL. 100'-0"
4"x4" GALV. EMBED PLATE @ 2' -0" O.C.
21"ø CONCRETE COLUMN ABOVE
4'-10"
11 1/2"
4 3/8"
1'-3"
3 1/8"
8 1/2"
2'-0 3/8"
FEATURE COLUMN
LAYOUT PLAN @ 142'-0
7
SCALE 1/4" = 1'-0"
OUTLINE OF COLUMN @ EL. 100'-0"
4"x4" GALV. EMBED PLATE @ 2' -0" O.C.
1'-3"
1'-3"
2'-10 1/2"
FEATURE COLUMN
LAYOUT PLAN @ 112'-0
6
SCALE 1/4" = 1'-0"
9
A306
1'-3 1/2"
6 1/2"
4 1/4"
2'-11 7/8"
FEATURE COLUMN
LAYOUT PLAN @ 108'-0
5
SCALE 1/4" = 1'-0"

22

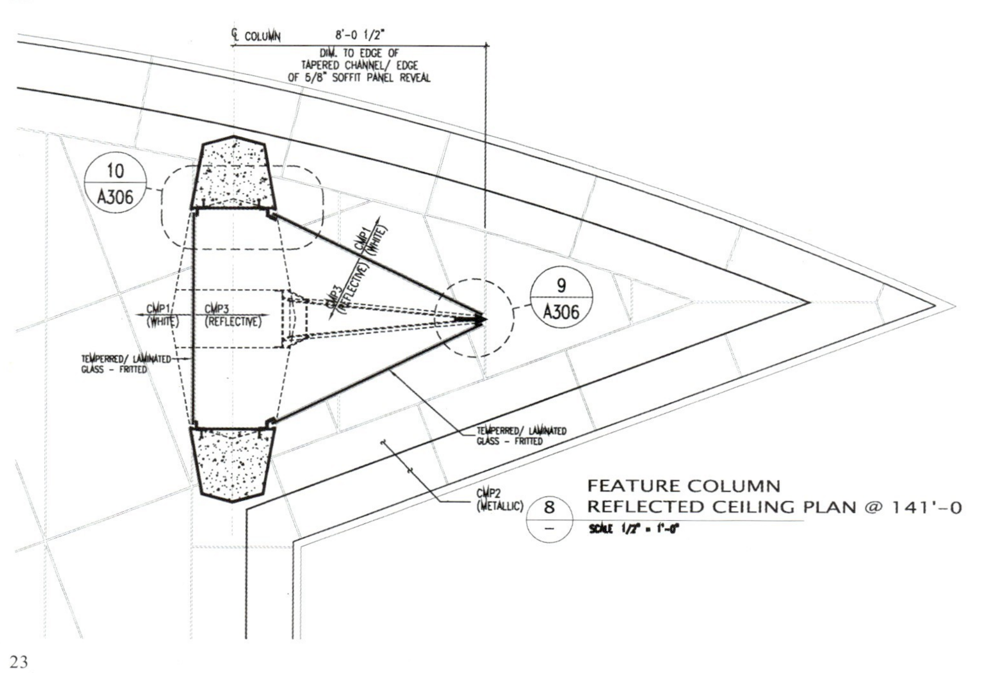

℄ COLUMN
8'-0 1/2"
DIM. TO EDGE OF TAPERED CHANNEL/ EDGE OF 5/8" SOFFIT PANEL REVEAL
10
A306
9
A306
CMP1 (WHITE)
CMP3 (REFLECTIVE)
TEMPERRED/ LAMINATED GLASS - FRITTED
CMP2 (METALLIC)
FEATURE COLUMN
REFLECTED CEILING PLAN @ 141'-0
8
SCALE 1/2" = 1'-0"

23

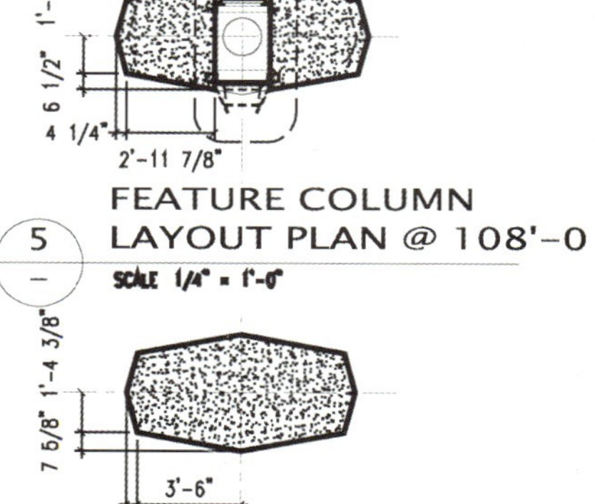

1'-4 3/8"
7 5/8"
3'-6"
4 1/2"
FEATURE COLUMN
LAYOUT PLAN @ 100'-0
2
SCALE 1/4" = 1'-0"

24

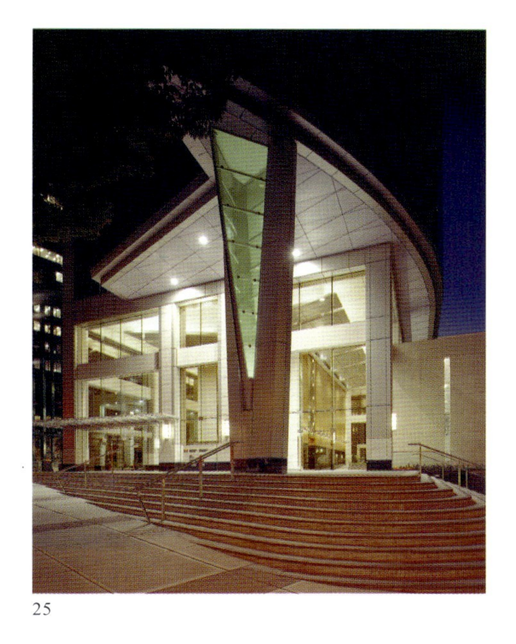

25

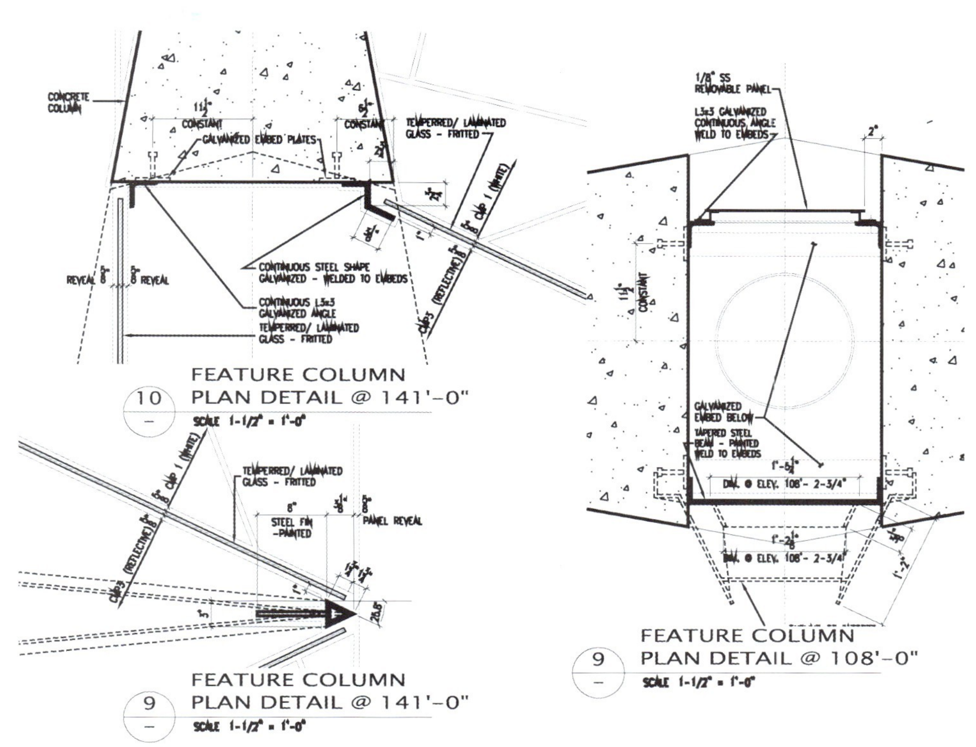

26

21~22　入口细部
23　入口顶棚平面详图
24　柱截面详图
25　入口夜景
26　V形柱构造详图
27　V形柱立面详图

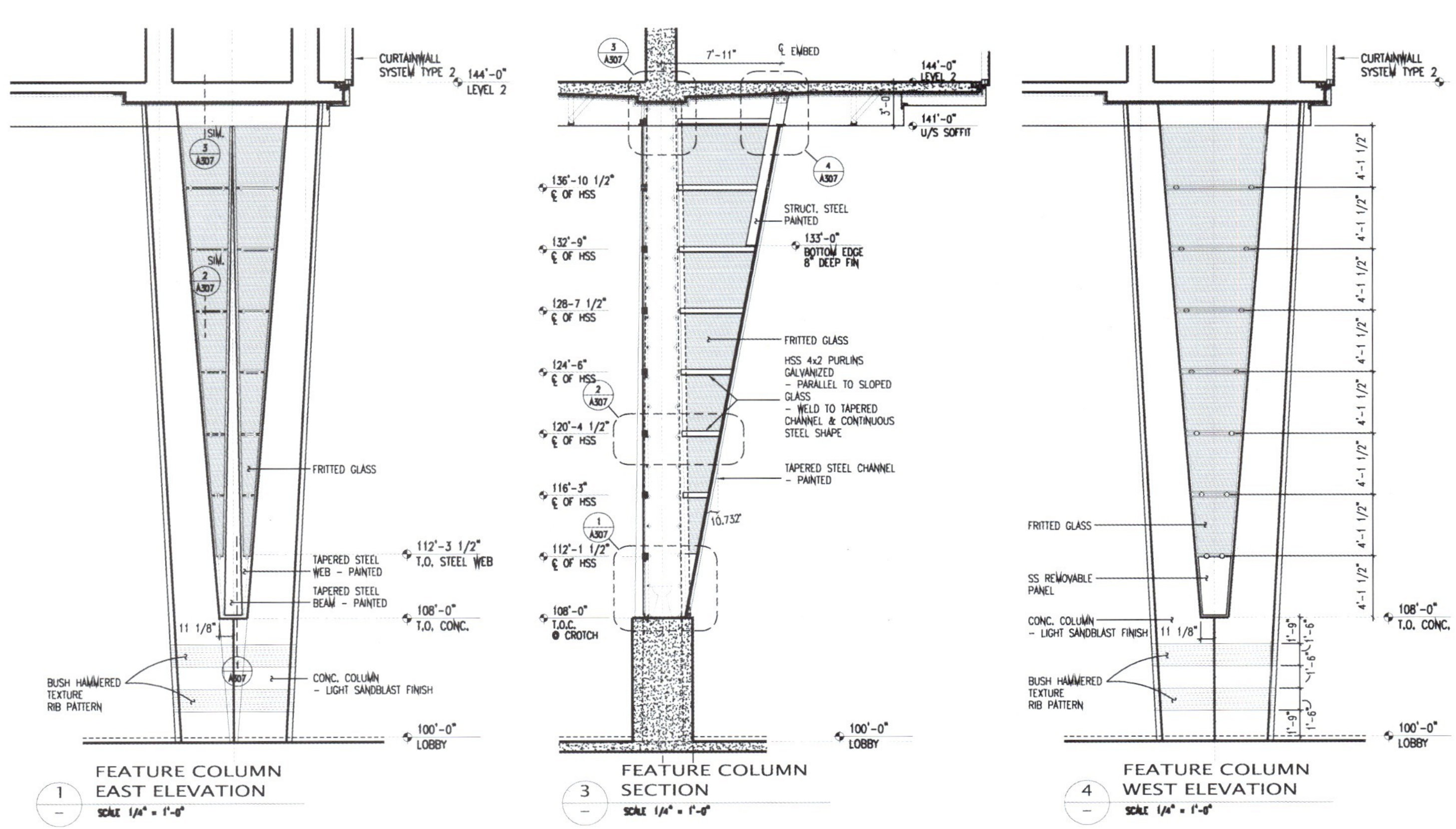

27

Liu Center for the Study of Global Issues
刘氏国际问题研究中心

业　　主：University of British Columbia
地　　点：温哥华
规　　模：1 750m²
完工时间：2000年
合作建筑师：亚瑟·埃里克森

The Liu Center has been successful in meeting a rigorous 'green' agenda. The Liu Center has two distinct elements: an office, research and administration wing for 72 participants; and a meeting wing including a case room, multi-purpose room and two seminar rooms. A rare Kadsura tree stands in the entry courtyard to the north, while the south courtyard opens to a Zen garden and the forest backdrop beyond. The Liu Center has been committed to a rigorous 'green' agenda: it was constructed within the footprint of its predecessor and makes extensive use of materials from that carefully deconstructed building; and it is sited with sensitivity to the quality of indoor light and its natural forest setting.

刘氏国际问题研究中心成功地达到了加拿大严谨的"绿色"建筑的要求。刘氏中心的建筑分为两个明显的部分：一个办公区，研究和接待侧翼有72个小间，和一个会议区，包括一个档案室、多功能厅和两个教室。一株稀有的南五味子树位于北侧出入口花园内，南部的院落面向一个禅宗花园，更远处是树林。刘氏中心设计是要求十分苛刻的"绿色"建筑：新建筑物的材质与原有建筑一致，尽量使用被小心的从旧建筑上拆下来的建筑材料，并精心设计房间采光和位于天然树林之中的场地。

1

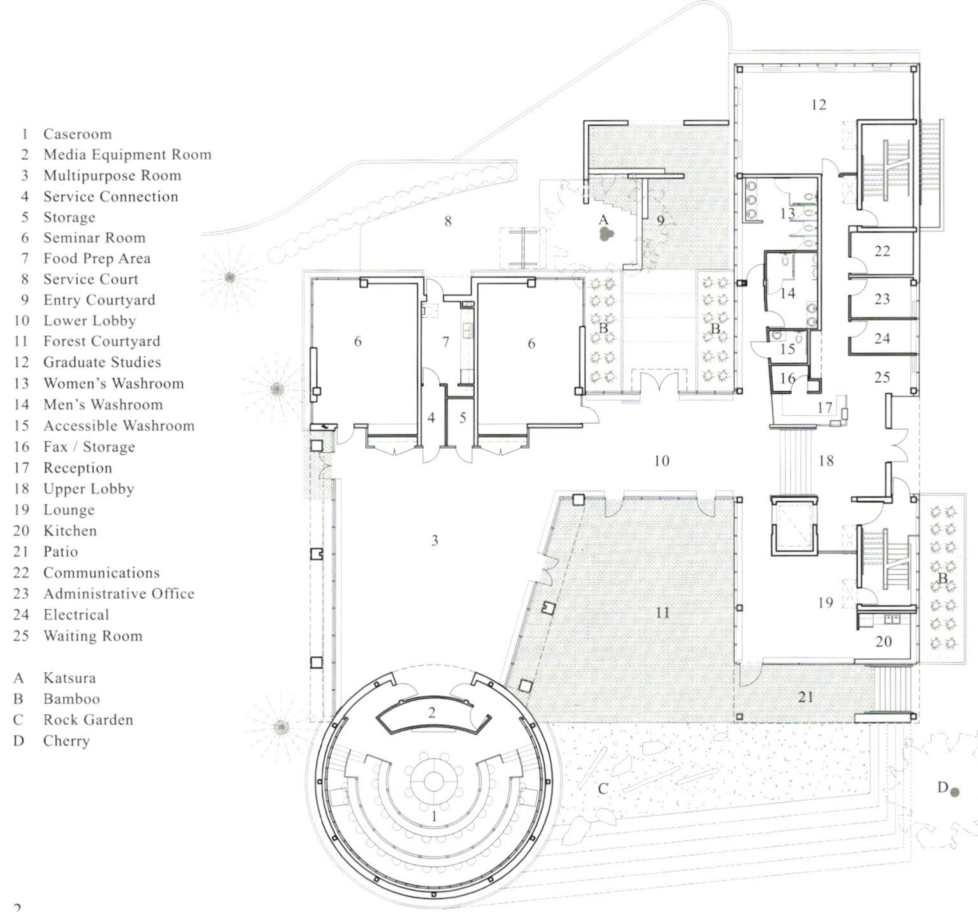

2

3

4

5

6

1 建筑模型
2 首层平面图
3 中心入口
4 入口庭院
5 圆形演讲厅
6 室内外的视线联系

7

8

9

1 Post Doctoral Studies
2 Research Office
3 Washroom
4 Communications
5 Electrical
6 Open Office / Copy Area
7 Graphic Work Station
8 Reading Room

10

7~8 内院
9 落地玻璃幕墙
10 二层平面图
11 中心入口
12 墙体细部
13~15 立面图

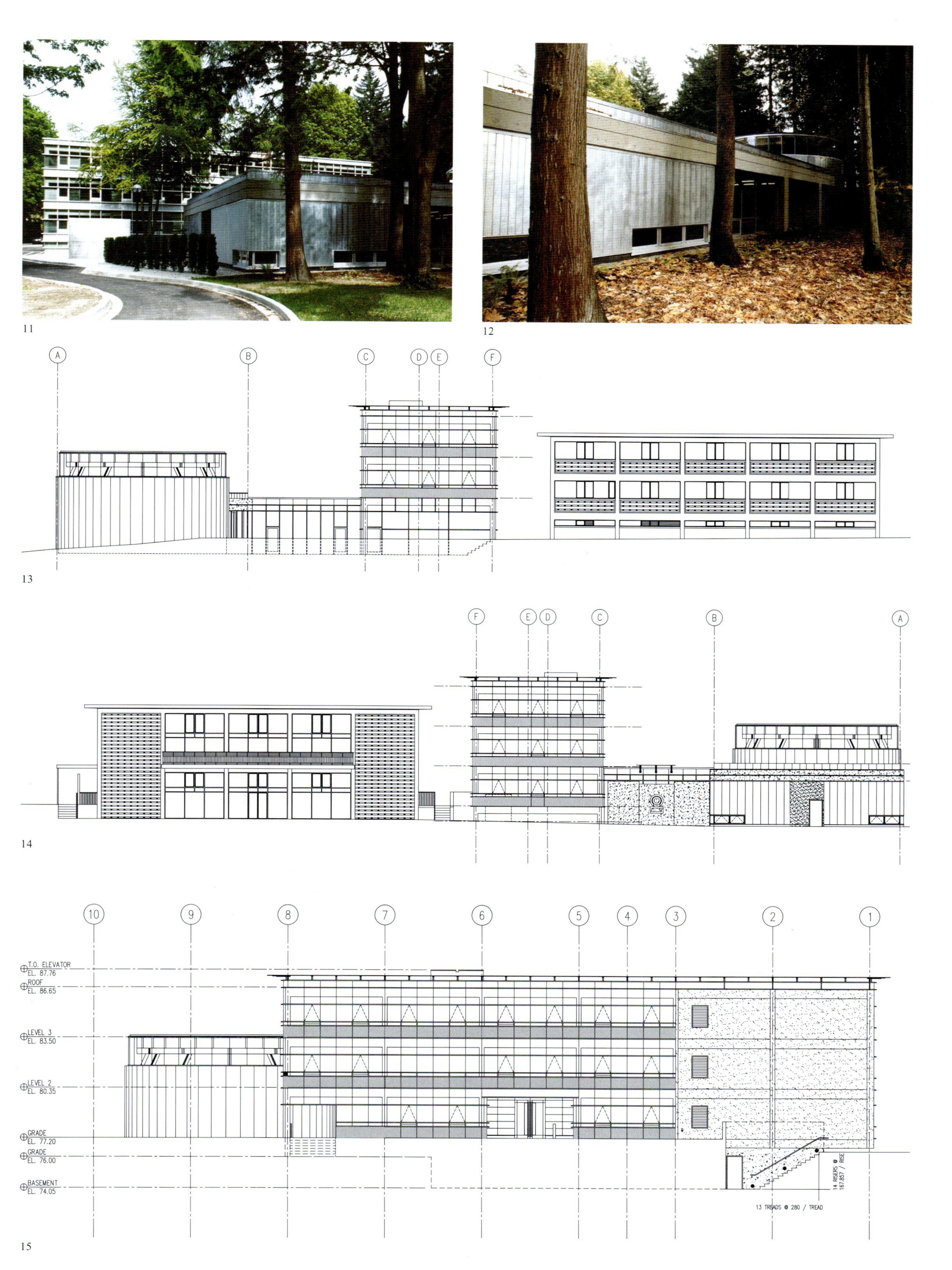
A
B
C
D
E
F
1
2
3
4
5
6
7
8
9
10
T.O. ELEVATOR
EL. 87.76
ROOF
EL. 86.65
LEVEL 3
EL. 83.50
LEVEL 2
EL. 80.35
GRADE
EL. 77.20
GRADE
EL. 76.00
BASEMENT
EL. 74.05
14 RISERS @ 167.857 / RISE
13 TREADS @ 280 / TREAD

11

12

13

14

15

16 剖面图
17 建筑物近景
18 过道细部
19 过道
20 会议室
21 演讲厅细部
22 演讲厅细部
23 演讲厅剖面
24 顶棚平面图

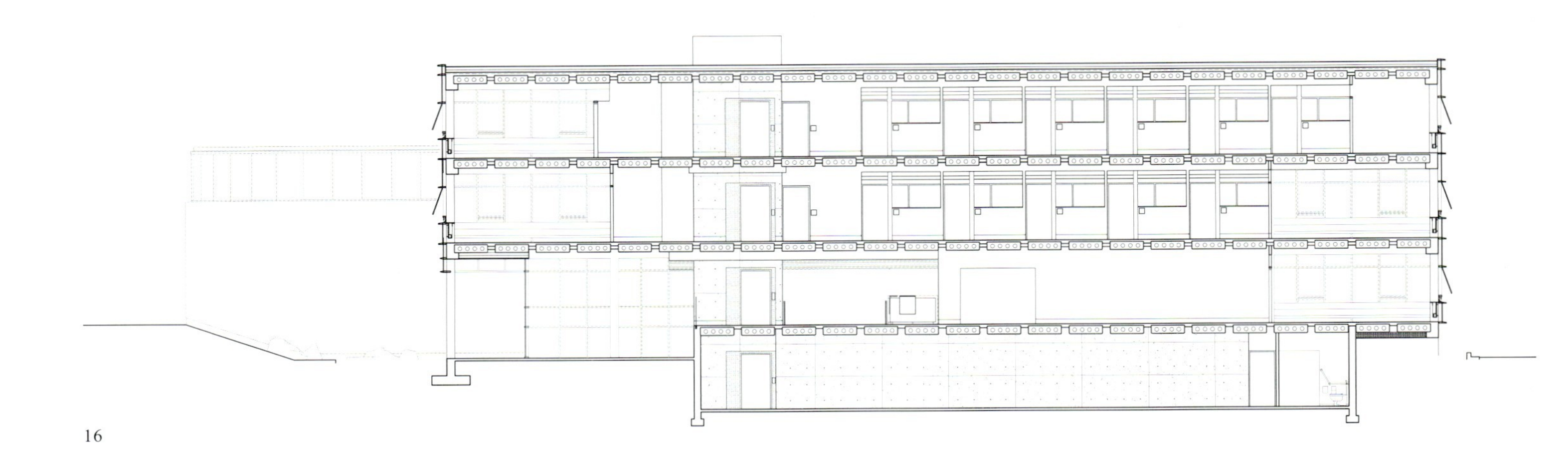

16

17

18

19

20

21

22

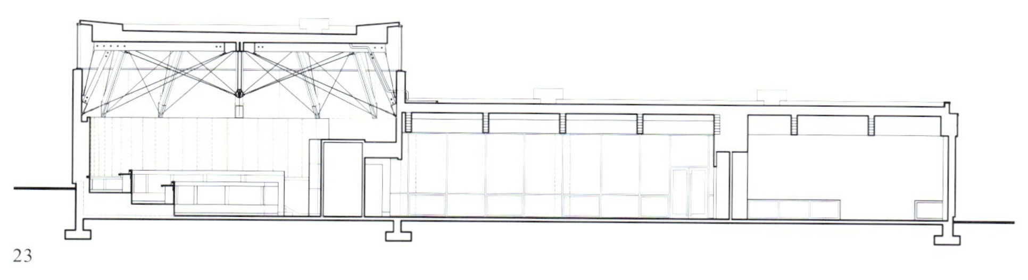
23

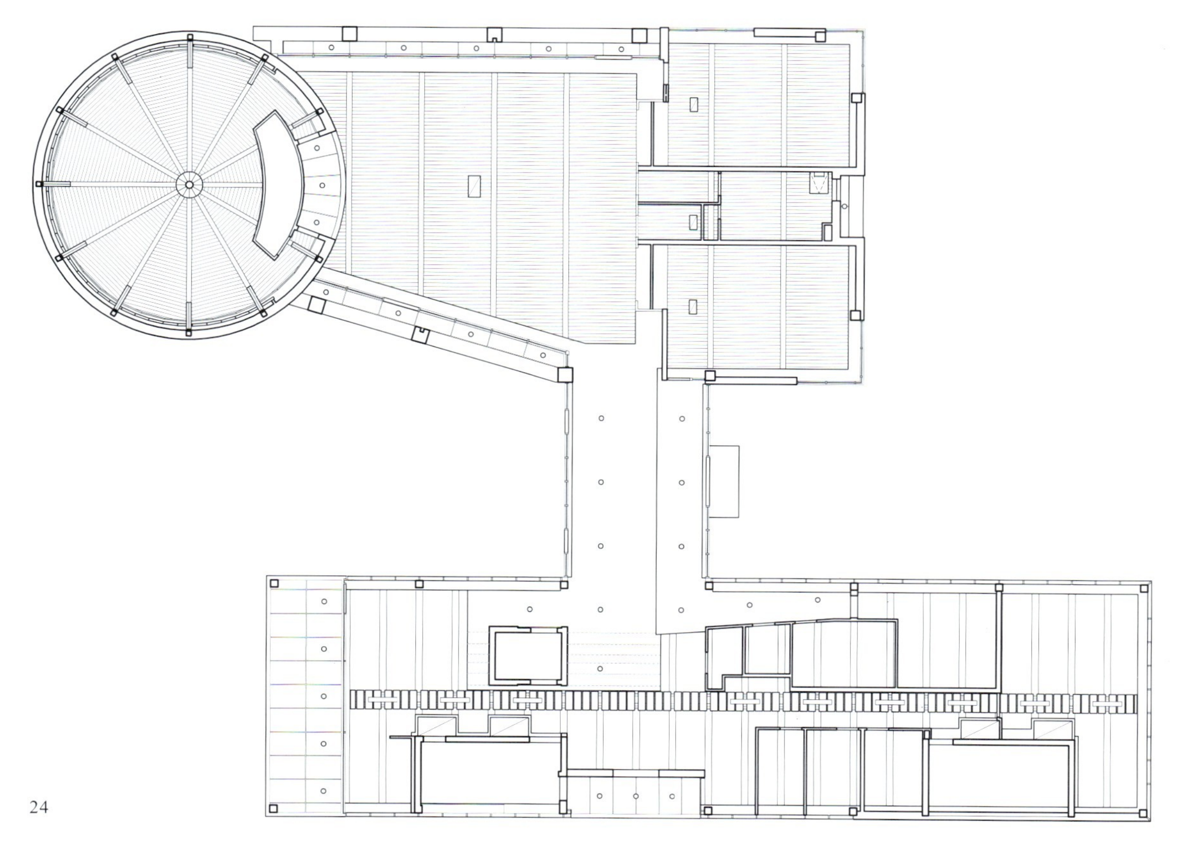
24

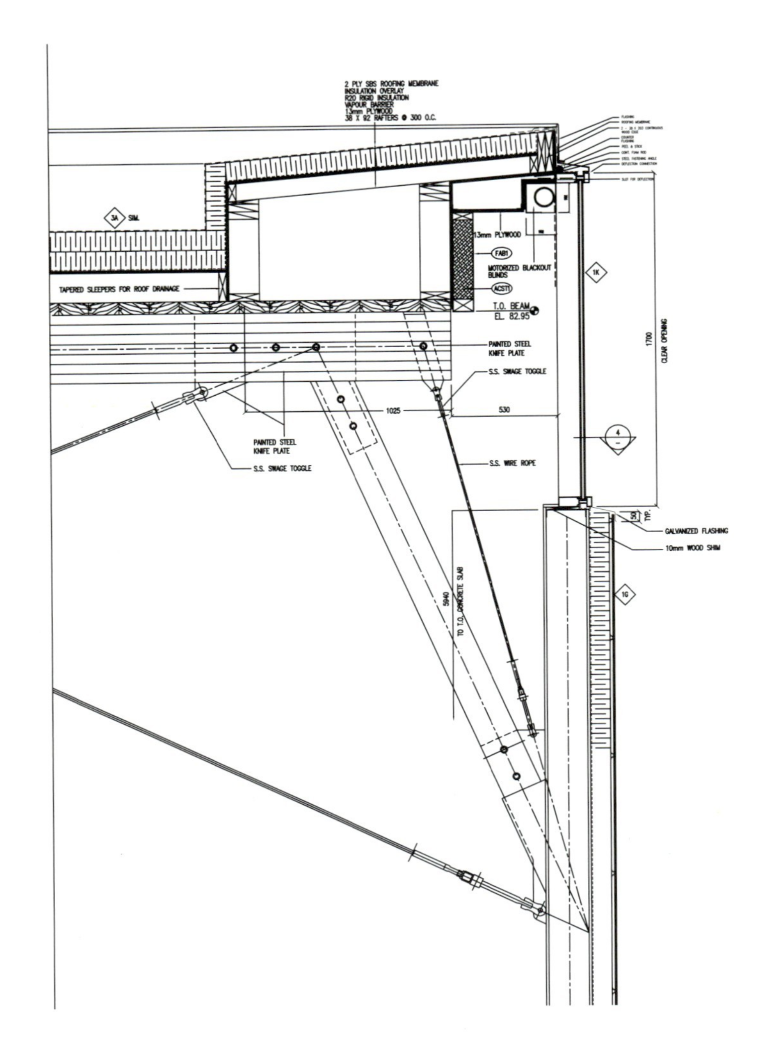
2 PLY SBS ROOFING MEMBRANE
INSULATION OVERLAY
R20 RIGID INSULATION
VAPOUR BARRIER
13mm PLYWOOD
38 X 92 RAFTERS @ 300 O.C.
3A SIM.
TAPERED SLEEPERS FOR ROOF DRAINAGE
13mm PLYWOOD
FAB1
MOTORIZED BLACKOUT BLINDS
ACST1
1K
T.O. BEAM
EL. 82.95
PAINTED STEEL KNIFE PLATE
S.S. SWAGE TOGGLE
1025
530
1700
CLEAR OPENING
PAINTED STEEL KNIFE PLATE
S.S. SWAGE TOGGLE
S.S. WIRE ROPE
4
50 TYP.
GALVANIZED FLASHING
10mm WOOD SHIM
5940
TO T.O. CONCRETE SLAB
1G

25

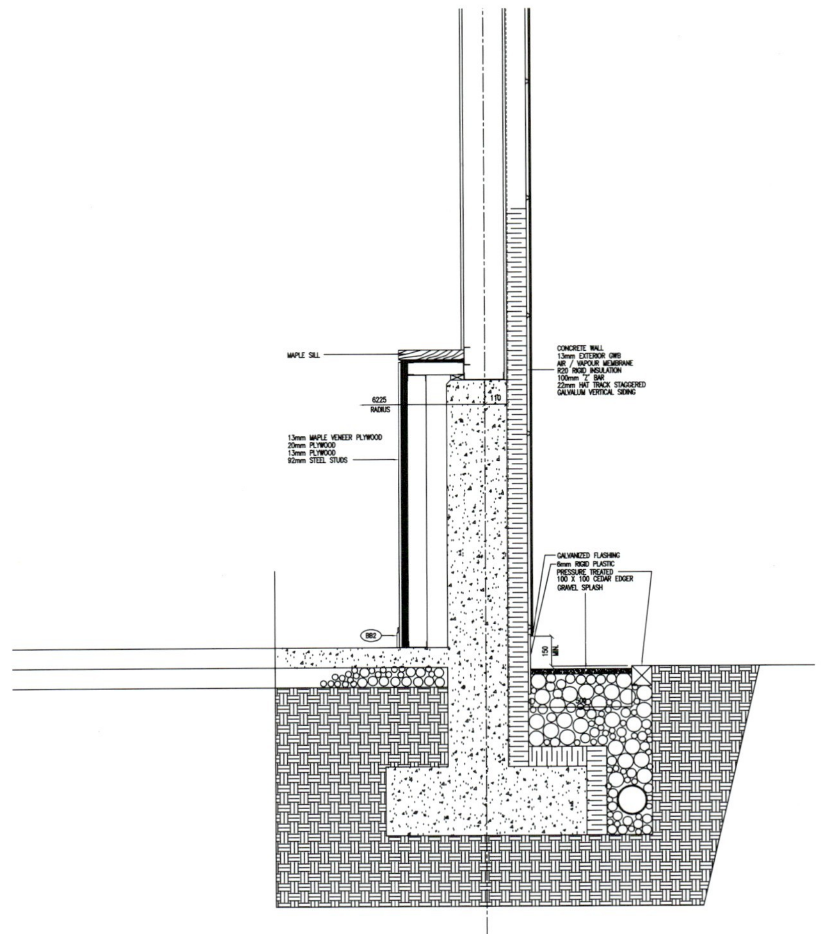
MAPLE SILL
CONCRETE WALL
13mm EXTERIOR GWB
AIR / VAPOUR MEMBRANE
R20 RIGID INSULATION
100mm "Z" BAR
22mm HAT TRACK STAGGERED
GALVALUM VERTICAL SIDING
6225
RADIUS
110
13mm MAPLE VENEER PLYWOOD
20mm PLYWOOD
13mm PLYWOOD
92mm STEEL STUDS
GALVANIZED FLASHING
6mm RIGID PLASTIC
PRESSURE TREATED
100 X 100 CEDAR EDGER
GRAVEL SPLASH
BB2
150 MIN.
500

26

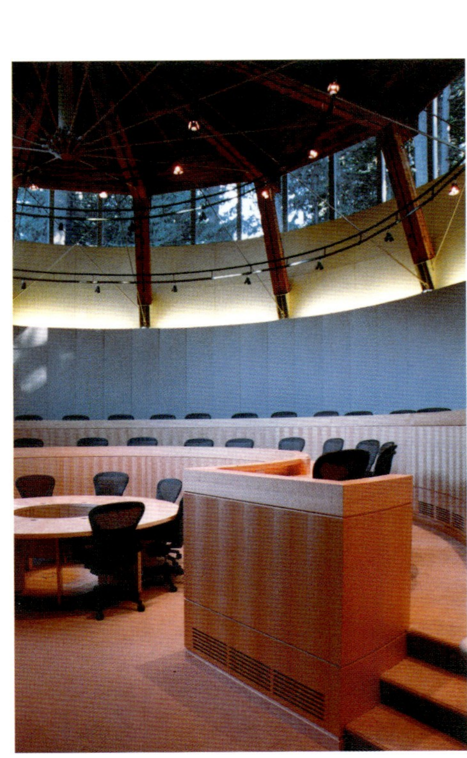

27

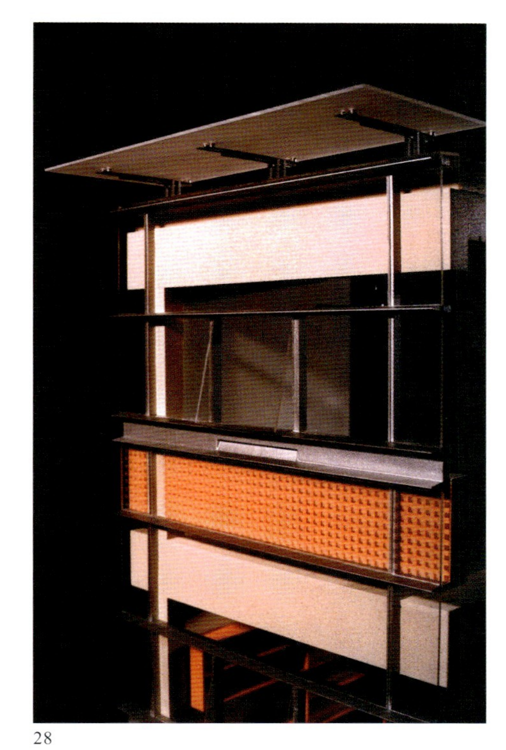

28

29

30

MDF SPACER
SECURED TO MULLION
MDF FACIA
MDF PAINTED
ROUTED SLOTS
FOR DIFFUSION
SYSTEM
SUPPORT
TYPE 2
HEATING
WATER LINES
SYSTEM
SUPPORT
TYPE 1
GYPSUM
FIBRE
PANEL
SYSTEM
SUPPORT
TYPE 1
ELECTRICAL
RACEWAY
FIN TUBE HOT
WATER HEATER
CONCRETE TOPPING
PRECAST CONCRETE SLABS
CONTINUOUS
NAILER
CONTINUOUS
NAILER
PLYWOOD SOFFIT
ELECTRICAL
CONDUITS AS
REQUIRED

31

25 顶棚构架构造详图
26 墙体详图
27 圆形演讲厅细部
28 窗户模型细部
29 楼板模型
30 窗户剖面详图
31 高能效供暖系统示意图

Trade & Technology Center
贸易和科技中心

业　　主：University College of the Cariboo
地　　点：坎卢普斯市
规　　模：10 000m²
完成时间：1998年

The Trade & Technology Center at Kamloops' University College of the Cariboo has been designed to provide appropriate space for existing technology programs, to accommodate new programs such as Robotics and Telecommunications, and to be adaptable to evolving program priorities.

These goals have been met with a rational design approach based on an effective use of limited budget resources ($1,100 psm) applied to an ambitious space program of 10,000 square meters. Flexibility and adaptability to future programs have been achieved by a primary structural system of clear-span three-dimensional steel trusses supported on a repetitive grid of concrete piers, which can be easily extended or re-partitioned. Exposed on both the interior and he exterior, the 20-metre long, curving trusses dramatically convey the technological purpose of the building.

Exposed structural elements and detailing that celebrate the assembly and connection of materials are common in the design of primary spaces of the building and in the resolution of stairs, canopies and sunscreens, providing an architectural expression that is highly appropriate to the technical and educational nature of the building.

坎卢普斯大学的贸易和科技中心的设计主题是为已有的科技学科提供教学空间，并满足新的学科专业——机器人、远程通信和其他优先学科的使用要求。

设计的挑战在于对造价的控制。由重复的混凝土柱网支撑的大跨度的三维钢桁架结构体系形成了灵活的、适应未来发展需要的空间。暴露于室内外的20m长的弧形屋架表现出科技建筑的特点。

为表现材料的组合与联系而暴露结构要素和细部的情况，在建筑的主要空间、楼梯处理和雨篷及遮阳篷上都十分普遍。这使得该建筑与教育及高科技建筑的特征非常吻合。

1

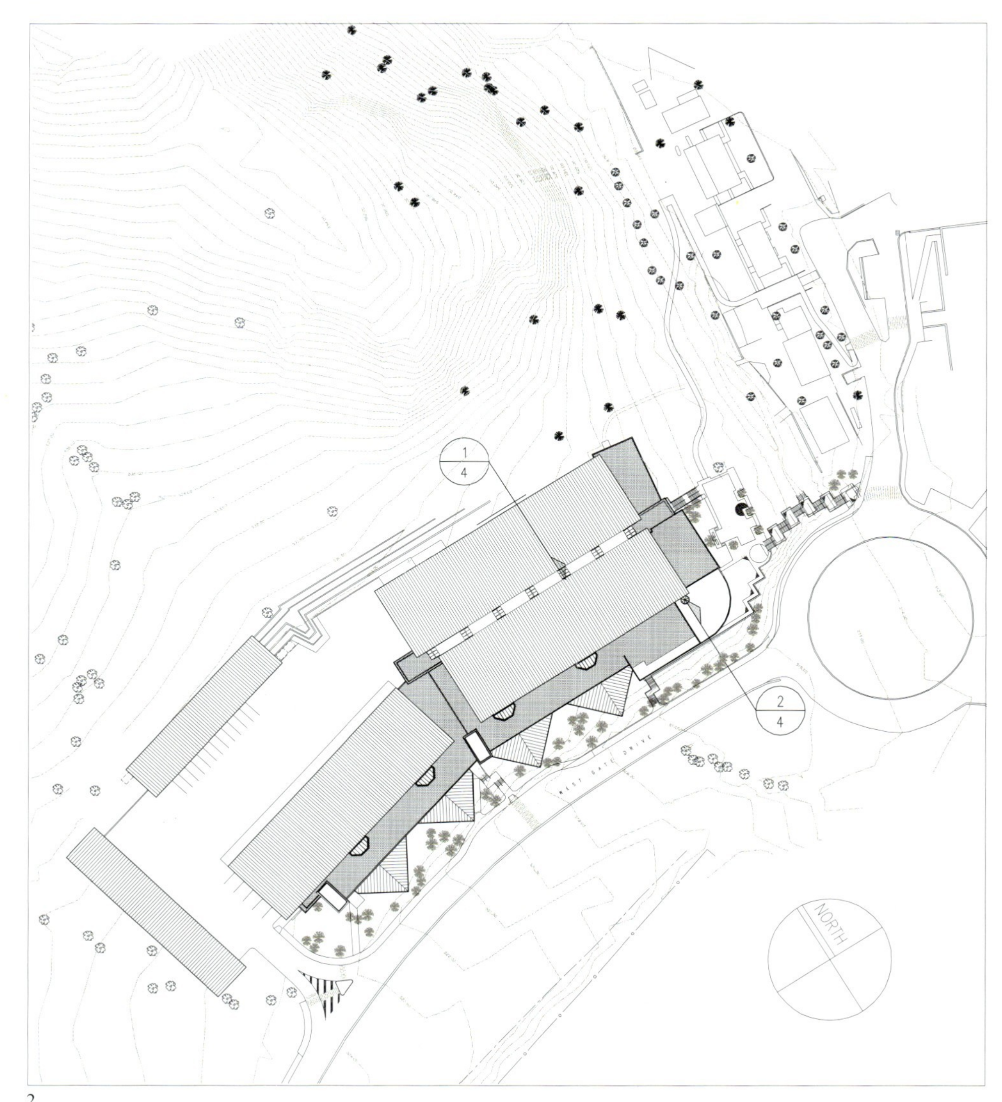

2

1　全景
2　总平面图
3　中心全景
4　二层平面图
5　首层平面图

3

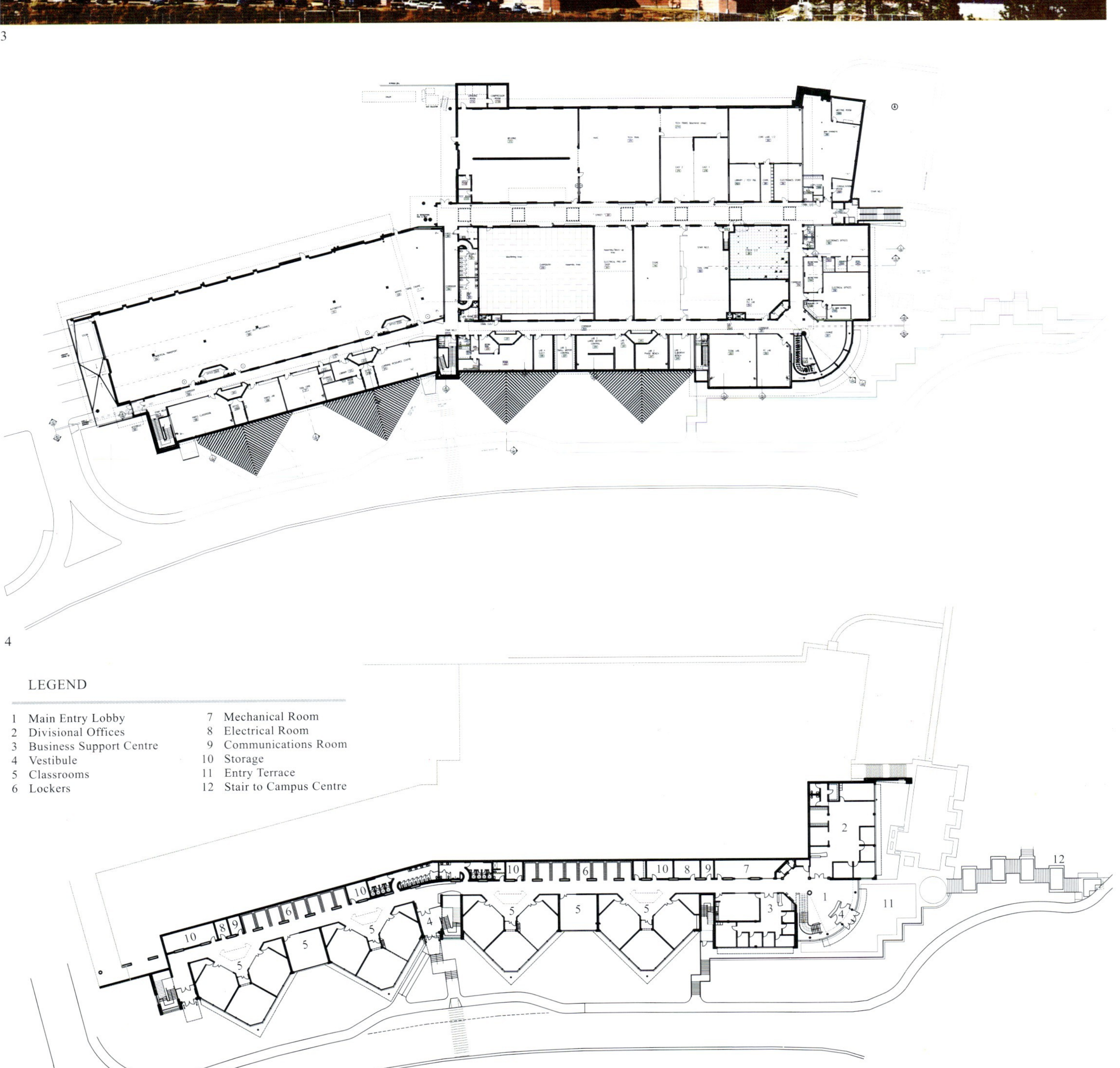
LEGEND
1 Main Entry Lobby
2 Divisional Offices
3 Business Support Centre
4 Vestibule
5 Classrooms
6 Lockers
7 Mechanical Room
8 Electrical Room
9 Communications Room
10 Storage
11 Entry Terrace
12 Stair to Campus Centre
0m 5m 10m 20m 50m

4

5

6~7 立面细部
8 东立面图
9 西立面图
10 南立面图
11~13 屋顶细部
14 顶棚平面图

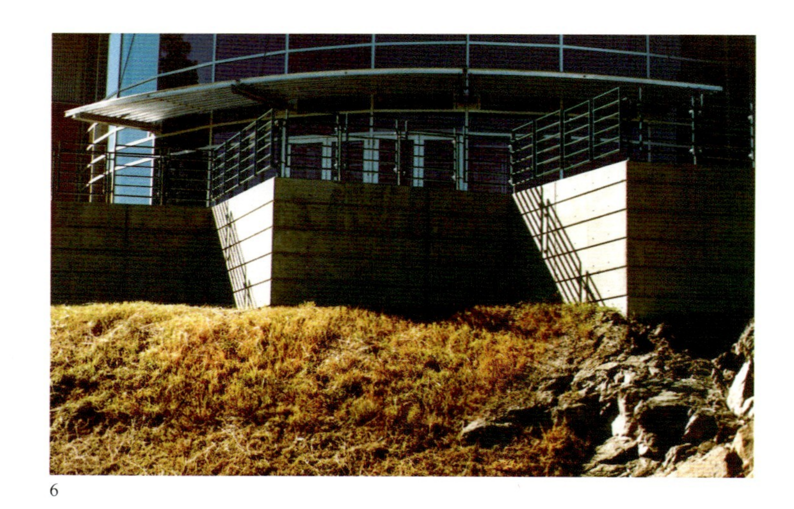
6

7

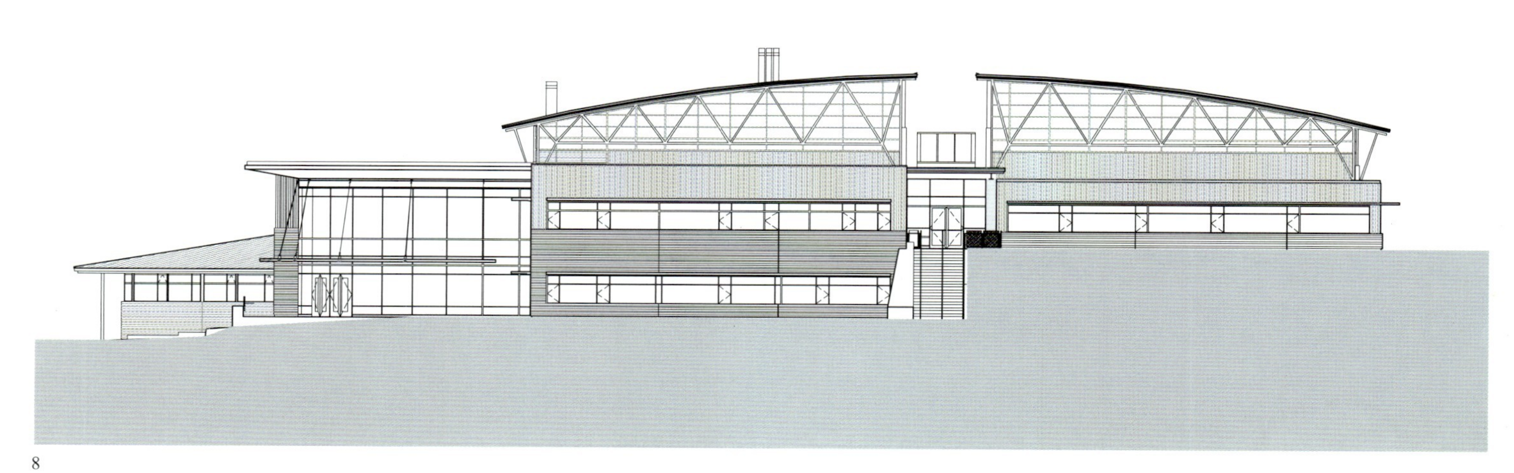
8

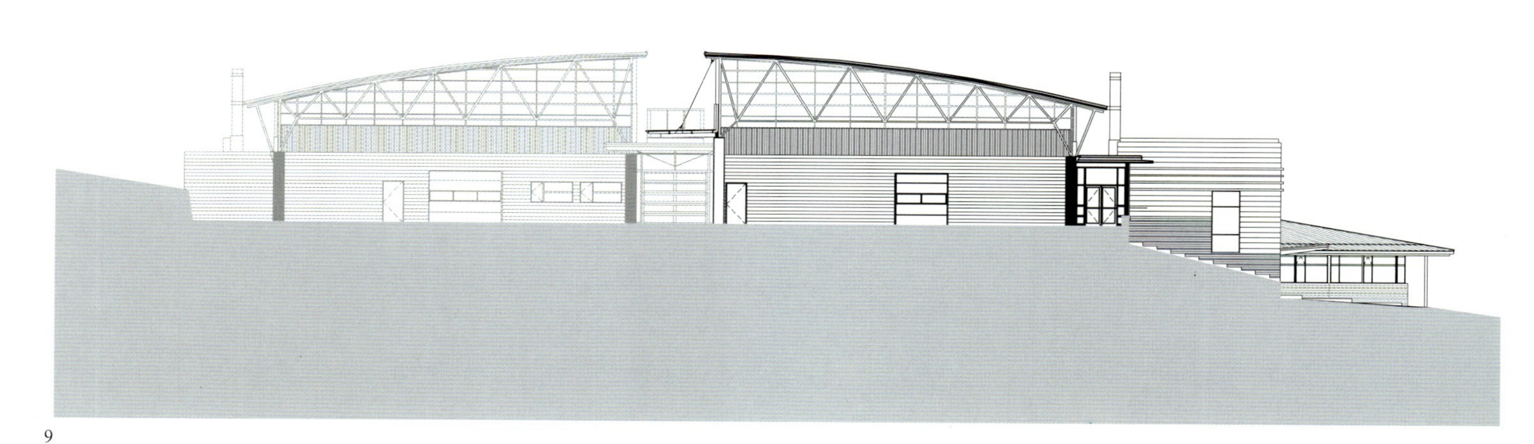
9

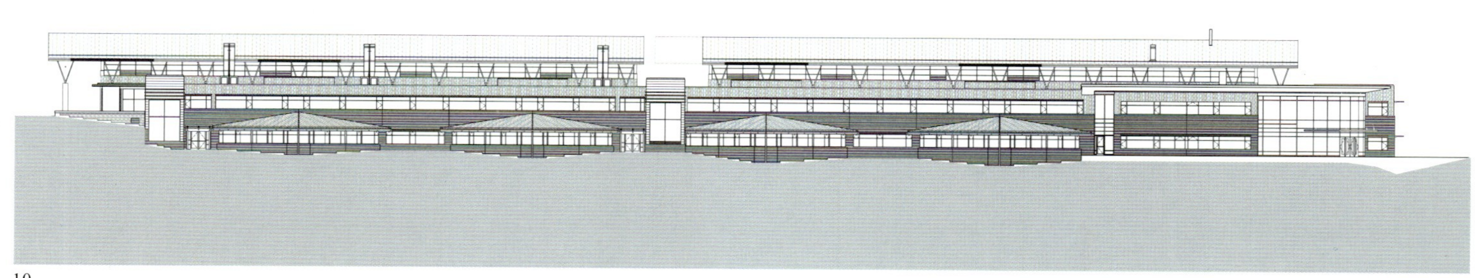
10

11

12

The building design responds to both the opportunity and challenges of a natural sloping site that establishes the current edge of the campus. The site slopes to the south east, affording spectacular views of the campus, the city of Kamloops and the Thompson River. The building and service compound are gently cut into the slope, cascading down the hillside in a series of roof forms that bring natural light into all levels of the building. The exterior street wall and projecting classroom bays are clad with a masonry base as outlined by campus design guidelines while the main entry conforms to the campus planning mandate to have all new building entries identified by a rotunda form.

Above the one-and-a-half story masonry base, the essential technological character of the building is expressed by the curved metal roof that floats above a skeleton of exposed trusses, with continuous clerestory glazing. At the easterly end of the street wall, the transparent double-height volume of the entry lobby and student lounge establishes the main point of entry and connection to the campus.

Steel and concrete were the prime construction materials in the project and used extensively throughout the building. The long-span, three- dimensional trusses, king post trusses, beams, columns, exposed structural acoustical desk, wall and roof claddings, stairways, guardrails, glazing support structures and wall studs were all constructed of steel. Cast in place concrete and concrete block were used extensively throughout the building, articulated precisely and left to view.

作为校园外缘的斜坡是中心建筑的表现的机遇，也是设计的挑战。面向东南的斜坡拥有全校园、坎卢普斯市和Thompson河的景色。中心建筑和服务部轻微切入山坡。沿山坡而设的一系列屋顶可以让自然光进入建筑的各层内部。外墙和保护教室隔间覆盖着石制的基座，并以其作为校园设计的主要语言，同时主入口也运用了这种校园规划语言，但是以圆形为标志物。

中心位于一层半的石制基座上，主要的技术特征是通过位于裸露的屋架高窗之上的弧形金属屋顶。在东部尽端的两层高透明的大堂和学生休息室是出入口，也是与校园的连接点。

钢和混凝土是中心建筑的主要建筑材料。大跨度立体屋架，主屋架，大梁，柱子，暴露的隔声板，墙面和屋顶装饰，楼梯，栏杆，玻璃支撑和墙支撑都是钢制成的。装配制混凝土和混凝土板广泛地使用在整个建筑中，但都精心设计使其不阻碍视线。

13

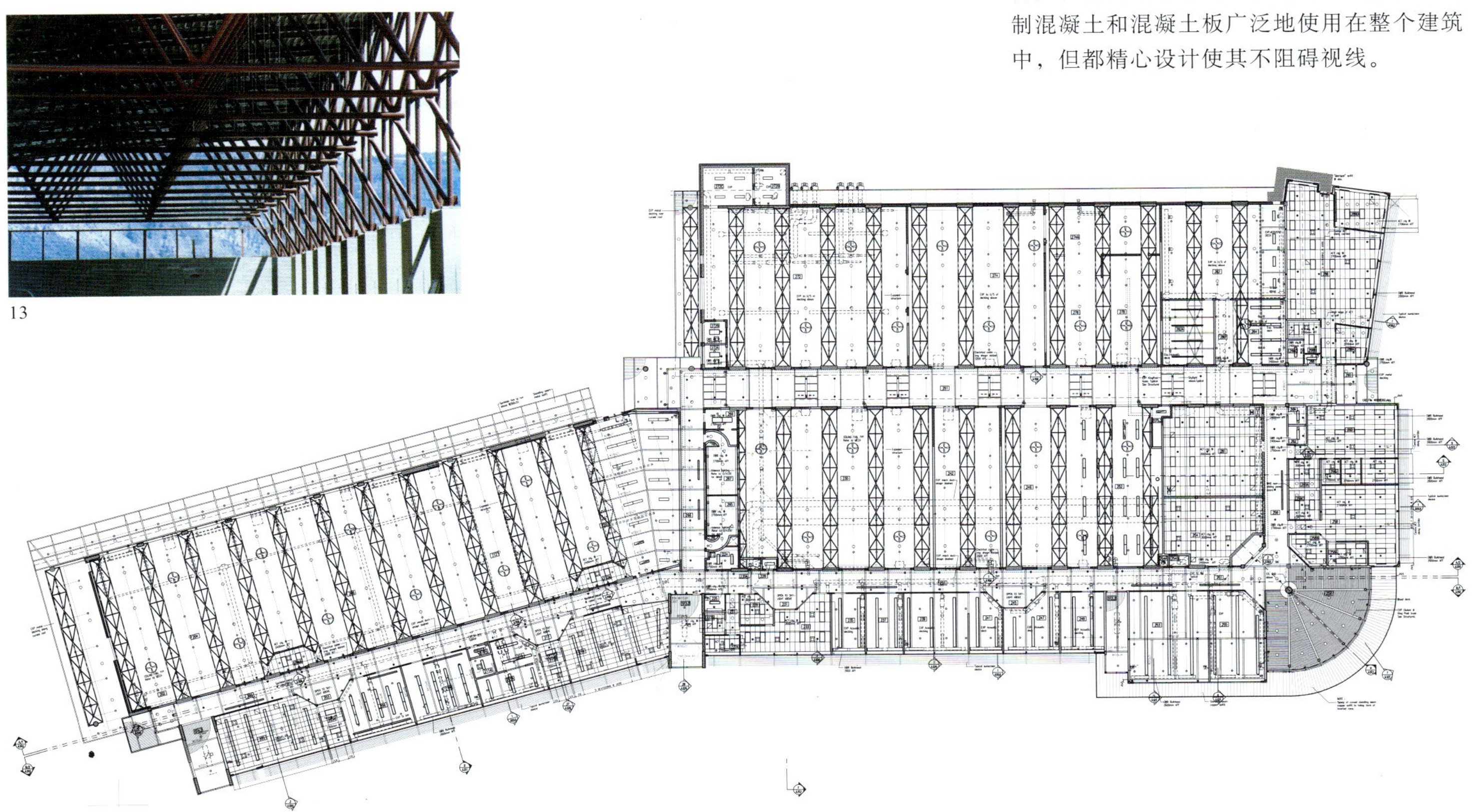

14

Many strategies were employed in the design of the building to contribute to its "eco-effective" success. The well-insulated and sealed "rain-screen" exterior walls have exterior aluminum louvered overhangs on south-facing windows that allow abundant natural light into the classrooms and labs. The reflective standing seam steel roofing that covers most of the building has broad thermally broken overhangs that provide shade to the clerestory glazing. The clerestory glazing over the shop spaces provides abundant natural light that in turn reduces electrical consumption. The lack of finishing veneers to the basic structural framing of the building fundamentally reduces the amount of resources required in its construction. High elevation motorized vents allow excessive heat build up to be naturally vented to reduce coolingloads. Most of the building materials utilized in the construction of this building are readily recyclable.

15

众多的措施用来减少对“生态效益”的冲击。在南向的窗上密闭的“雨幕”外墙外设有外露的铝制百叶悬垂物，以保证有足够的阳光进入教室和实验室内。有反光的、竖立有外框的钢屋顶覆盖着建筑的大部分。可推开的悬垂物可以给高窗提供足够的阴影。商店上面的高窗提供充足的阳光，可以减少对能源的消耗。位于高处的通风口可以减少对通风管的要求。建筑的大多数材料都是可以循环使用的。

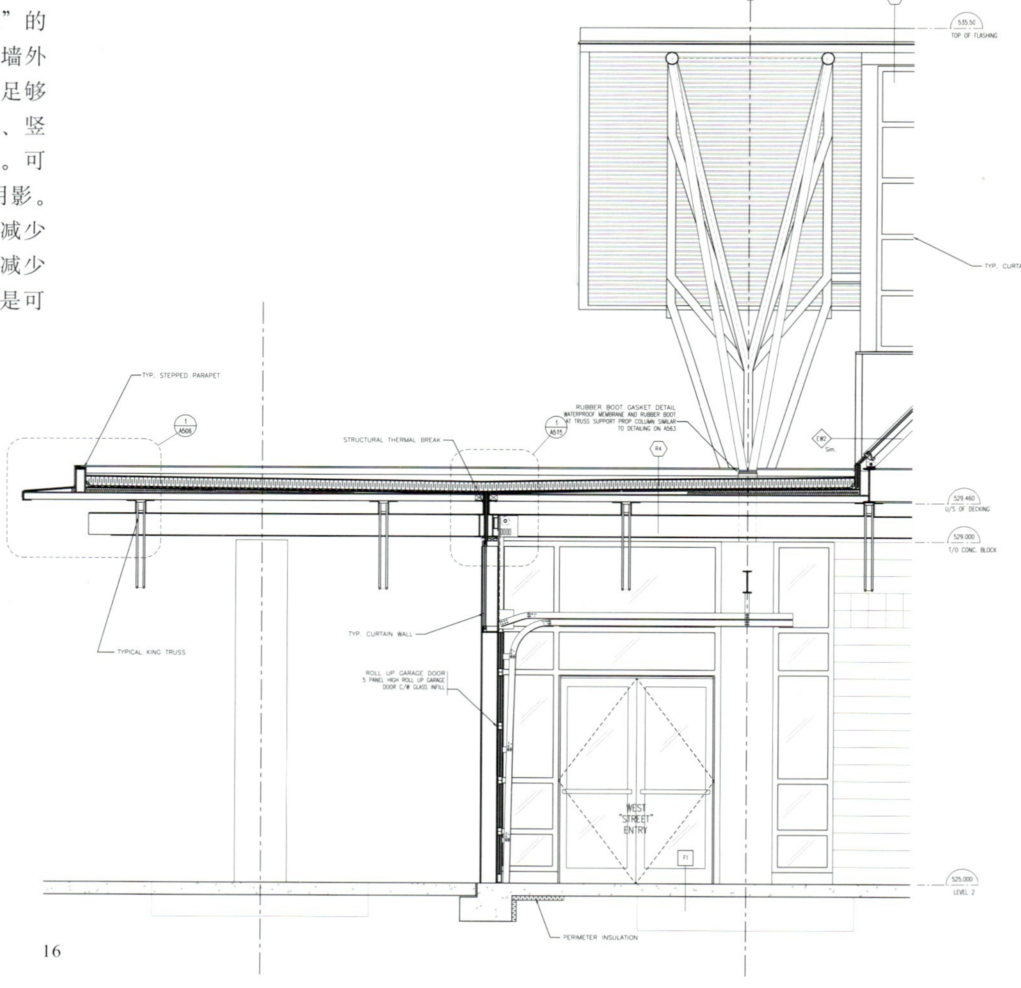

16

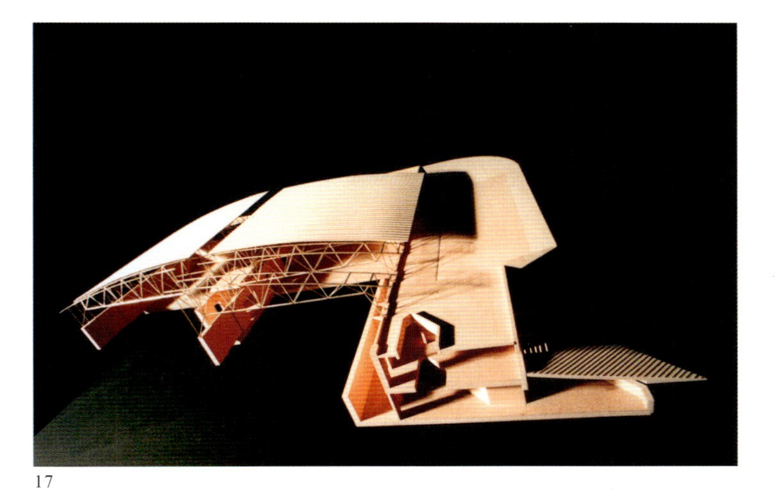

17

18

19

14

16

TYPICAL CURTAIN WALL ASSEMBLY

TALL PARAPET CONST.

TYPICAL SHADING DEVICE

COMP. TECH LAB

CORRIDOR

MEETING ROOM

ELECTRONICS OFFICES

UNEXCAVATED

COPY/ STORAGE

DIVISION OFFICES

B WALL SECTION
A202 A204

20

15 檐口细部
16 立面详图
17 建筑模型
18 屋面细部
19 过道
20 剖面详图

21

22

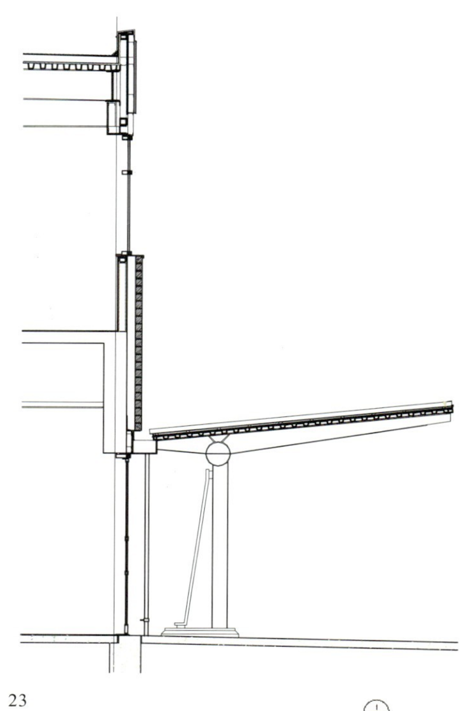

23

24

25

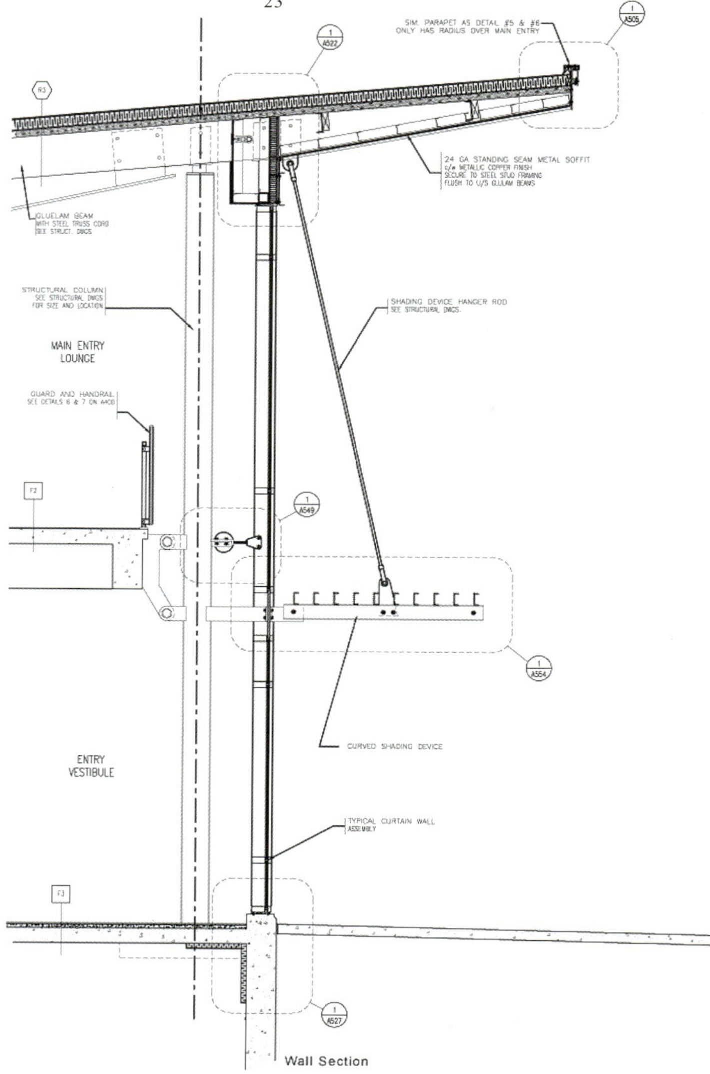

26

21~22　入口细部
23　入口剖面详图
24　室外细部
25　室内顶棚细部
26　剖面详图
27　室内细部
28　过道细部
29　楼梯西部
30　楼梯立面详图

27

28

29

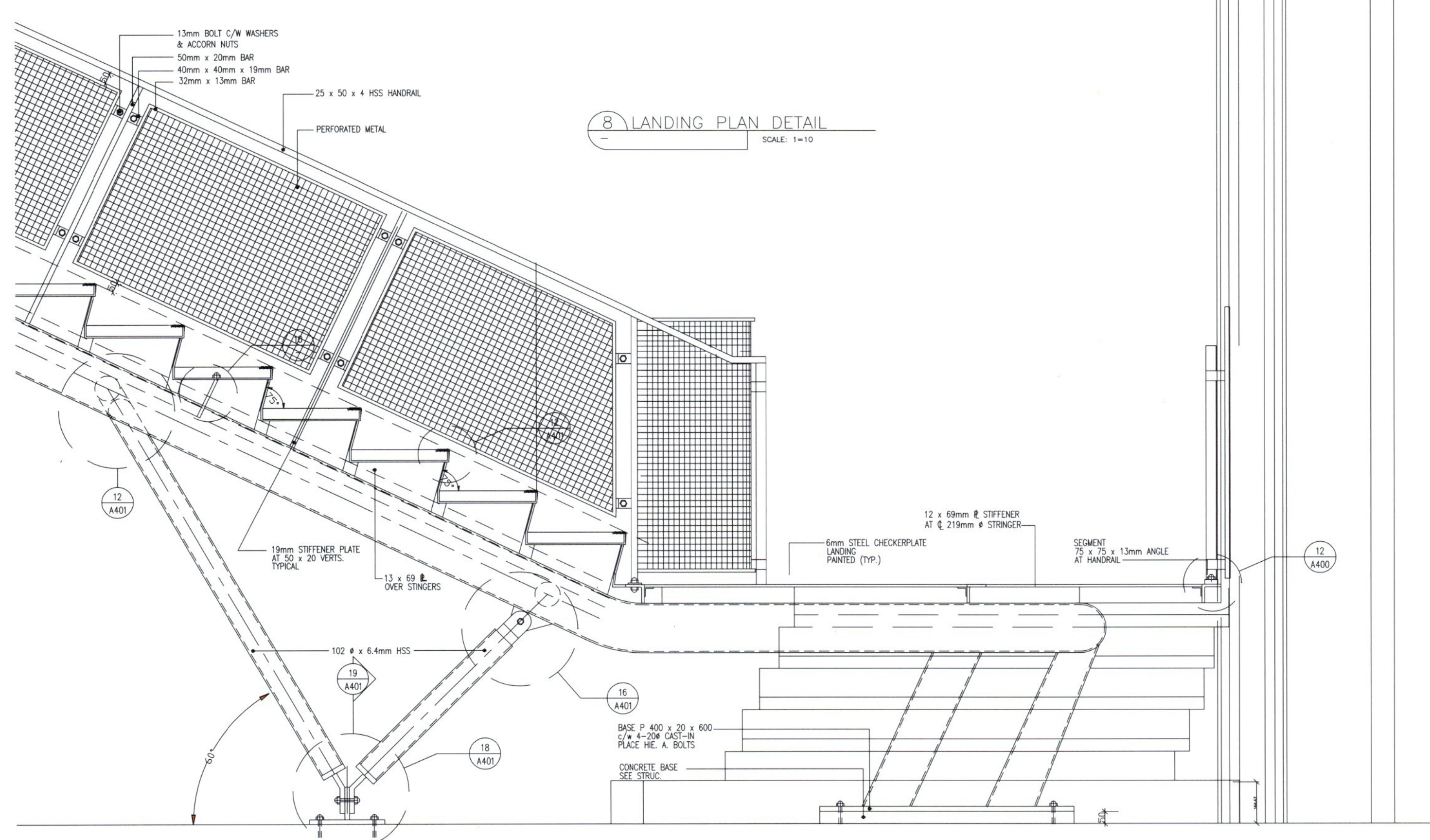

30

SFU Morris J. Wosk Center for Dialogue
西蒙菲沙大学 Morris J.Wosk 交流中心

地　　点：温哥华市
业　　主：Simon Fraser University
规　　模：3 700m²
完成时间：2000 年

Stantec Architecture was commissioned to provide full architectural and interior design services to create an adaptive re-use design for this Class A Heritage structure. The Simon Fraser University Center for Dialogue is designed to provide world-class conference and meeting facilities for the adjacent SFU Downtown Campus, the facility is both sensitive to its heritage lineage and an impressive display of contemporary design. In addition to a wide range of fully networked meeting facilities, the Conference Center contains two unique major assembly spaces. The Dialogue Hall is a unique meeting area for 150, with a circular seating format, designed for face-to-face discourse. The four-story entry atrium that links old and new is fully interconnected with the Delta Hotel next door.

Stantec承接了这座A级历史保护建筑的全部建筑单体及室内设计任务。Morris J. Wosk 交流中心位于西蒙菲沙大学市中心校园内，其主要用途为提供会议设施。Morris J. Wosk交流中心的建筑设计不仅反映其历史性建筑的特点，而且运用现代建筑语言。会议中心包括两个大的多功能空间和一系列完全由网络连接的会议设施。交流中心大厅是一个独特的环形，可以提供面对面交流的150 座大空间。把新建筑与原有老建筑联系在一起的 4 层的入口大厅与紧邻的三角洲酒店的建筑形态的对话十分亲密。

1

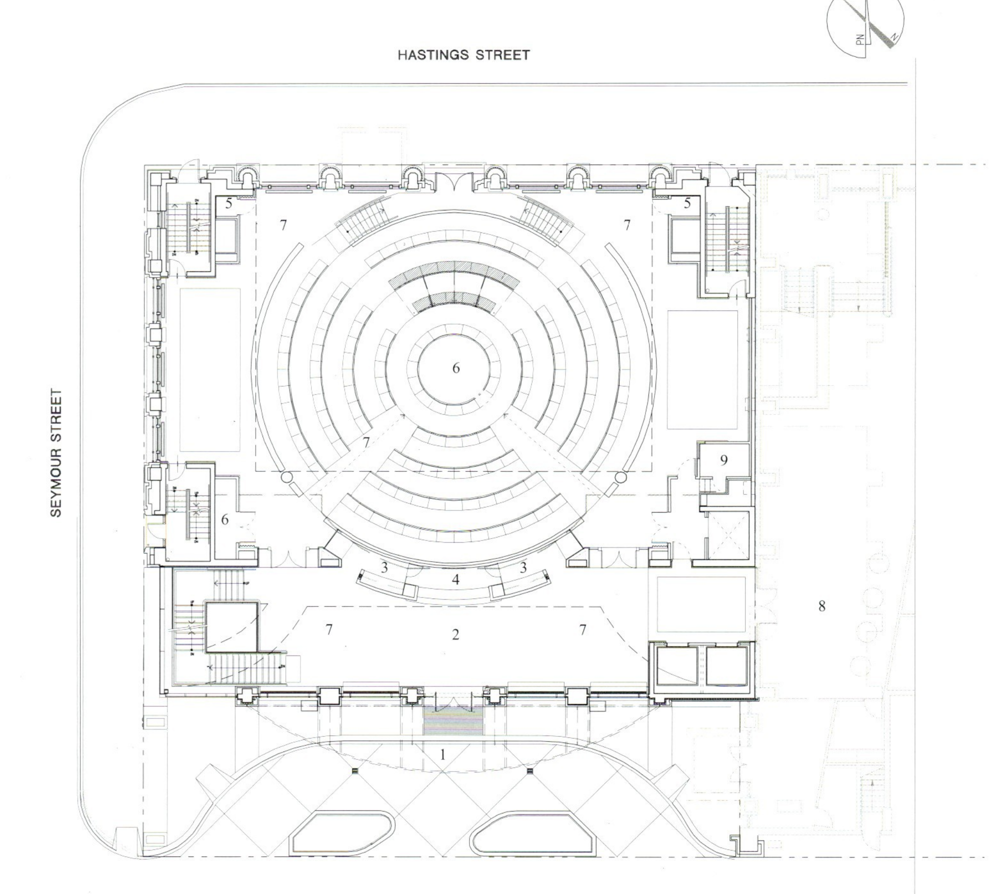

2

1 圆形会议室
2 首层平面图
3 建筑西北角

3

4

5

6

7

8

9

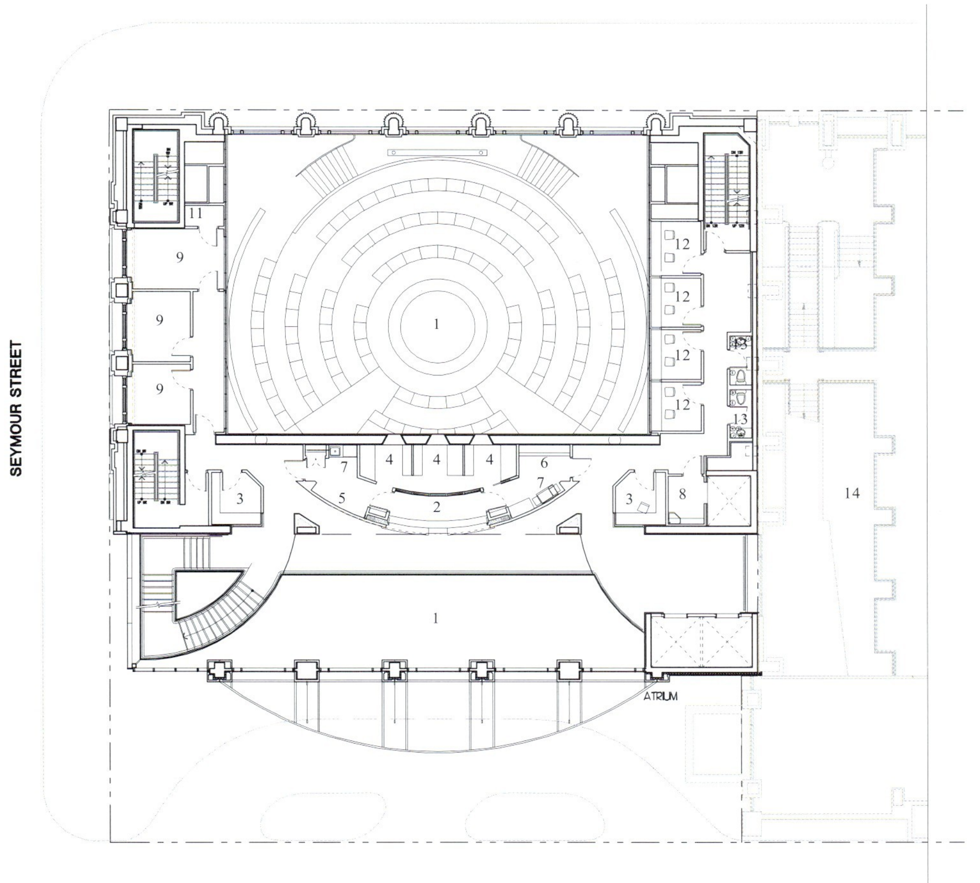

10

4 会议室
5 门的细部
6 具有传统风格的建筑立面
7 会议室细部
8 圆形会议室室内
9 大会议厅室内
10 二层平面图

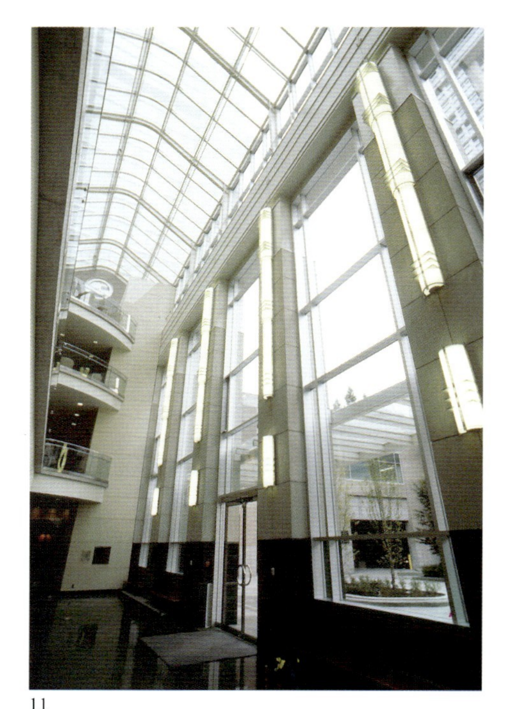

11

12

13

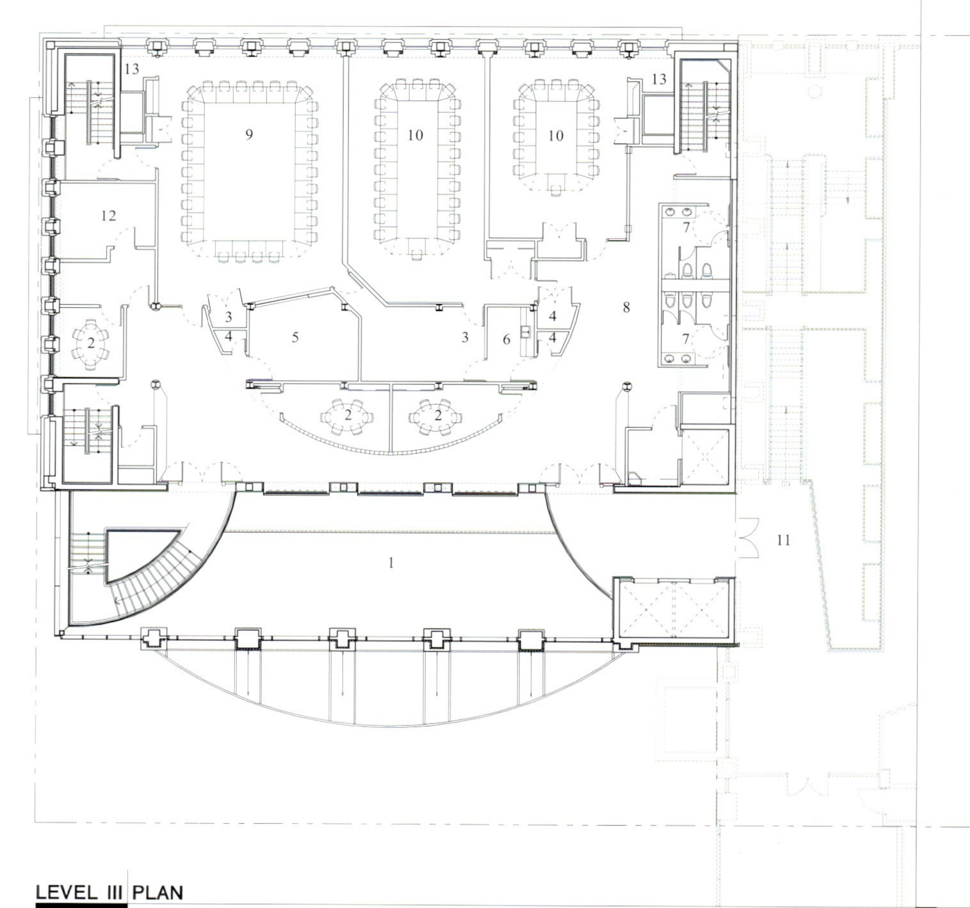

1 Open to Below
2 Meeting Room
3 Storage
4 Telephone
5 Projection Room
6 Food Prep.
7 Washrooms
8 Vestibule
9 Boardroom
10 Conference Room
11 Adjacent Hotel
12 Office
13 Coats

14

11 入口
12 入口接待室
13 入口厅仰视
14 三层平面图
15 墙体细部
16 楼梯细部
17 四层平面图
18 楼梯细部

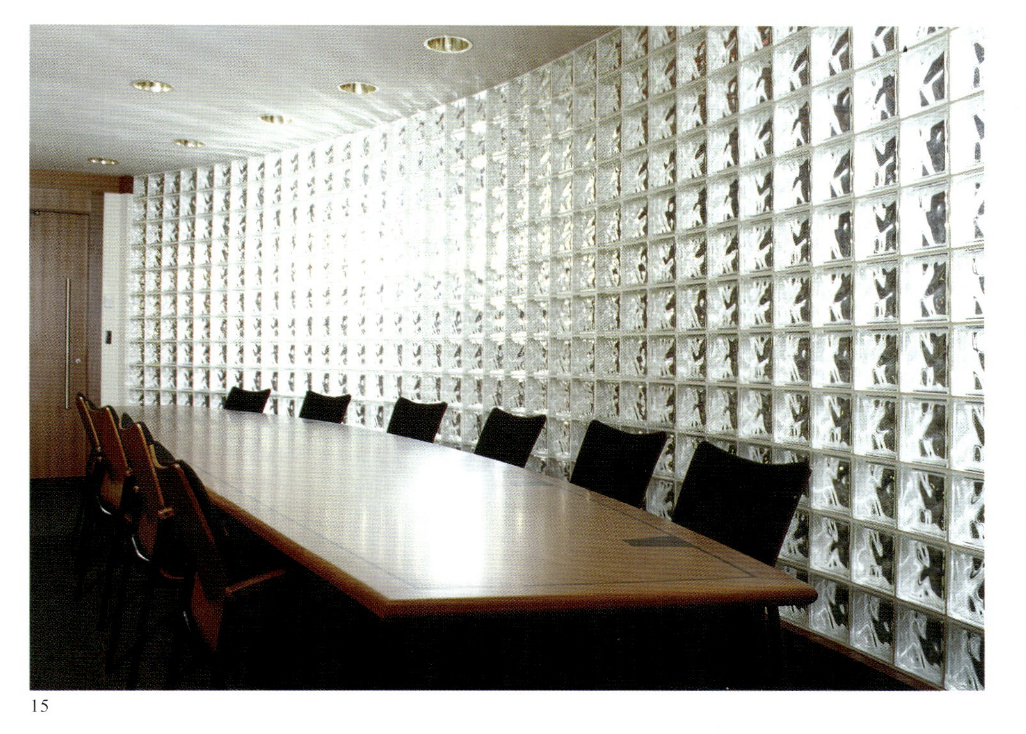

15

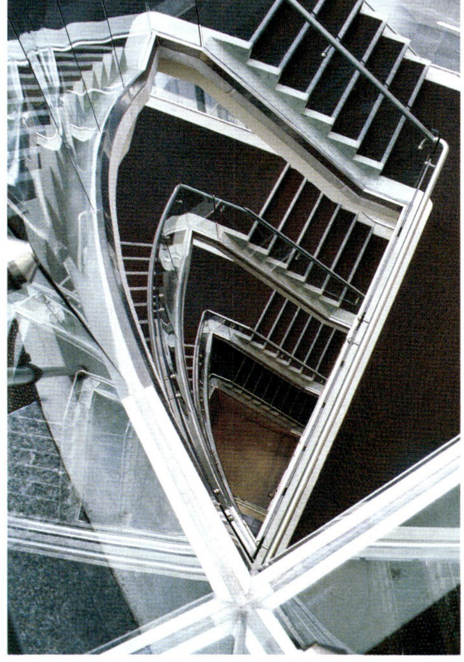

16

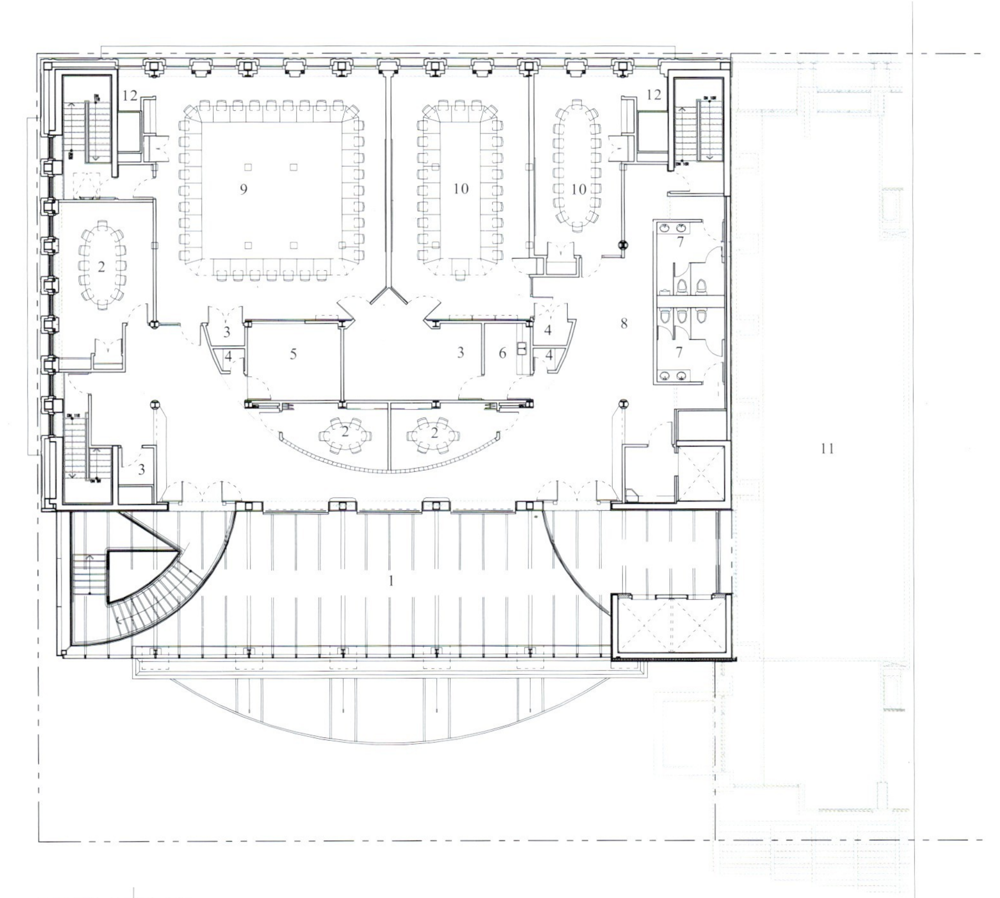

LEVEL IV PLAN

LEGEND

1 Open to Below
2 Meeting Room
3 Storage
4 Telephone
5 Projection Room
6 Food Prep.
7 Washrooms
8 Vestibule
9 Boardroom
10 Conference Room
11 Adjacent Hotel
12 Coats

17

18

19

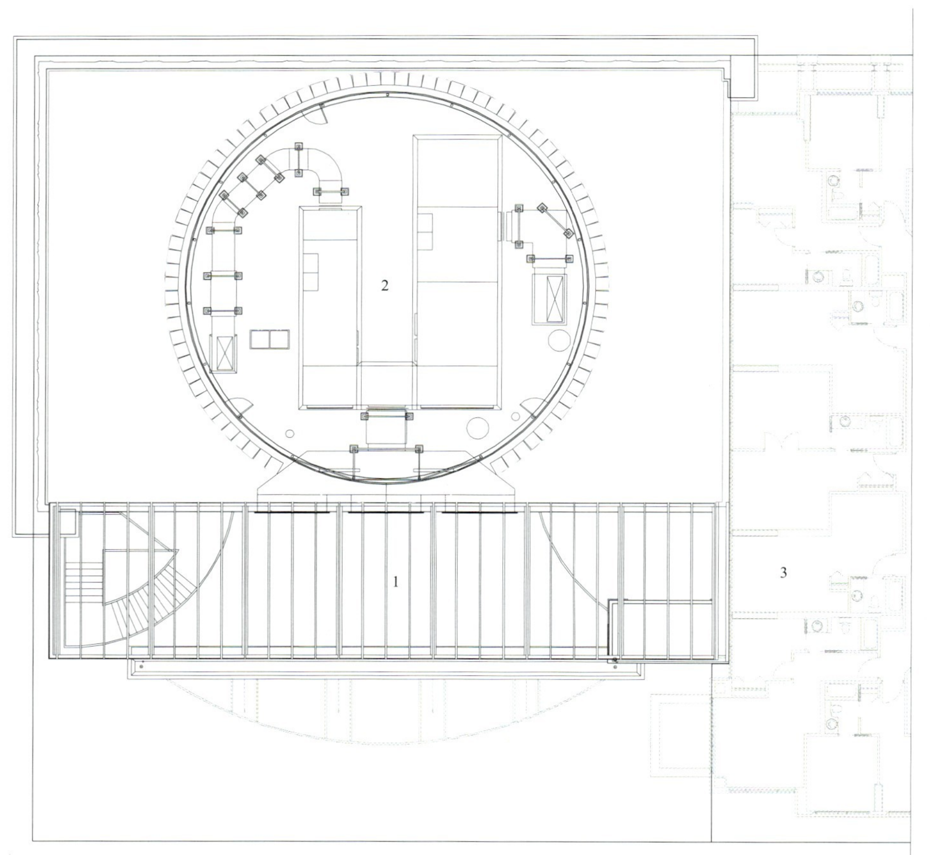

21

ROOF PLAN

LEGEND

1 Skylight
2 Mechanical
3 Adjacent Hotel

20

19　顶棚细部
20　屋顶平面图
21　吊灯细部
22　沿街立面
23　北立面图

22

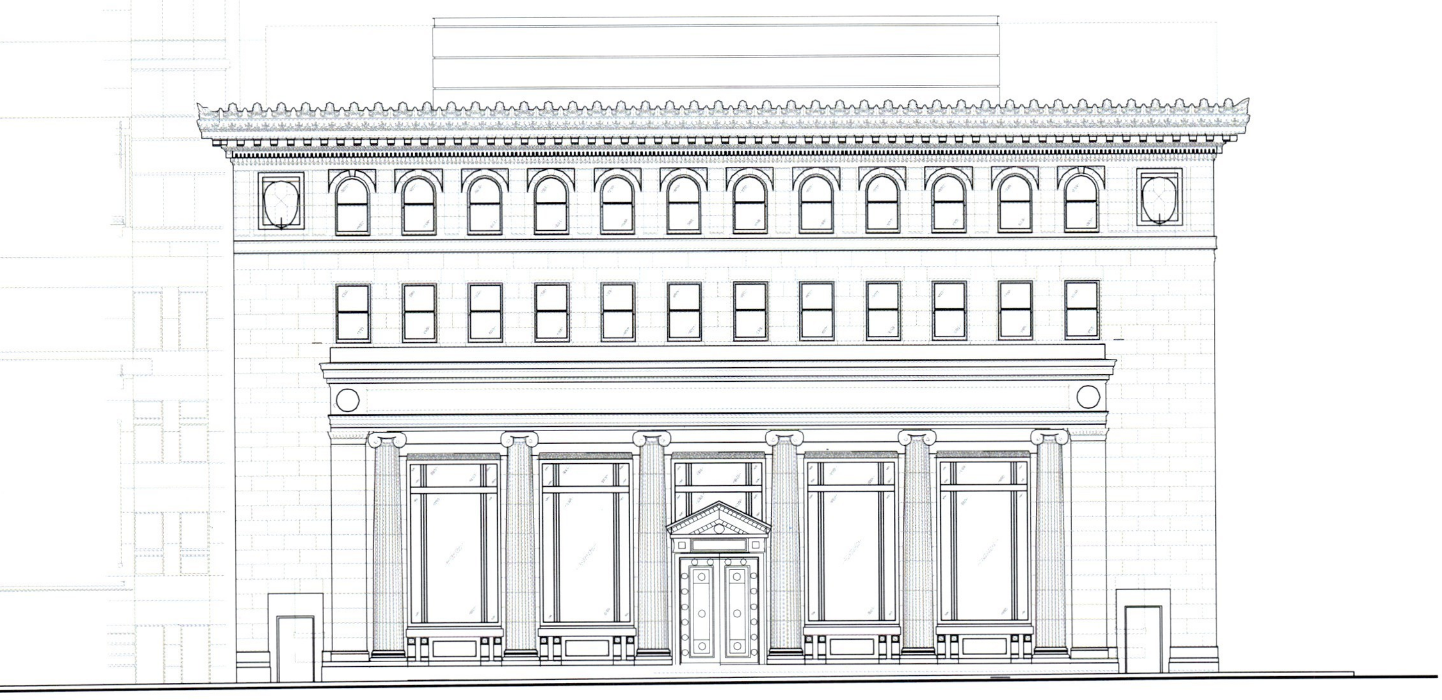

23

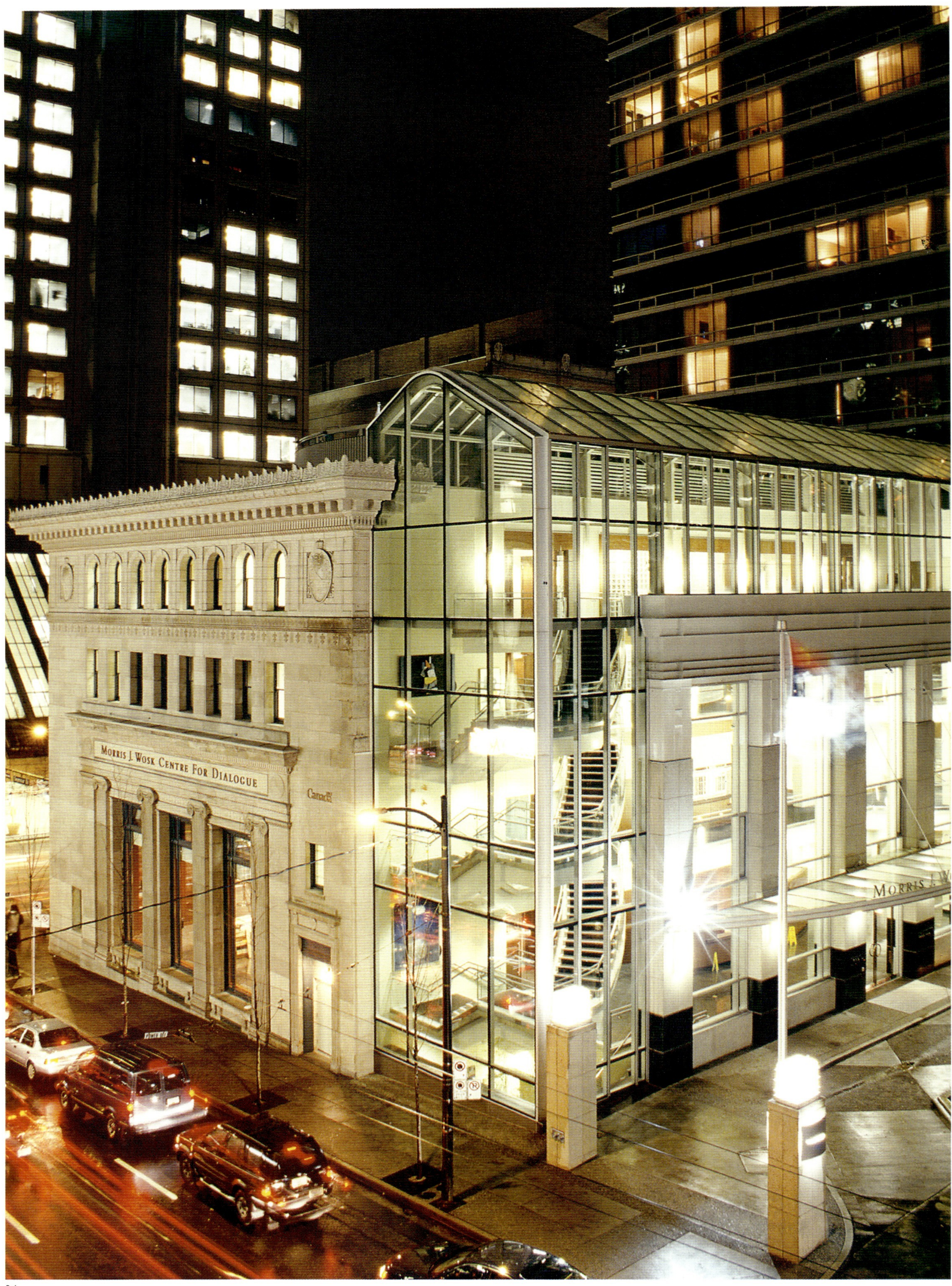

24

24　建筑夜景
25　立面细部
26　立面细部
27　西立面图

25

26

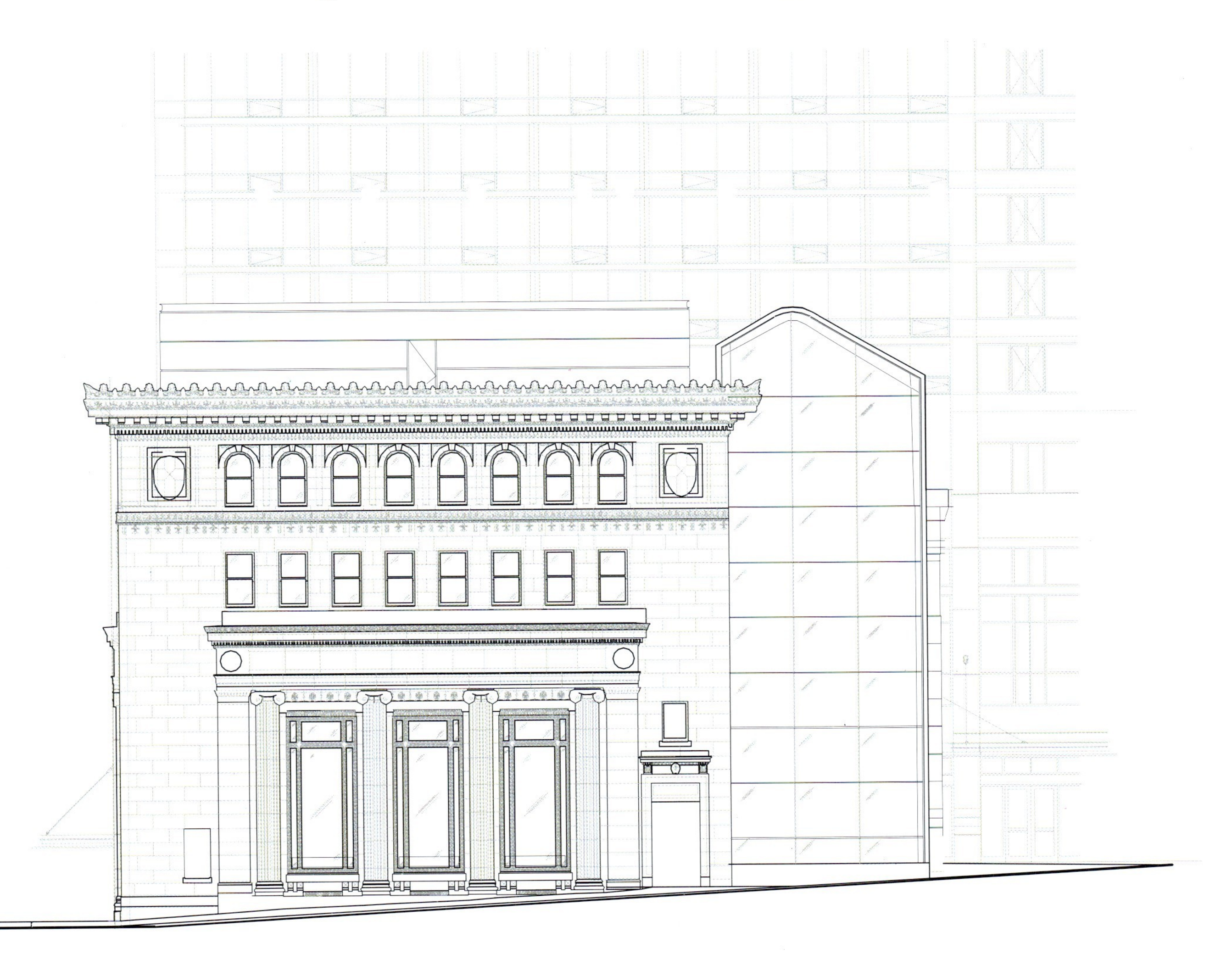

27

28

29

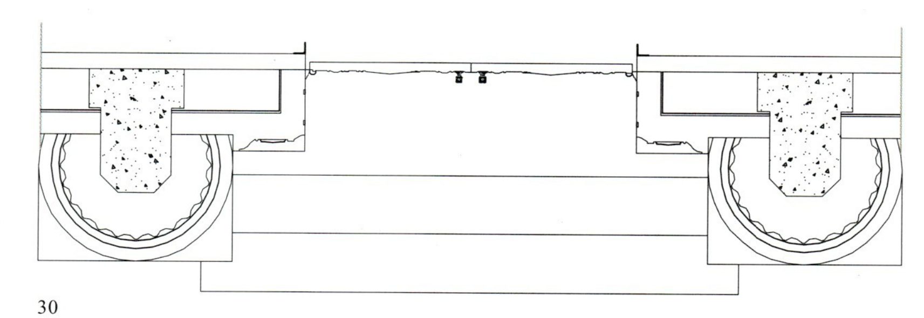

30

31

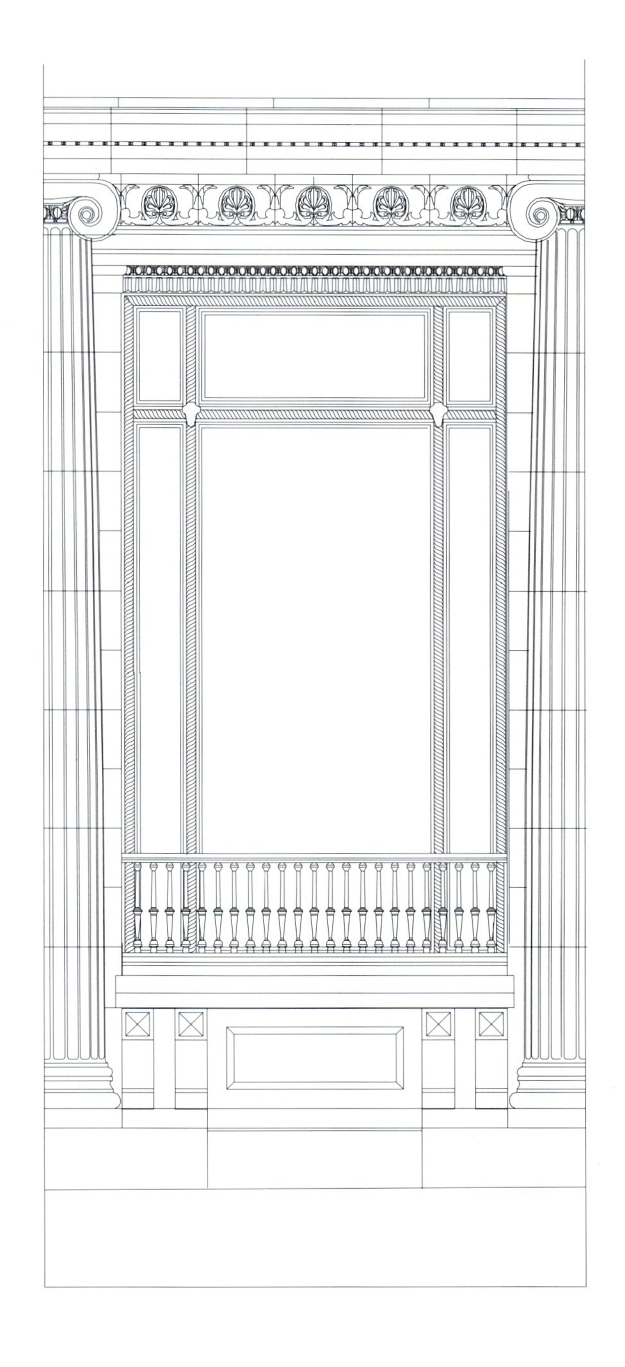

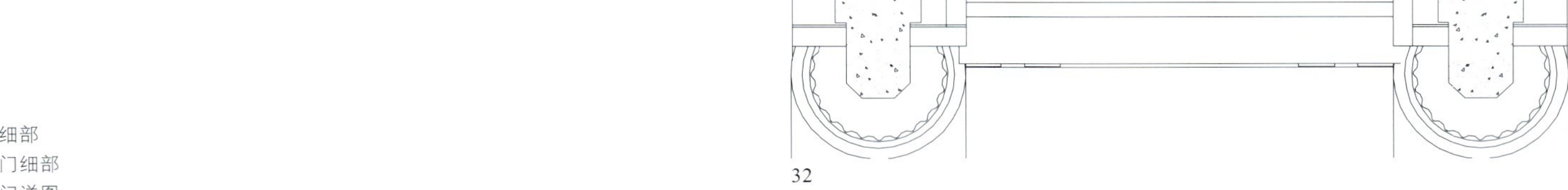
32

28 细部
29 门细部
30 门详图
31 中心秋景
32 窗详图

Teeple Architects Inc.

Teeple 建筑设计事务所

From its inception in 1989, Teeple Architects Inc. has built a reputation for innovative design and exceptional service. The firm established this reputation through a broad range of institutional, commercial and residential projects. The firm is known for designing projects of exceptional material and spatial quality, with a strong conceptual basis derived from the specific needs and aspirations of each client.

The goal of Teeple Architects is to create innovative design projects in which the architectural concept is intimately linked to the day-to-day use and inhabitation of the building. An ability to respond creatively to the dictates of the site, context, budget and client requirements has characterized the work. In this era of specialization, the firm believes in the continuing value of the general architectural practice and to this end, has pursued work in a variety of fields at a wide range of scales.

The firm believes that a high level of caring, diligent architectural services must be combined with innovative, creative design abilities to achieve true design excellence. Successful architecture results from dedication to the entire design and construction process. To this end, at each stage of the process, they seek to ensure that the detailed concerns of the client and user are fully incorporated into the building.

Teeple 建筑设计事务所创建于 1989 年，以创新化的设计和出众的服务闻名于加拿大建筑界。Teeple 建筑设计事务所的业务范围广泛，包括文化建筑、商业建筑和居住建筑。每一个业主的个性化要求和灵感是Teeple建筑设计事务所设计最初考虑的方向。

Teeple建筑设计事务所的目标是创造革新化的建筑，可以充分体现日常使用和建筑现场的特点。现场的特点、建筑的内容、预算和客户的要求是建筑创新的基础。正因为如此，Teeple 建筑设计事务所获得了众多类型的、规模各异的建筑任务。

Teeple建筑设计事务所相信高层次的优质设计服务必须包括设计创新的能力，成功的建筑设计贯穿设计和建造的全过程。因此，Teeple 建筑设计事务所试图确保在建筑设计全过程的每一个阶段与业主和使用者有关的每一个细节都能与整个建筑完全融合。

East End Community Health Centre
东端社区健康中心

业　　主：东端社区健康中心
地　　点：多伦多市
规　　模：约 1 580m²
完成时间：2003 年 8 月

This centre is designed as a "healthy building" - a building that promotes health through the use of sustainable materials and that feels healthy and welcoming. To this end, ones principal experience of the building is one of nature, the principle focus of the design. The center piece of the design is a small-protected courtyard. From within the facility one will view the seasonal changes of the natural environment, while at the same time the court will serve as a soothing space. Its protected climate will ensure that it can be used over much of the year.

The health centre provides three distinct types of care services - community, mental and physical health. Each of these departments is signified by a brightly colour light well, that animates the interior with natural light and colour. The building includes outreach facilities such as children's playroom, a community kitchen and community room, consultation rooms, examination and procedure rooms as well as laboratories and administrative spaces.

Sustainable design features include: the use of natural, low VOC materials, free mechanical units, heat recovery, natural ventilation and energy efficient envelope design and lighting.

1

2

3

东端社区健康中心设计为一个健康的建筑——使用可持续化的建筑材料和外表亲切的建筑造型。东端社区健康中心的两个设计理念：一个是自然健康，另一个是注重设计。东端社区健康中心的重点是一个小小的受保护的院落。由室内的角度来看，院落的景色是随四季的变换而改变的，同时院落还是一个放松精神的空间。独特的设计使其在全年的大多数时间内都可以使用。

东端社区健康中心提供三种不同的医疗服务：社区医疗、精神健康和理疗护理。每一个部门由不同的色彩艳丽的采光天井相互区分。也是这些采光天井赋予室内以自然光线和缤纷的色彩。中心内设有一些服务设施，包括儿童游戏室、社区厨房、社区多功能室、咨询室、检测室、等候室、实验室和行政办公室。

可持续化设计的特点包括：使用天然、低VOC建筑材料，无空调房间，热能回收，自然通风，低能耗外墙设计和采光设计。

1 健康中心庭院
2 入口夜景
3 健康中心庭院夜景

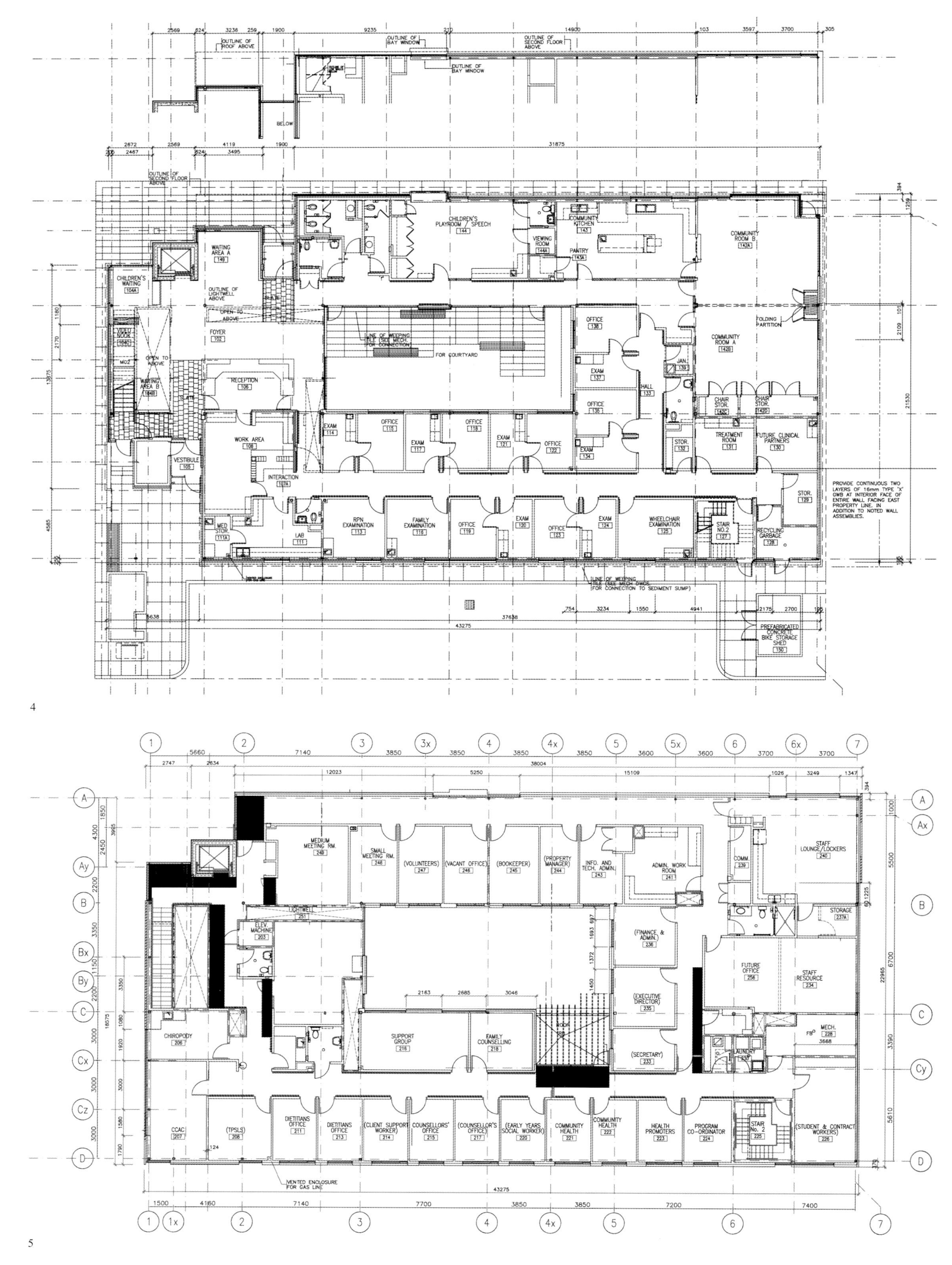

OUTLINE OF ROOF ABOVE
OUTLINE OF BAY WINDOW
OUTLINE OF SECOND FLOOR ABOVE
BELOW
CHILDREN'S PLAYROOM / SPEECH 144
COMMUNITY KITCHEN 143
VIEWING ROOM 144A
PANTRY 143A
COMMUNITY ROOM B 142A
WAITING AREA A 149
CHILDREN'S WAITING 104A
OUTLINE OF LIGHTWELL ABOVE
OPEN TO ABOVE
FOYER 102
VIDEO NOOK 104C
OPEN TO ABOVE
WAITING AREA B 104B
RECEPTION 106
FOR COURTYARD
OFFICE 138
EXAM 137
OFFICE 135
HALL 133
JAN. 139
COMMUNITY ROOM A 142B
FOLDING PARTITION
CHAIR STOR. 142C
CHAIR STOR. 142D
WORK AREA 108
VESTIBULE 105
INTERACTION 107A
EXAM 114
OFFICE 115
EXAM 117
OFFICE 118
EXAM 121
OFFICE 122
EXAM 134
STOR. 132
TREATMENT ROOM 131
FUTURE CLINICAL PARTNERS 130
STOR. 129
MED STOR. 111A
LAB 111
RPN EXAMINATION 113
FAMILY EXAMINATION 116
OFFICE 119
EXAM 120
OFFICE 123
EXAM 124
WHEELCHAIR EXAMINATION 125
STAIR NO.2 127
RECYCLING GARBAGE 128
PROVIDE CONTINUOUS TWO LAYERS OF 16mm TYPE 'X' GWB AT INTERIOR FACE OF ENTIRE WALL FACING EAST PROPERTY LINE, IN ADDITION TO NOTED WALL ASSEMBLIES.
LINE OF WEEPING TILE (SEE MECH DWGS FOR CONNECTION TO SEDIMENT SUMP)
PREFABRICATED CONCRETE BIKE STORAGE SHED 150
4
MEDIUM MEETING RM. 249
SMALL MEETING RM. 248
(VOLUNTEERS) 247
(VACANT OFFICE) 246
(BOOKKEEPER) 245
(PROPERTY MANAGER) 244
INFO. AND TECH. ADMIN. 243
ADMIN. WORK ROOM 241
COMM. 239
STAFF LOUNGE/LOCKERS 240
STORAGE 237A
LIGHTWELL 251
ELEV MACHINE 203
(FINANCE & ADMIN.) 236
FUTURE OFFICE 256
STAFF RESOURCE 234
(EXECUTIVE DIRECTOR) 235
CHIROPODY 206
SUPPORT GROUP 216
FAMILY COUNSELLING 218
(SECRETARY) 233
MECH. 228
LAUNDRY
CCAC 207
(TPSLS) 208
DIETITIANS OFFICE 211
DIETITIANS OFFICE 213
(CLIENT SUPPORT WORKER) 214
COUNSELLORS' OFFICE 215
(COUNSELLOR'S OFFICE) 217
(EARLY YEARS SOCIAL WORKER) 220
COMMUNITY HEALTH 221
COMMUNITY HEALTH 222
HEALTH PROMOTERS 223
PROGRAM CO-ORDINATOR 224
STAIR No. 2 225
(STUDENT & CONTRACT WORKERS) 226
VENTED ENCLOSURE FOR GAS LINE
5

6

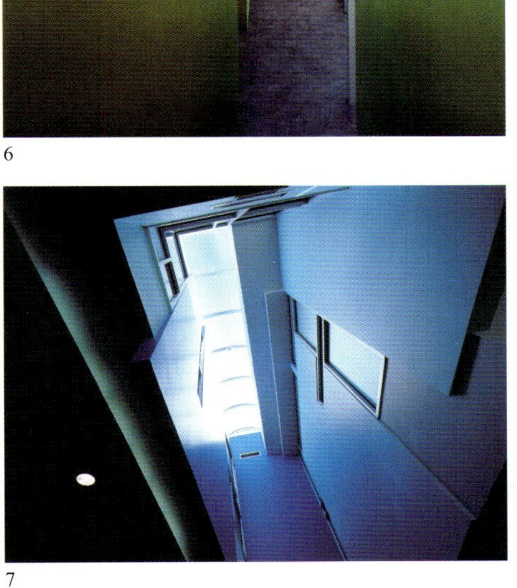

7

EXIT

8

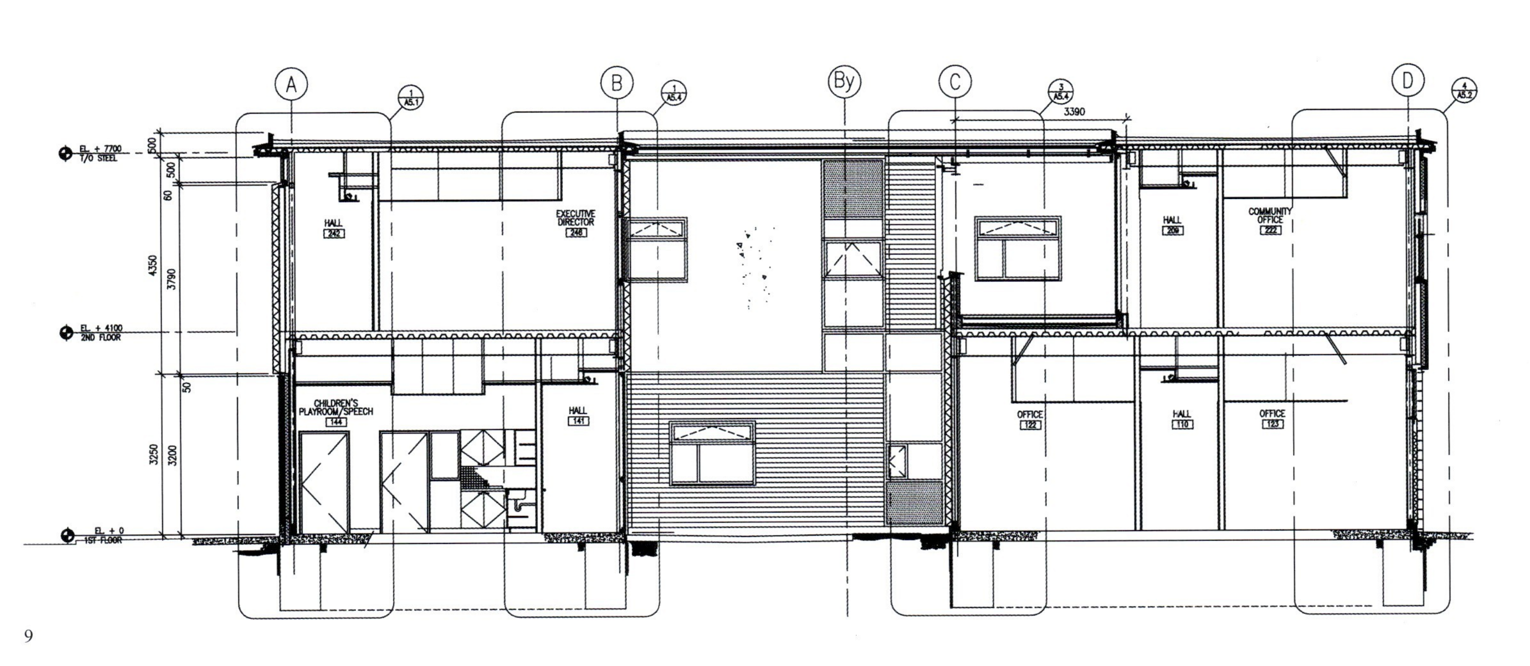
A
B
By
C
D
3390
EL + 7700
T/O STEEL
EL + 4100
2ND FLOOR
EL + 0
1ST FLOOR
4350
3790
3250
3200
HALL
EXECUTIVE DIRECTOR
HALL
COMMUNITY OFFICE
CHILDREN'S PLAYROOM/SPEECH
HALL
OFFICE
HALL
OFFICE

9

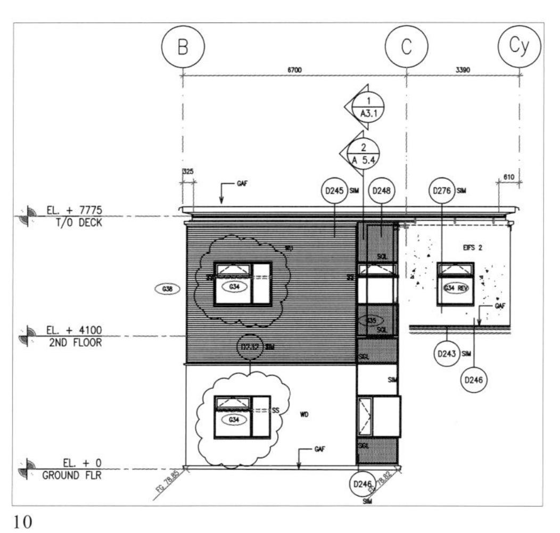

10

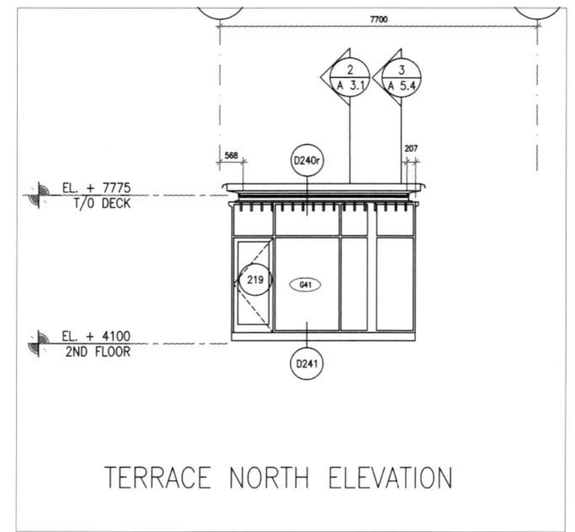

4　首层平面图
5　二层平面图
6　室内采光细部——以不同色彩划分区域
7　室内采光细部
8　室内采光天井及庭院
9　剖面图
10~16　墙体立面详图

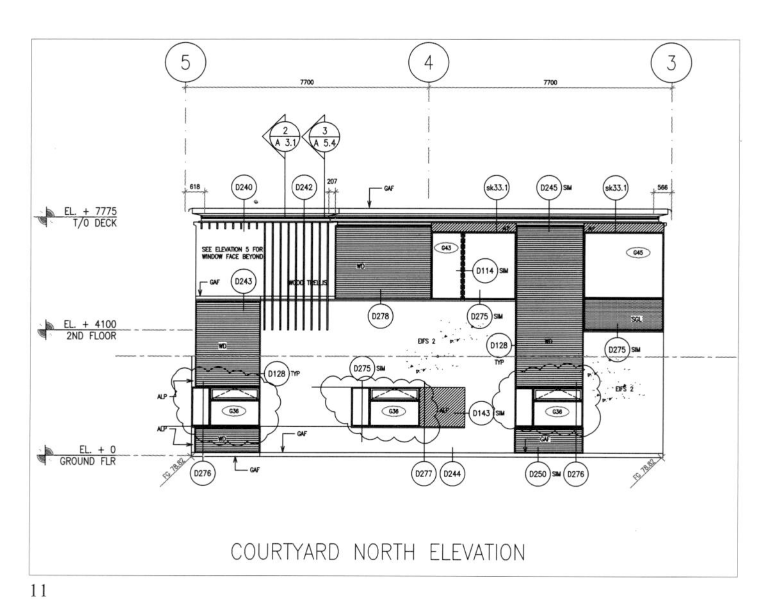

11

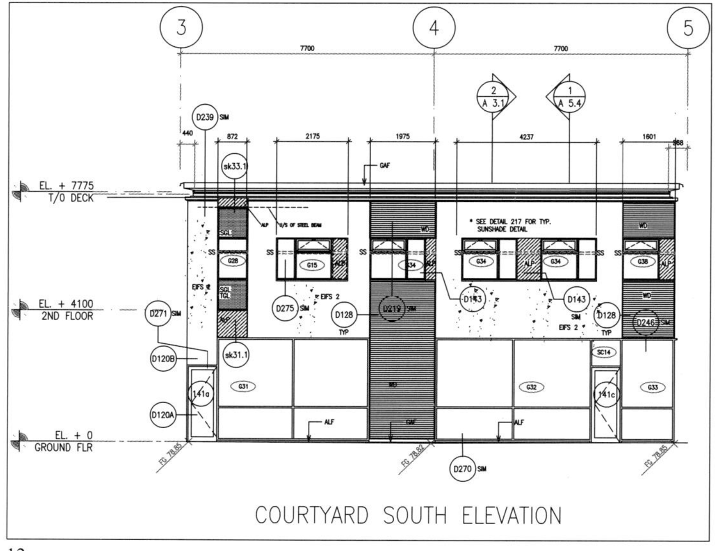

12

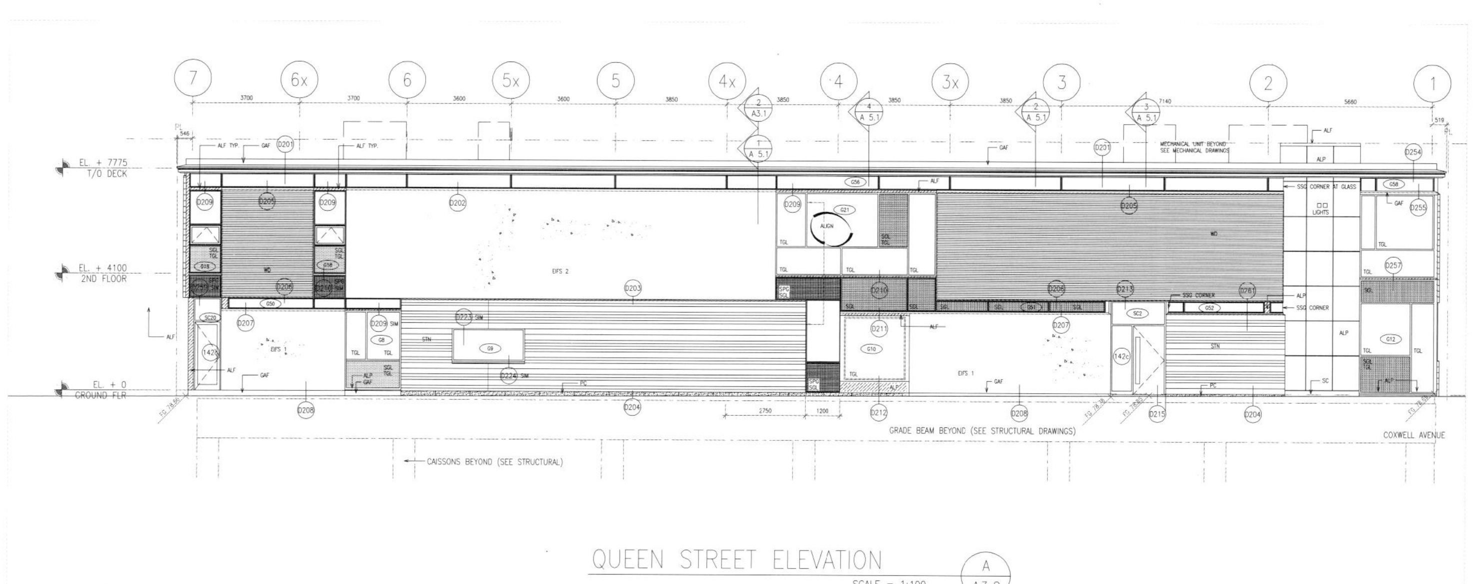

13

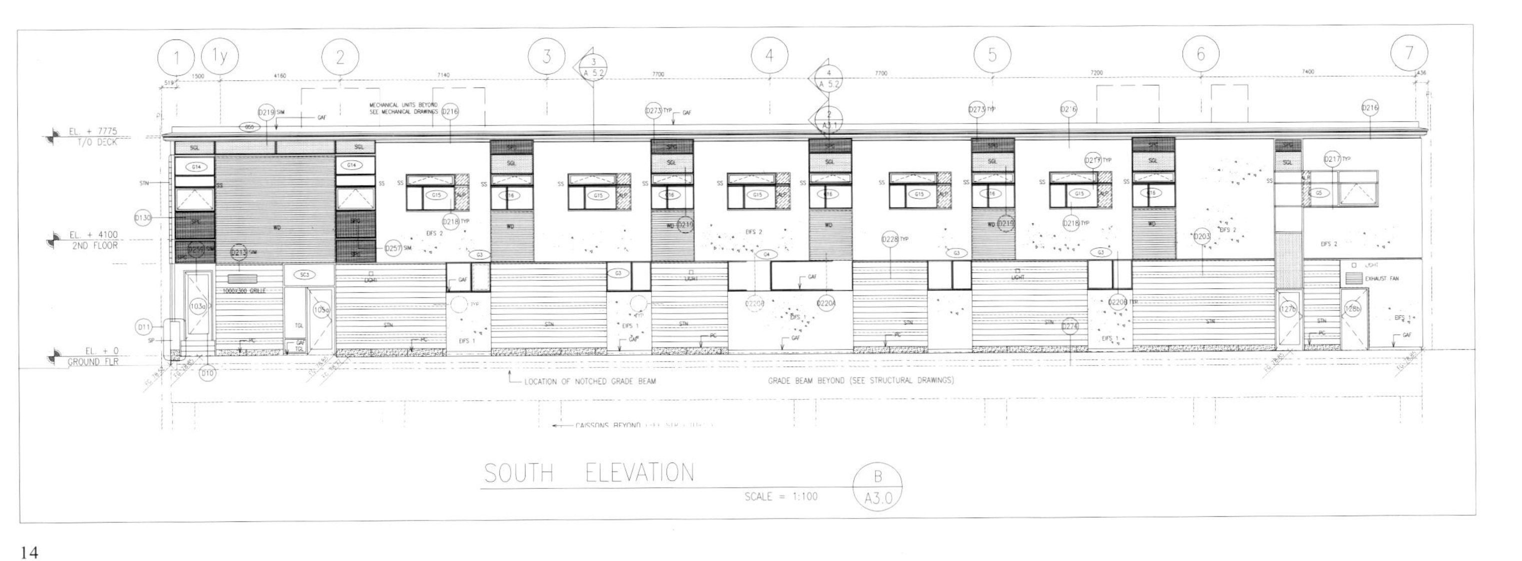

14

COXWELL AVENUE ELEVATION

SCALE = 1:100

C
A3.0

LEGEND

EIFS 1	EIFS WITH HEAVY DUTY MESH
EIFS 2	EIFS WITH REGULAR MESH
STN	STONE
WD	WOOD SIDING
GAF	"GALVALUME" FLASHING
ALF	ALUMINUM FLASHING
ALP	ALUMINUM PANELS
PC	PRECAST CONCRETE CURB
SS	SUNSCREEN – SEE D217 TYP
SGL	SAND BLASTED GLASS
SPG	SAND BLASTED GLASS SPANDREL PAN

CEDAR WOOD SIDING

STONE (EBBLE STONE)

EIFS (STUCCO)

EL. + 7775 T/O DECK

EL. + 4100 2ND FLOOR

EL. + 0 GROUND FLR

QUEEN STREET

LOCATION OF NOTCHED GRADE BEAM

GRADE BEAM BEYOND (SEE STRUCTURAL DRAWINGS)

CAISSONS BEYOND

15

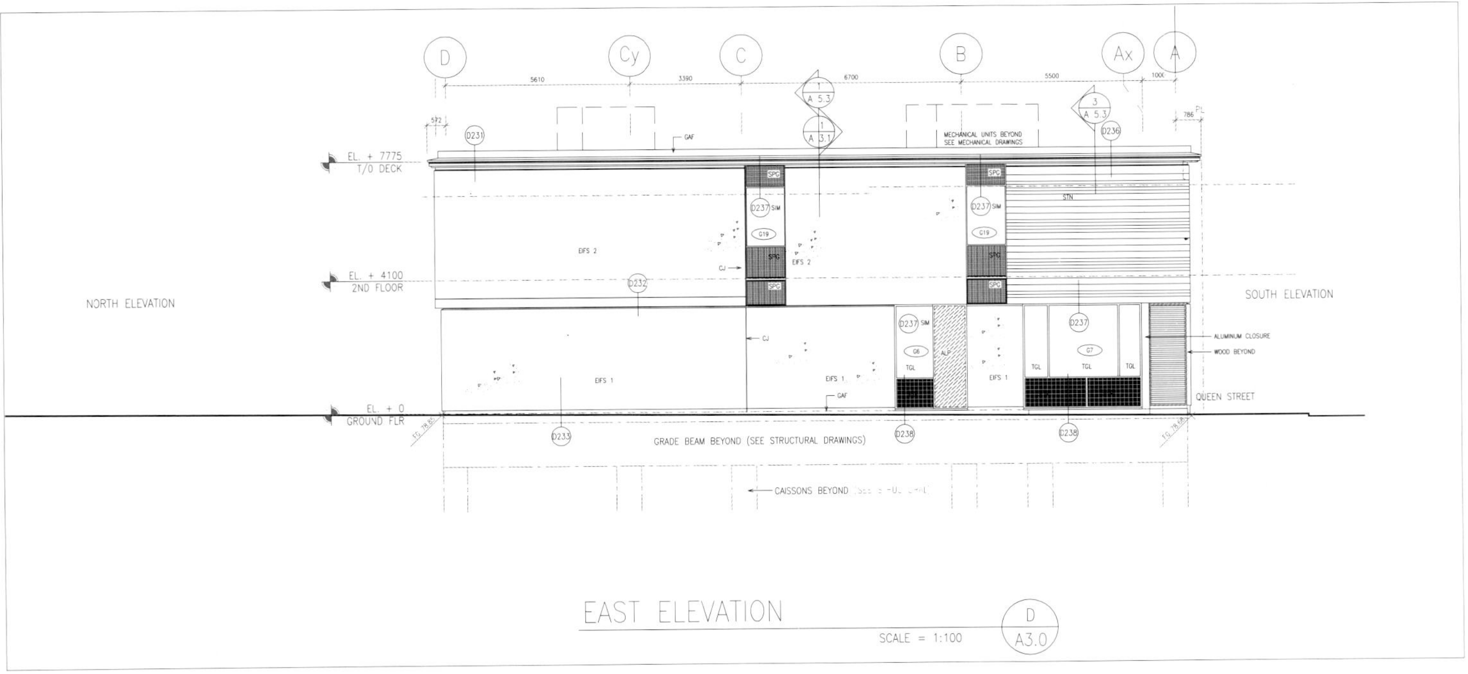

16

EAST END COMMUNITY HEALTH CENTRE

17

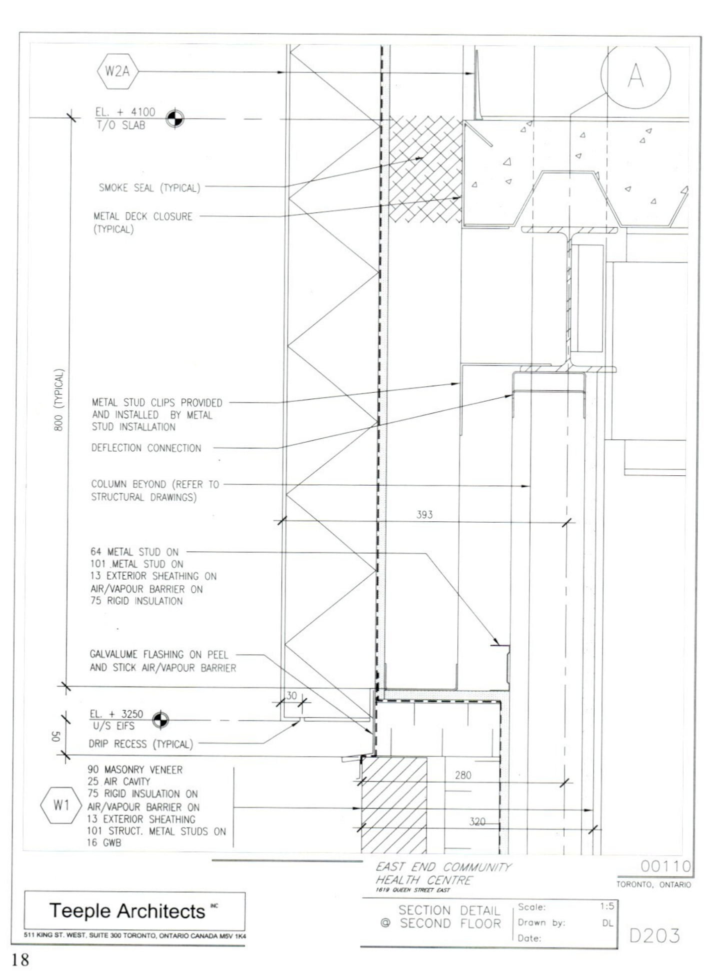
W2A
EL. + 4100
T/O SLAB
A
SMOKE SEAL (TYPICAL)
METAL DECK CLOSURE (TYPICAL)
800 (TYPICAL)
METAL STUD CLIPS PROVIDED AND INSTALLED BY METAL STUD INSTALLATION
DEFLECTION CONNECTION
COLUMN BEYOND (REFER TO STRUCTURAL DRAWINGS)
393
64 METAL STUD ON
101 METAL STUD ON
13 EXTERIOR SHEATHING ON
AIR/VAPOUR BARRIER ON
75 RIGID INSULATION
GALVALUME FLASHING ON PEEL AND STICK AIR/VAPOUR BARRIER
30
EL. + 3250
U/S EIFS
50
DRIP RECESS (TYPICAL)
90 MASONRY VENEER
25 AIR CAVITY
75 RIGID INSULATION ON
AIR/VAPOUR BARRIER ON
13 EXTERIOR SHEATHING
101 STRUCT. METAL STUDS ON
16 GWB
W1
280
320
EAST END COMMUNITY HEALTH CENTRE
1619 QUEEN STREET EAST
00110
TORONTO, ONTARIO
Teeple Architects INC
511 KING ST. WEST, SUITE 300 TORONTO, ONTARIO CANADA M5V 1K4
SECTION DETAIL @ SECOND FLOOR
Scale: 1:5
Drawn by: DL
Date:
D203

18

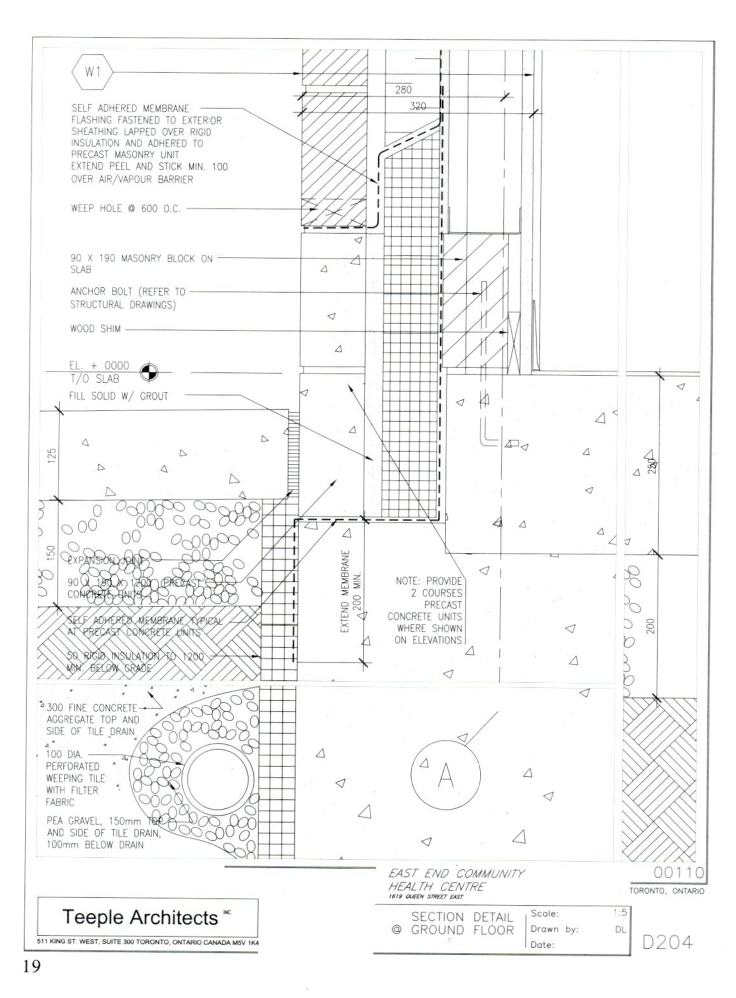
W1
280
320
SELF ADHERED MEMBRANE FLASHING FASTENED TO EXTERIOR SHEATHING LAPPED OVER RIGID INSULATION AND ADHERED TO PRECAST MASONRY UNIT EXTEND PEEL AND STICK MIN. 100 OVER AIR/VAPOUR BARRIER
WEEP HOLE @ 600 O.C.
90 X 190 MASONRY BLOCK ON SLAB
ANCHOR BOLT (REFER TO STRUCTURAL DRAWINGS)
WOOD SHIM
EL. + 0000
T/O SLAB
FILL SOLID W/ GROUT
125
150
EXPANSION JOINT
90 X 180 X 1200 PRECAST CONCRETE UNITS
SELF ADHERED MEMBRANE TYPICAL AT PRECAST CONCRETE UNITS
50 RIGID INSULATION TO 1200 MIN. BELOW GRADE
EXTEND MEMBRANE 200 MIN.
NOTE: PROVIDE 2 COURSES PRECAST CONCRETE UNITS WHERE SHOWN ON ELEVATIONS
280
200
300 FINE CONCRETE AGGREGATE TOP AND SIDE OF TILE DRAIN
100 DIA. PERFORATED WEEPING TILE WITH FILTER FABRIC
PEA GRAVEL, 150mm TOP AND SIDE OF TILE DRAIN, 100mm BELOW DRAIN
A
EAST END COMMUNITY HEALTH CENTRE
1619 QUEEN STREET EAST
00110
TORONTO, ONTARIO
Teeple Architects INC
511 KING ST. WEST, SUITE 300 TORONTO, ONTARIO CANADA M5V 1K4
SECTION DETAIL @ GROUND FLOOR
Scale: 1:5
Drawn by: DL
Date:
D204

19

17　沿城市快速车道立面
18~19,23~24　墙体详图
20~22　门节点详图

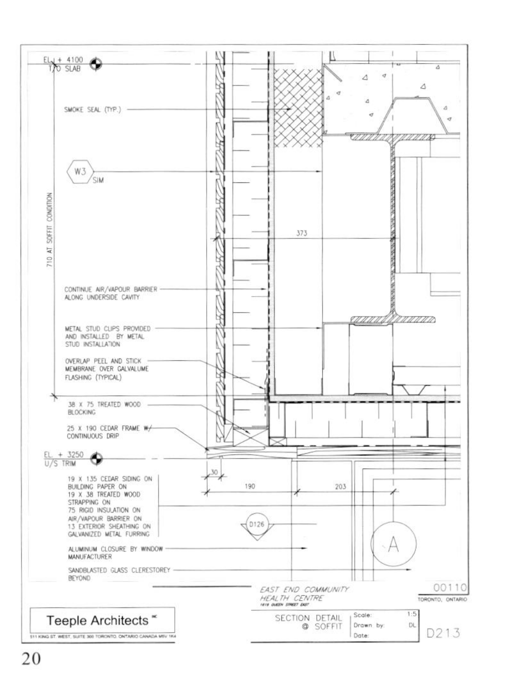

20

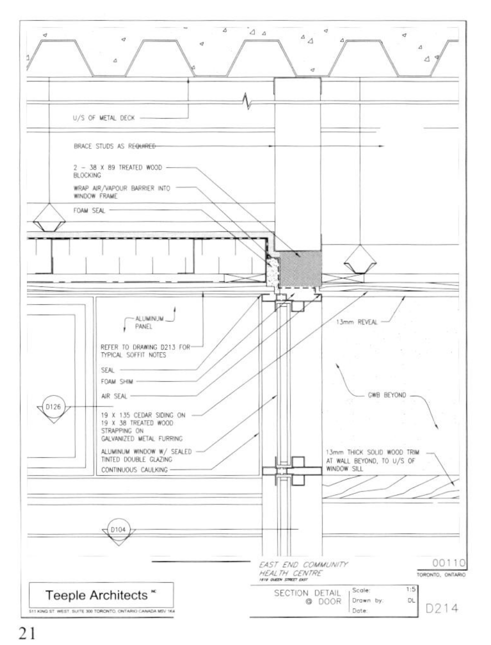

21

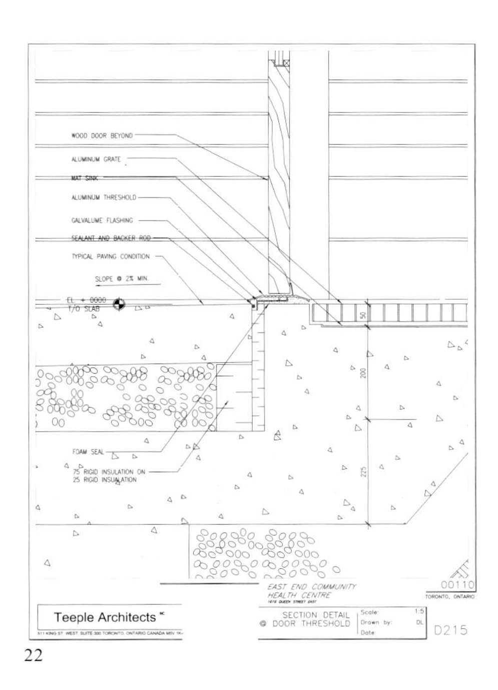

22

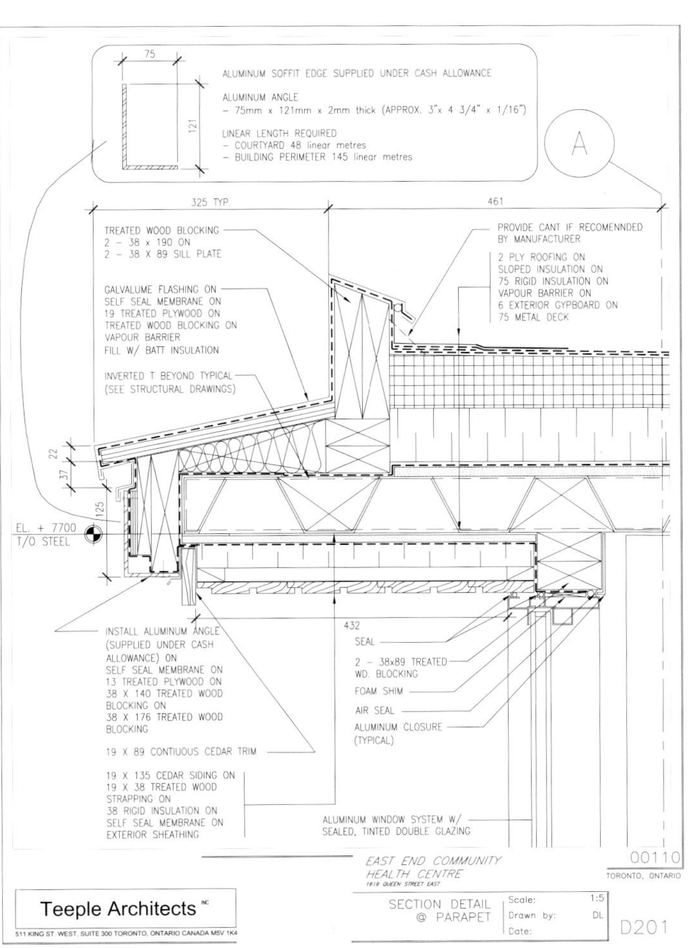

23

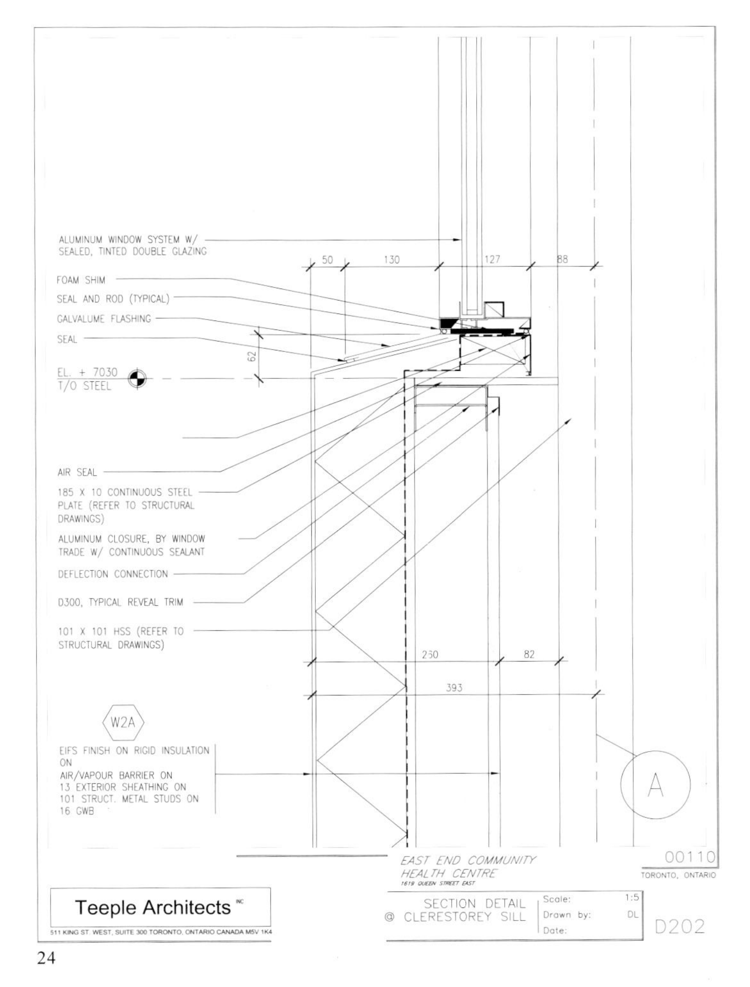

24

Eatonville Library
Eatonville 图书馆

业　　主：多伦多公共图书馆
地　　点：Etobicoke
规　　模：约 1 100m^2
完成时间：2001 年 7 月

The Eatonville Library replaced an existing facility that was outgrown by the local community. The library is situated at the edge of the city, at the intersection of a major city street and a regional highway leading to Toronto's international airport. It is at a juncture where the city's urbanity ends and the dynamic movement of the highway begins. The nature of this position is captured in the space of the library. A zinc band, reflecting the speed and dynamism of the highway, is stretched between stonewalls that ground the building to the urbanity of the street creating a tension between the local site and the larger environs.

The interior space of the library also reflects this tension. The interior ceilings are stretched between the inner surfaces of the exterior walls, pulling away from the walls to reveal the structure of the building. The facility includes fully networked public areas, a large children's area, study and lounge areas, as well as a community room.

1

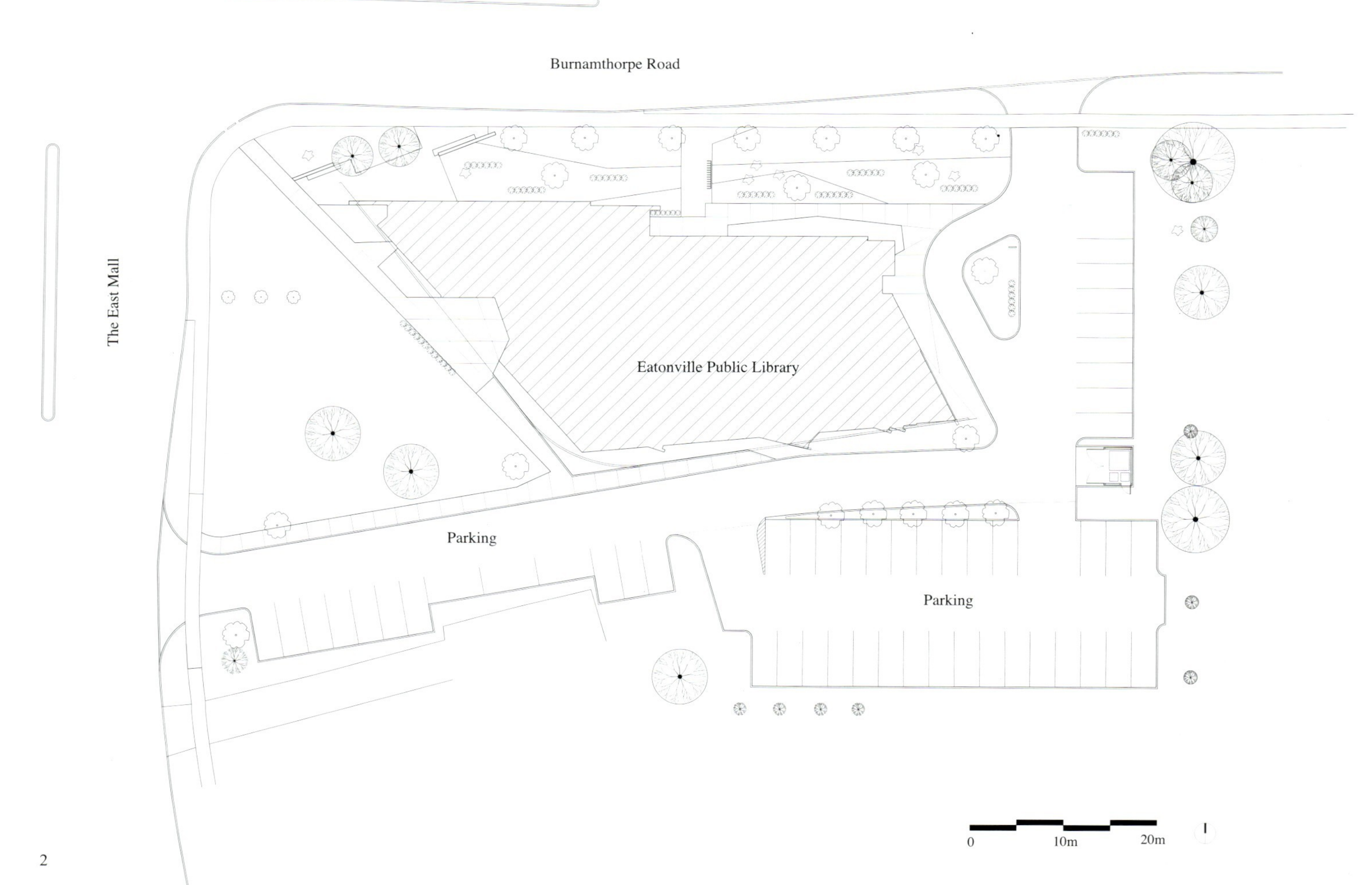

2

1 图书馆全景
2 总平面图
3 图书馆入口夜景
4 图书馆全景

Eatonville图书馆是新建的，以代替原有因当地社区成长而不敷使用的旧图书馆。Eatonville图书馆位于城市边缘，在城市干道和通向机场的高速公路的交叉口。这里是城市边缘和大规模高速公路交通起点的交会点，这些用地的特点影响建筑空间的设计。能够反射高速公路速度和活力的锌板伸出于石制的墙面。石制的墙面使建筑感觉上根植于城市街道的地面，而且营造出与社区和更大城市环境的联系。

建筑的室内同样反映出这种联系。由室外墙面一直延伸到室内屋顶，并裸露出建筑的构造。Eatonville图书馆设有网络化的公共空间、儿童室、学习休息区和公共活动室。

3

4

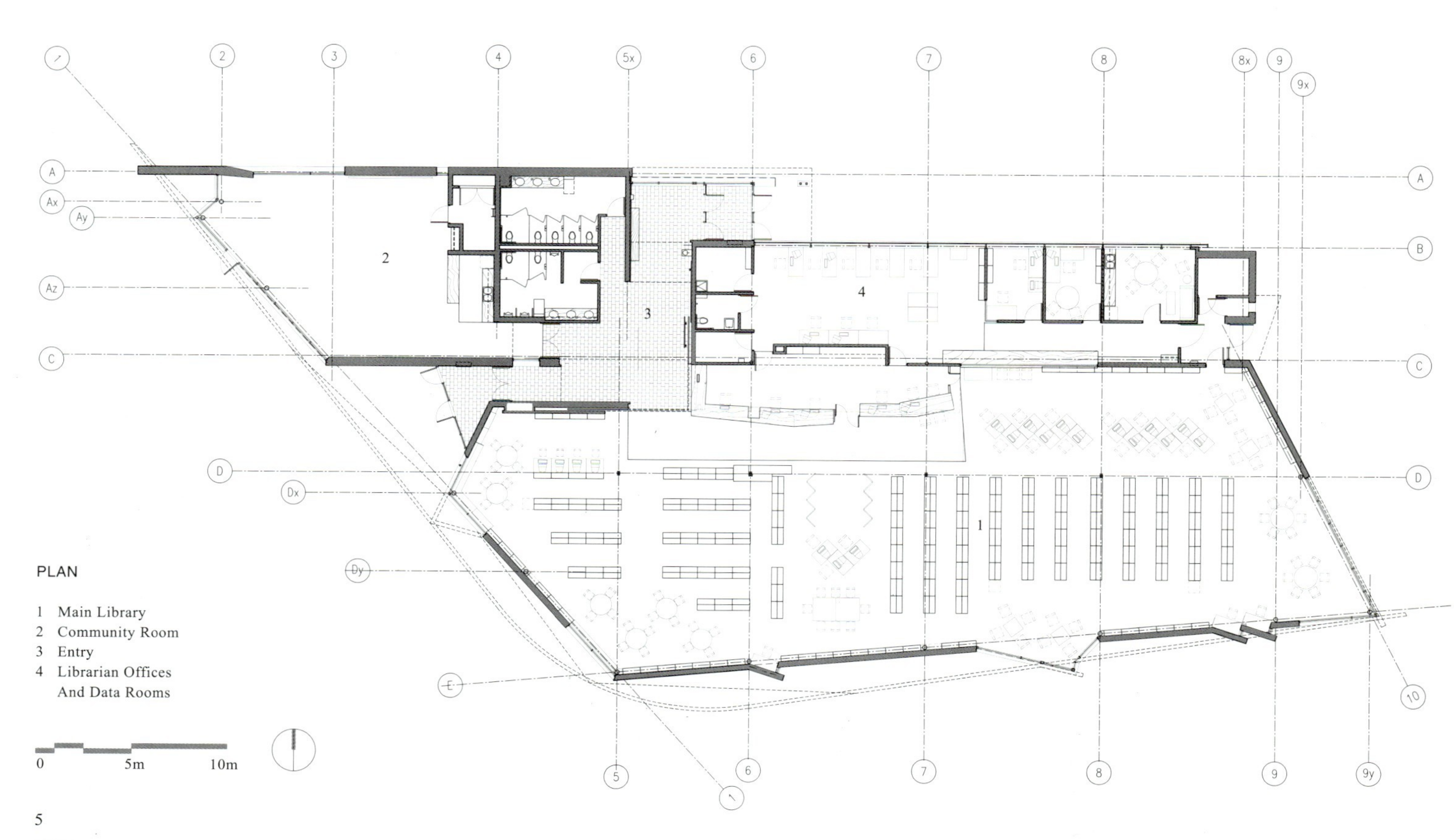
PLAN

1 Main Library
2 Community Room
3 Entry
4 Librarian Offices
And Data Rooms

0 5m 10m

5

6

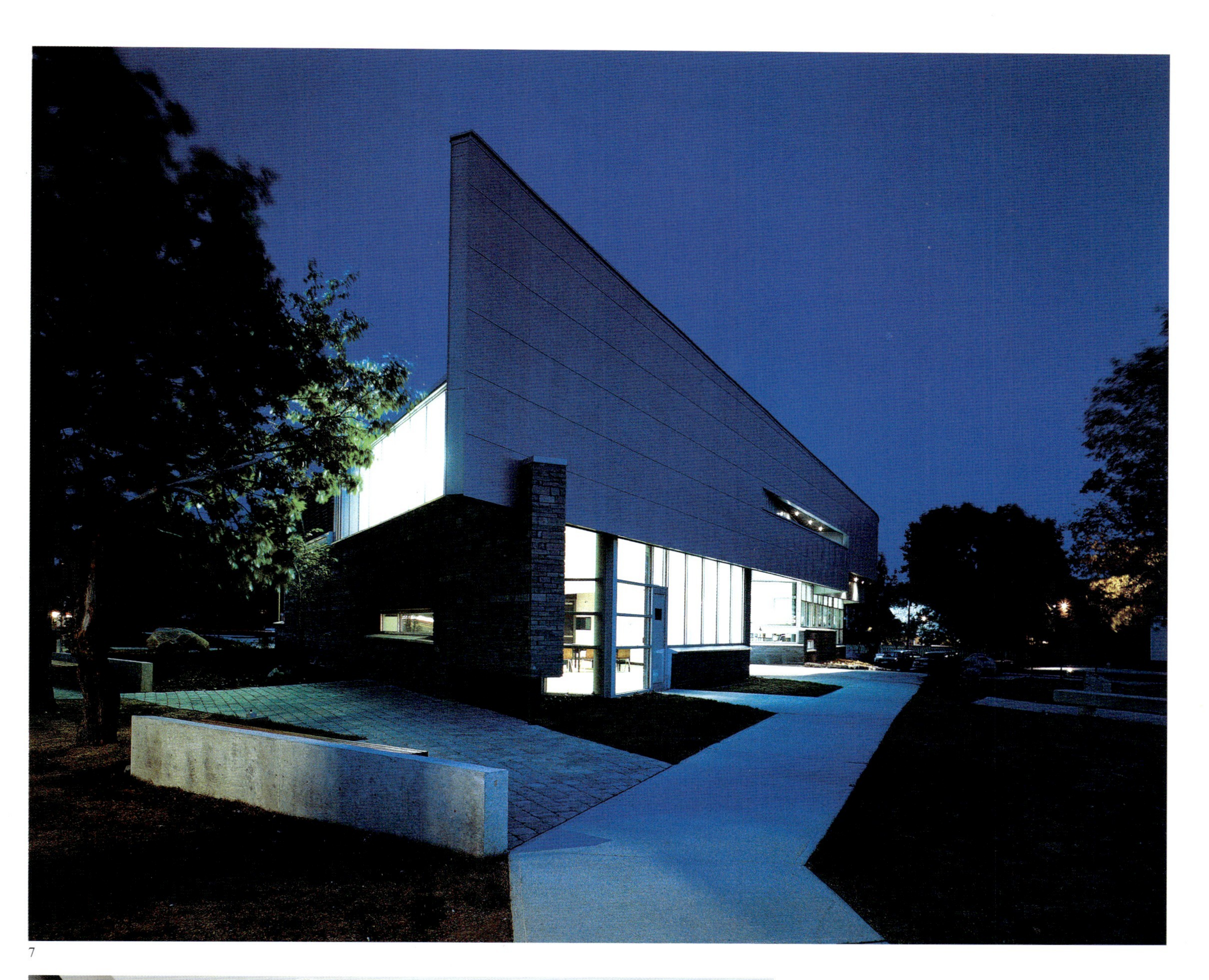

7

8

5　首层平面图
6　图书馆室内
7　图书馆夜景
8　图书馆室内

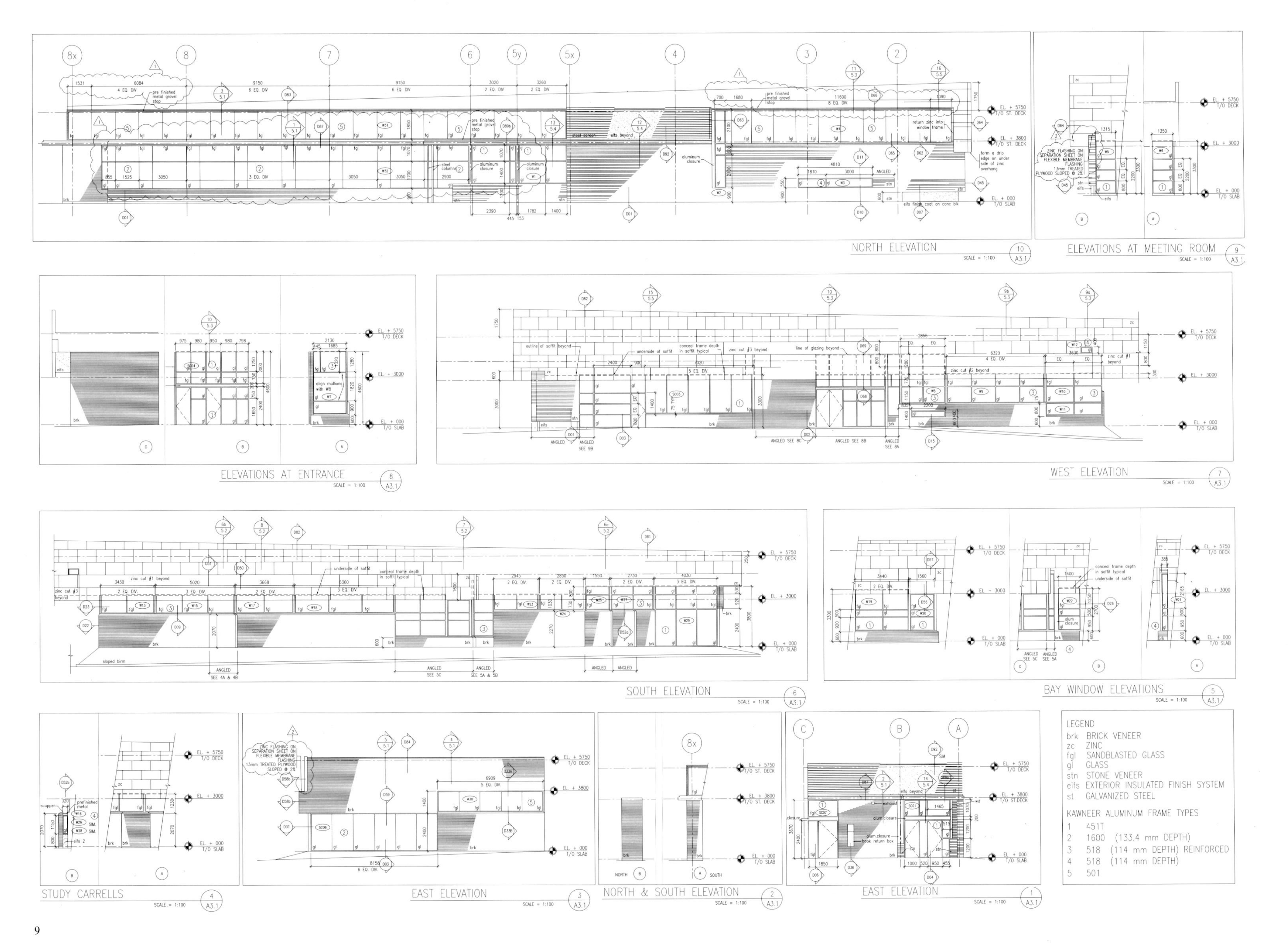
NORTH ELEVATION
SCALE = 1:100
ELEVATIONS AT MEETING ROOM
SCALE = 1:100
ELEVATIONS AT ENTRANCE
SCALE = 1:100
WEST ELEVATION
SCALE = 1:100
SOUTH ELEVATION
SCALE = 1:100
BAY WINDOW ELEVATIONS
SCALE = 1:100
STUDY CARRELLS
SCALE = 1:100
EAST ELEVATION
SCALE = 1:100
NORTH & SOUTH ELEVATION
SCALE = 1:100
EAST ELEVATION
SCALE = 1:100
LEGEND
brk BRICK VENEER
zc ZINC
fgl SANDBLASTED GLASS
gl GLASS
stn STONE VENEER
eifs EXTERIOR INSULATED FINISH SYSTEM
st GALVANIZED STEEL
KAWNEER ALUMINUM FRAME TYPES
1 451T
2 1600 (133.4 mm DEPTH)
3 518 (114 mm DEPTH) REINFORCED
4 518 (114 mm DEPTH)
5 501

9

9　立面详图
10　墙体细部
11　墙体剖面图

10

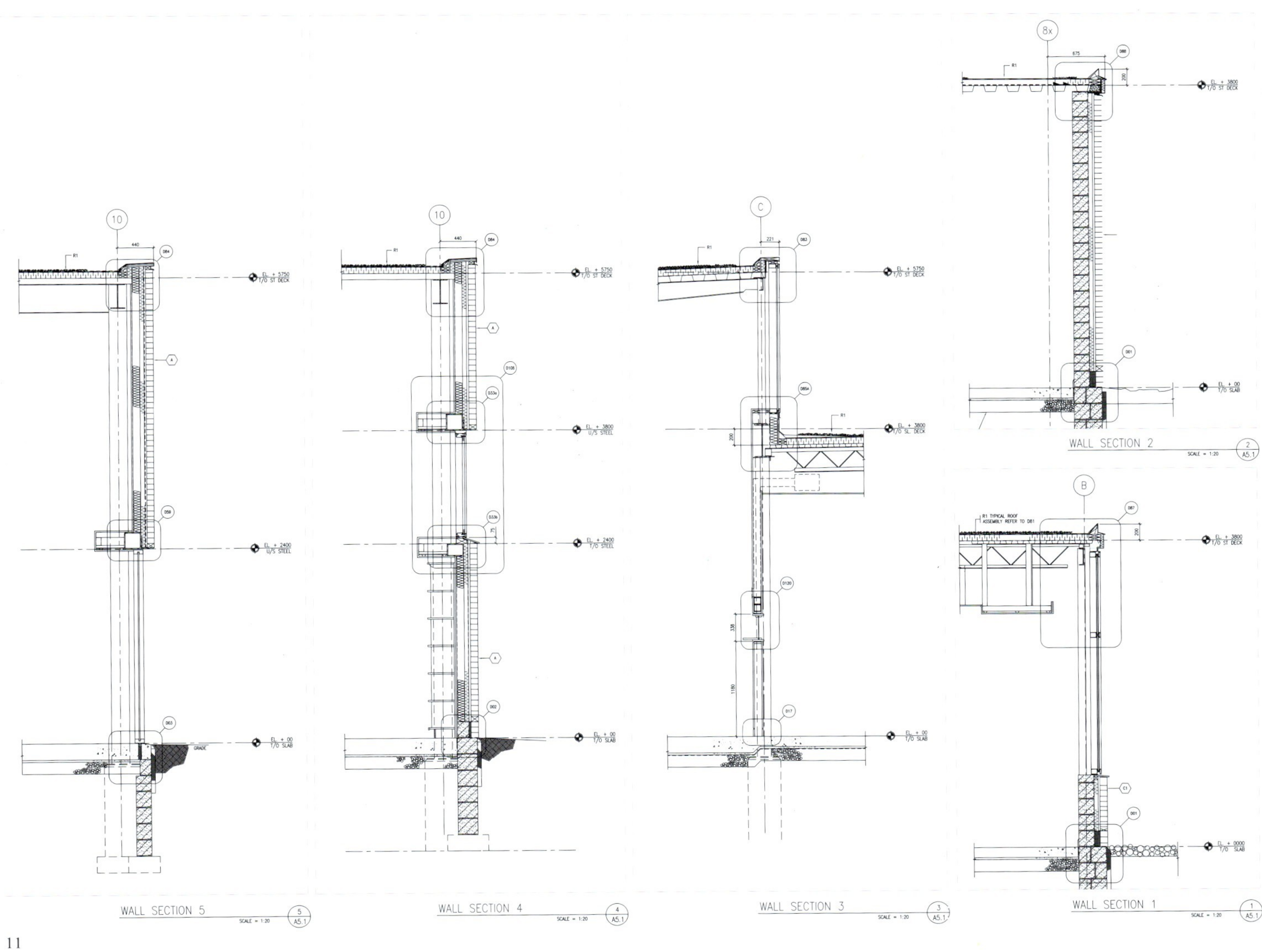

11

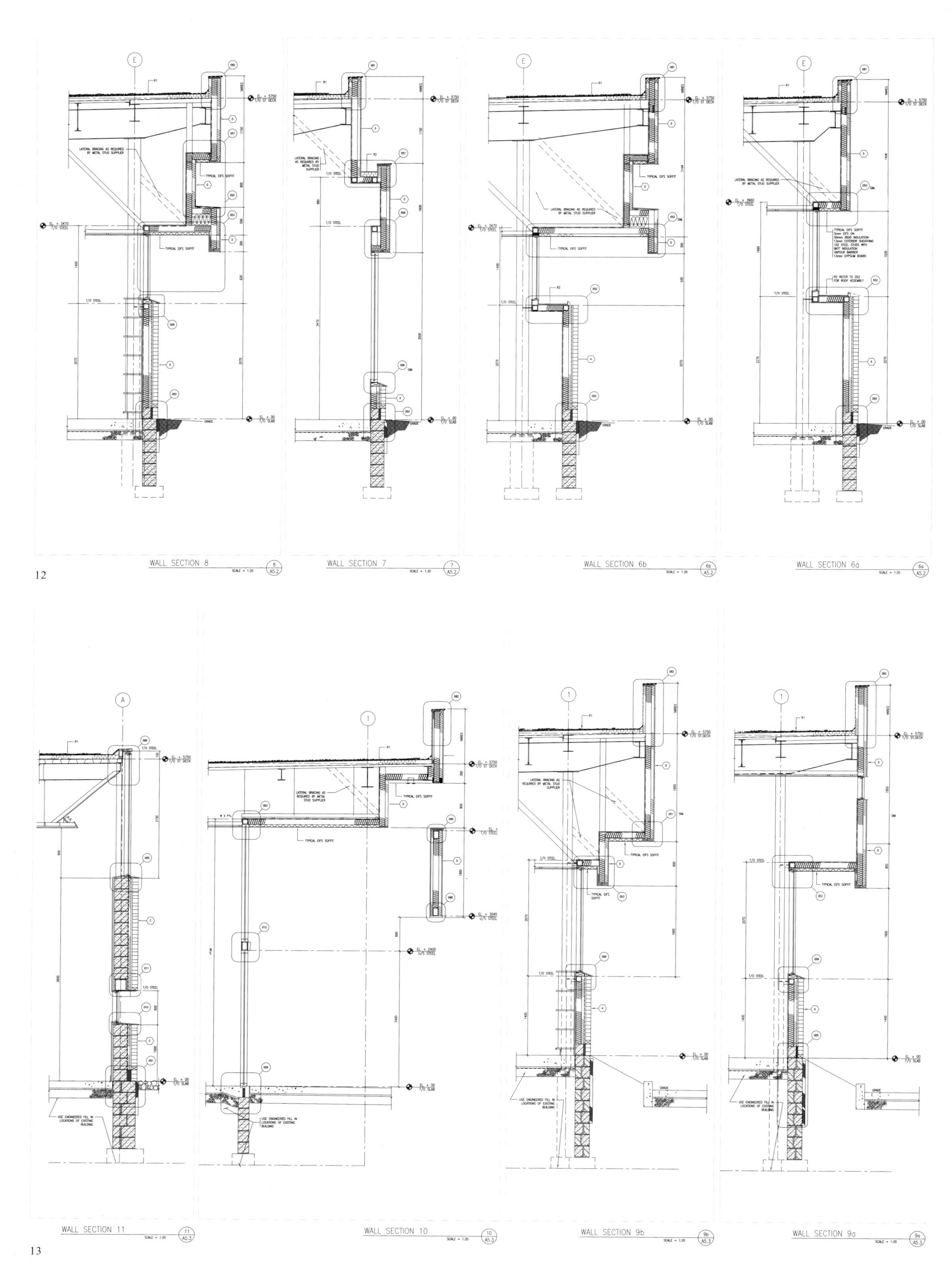
WALL SECTION 8
WALL SECTION 7
WALL SECTION 6b
WALL SECTION 6a
WALL SECTION 11
WALL SECTION 10
WALL SECTION 9b
WALL SECTION 9a

12

13

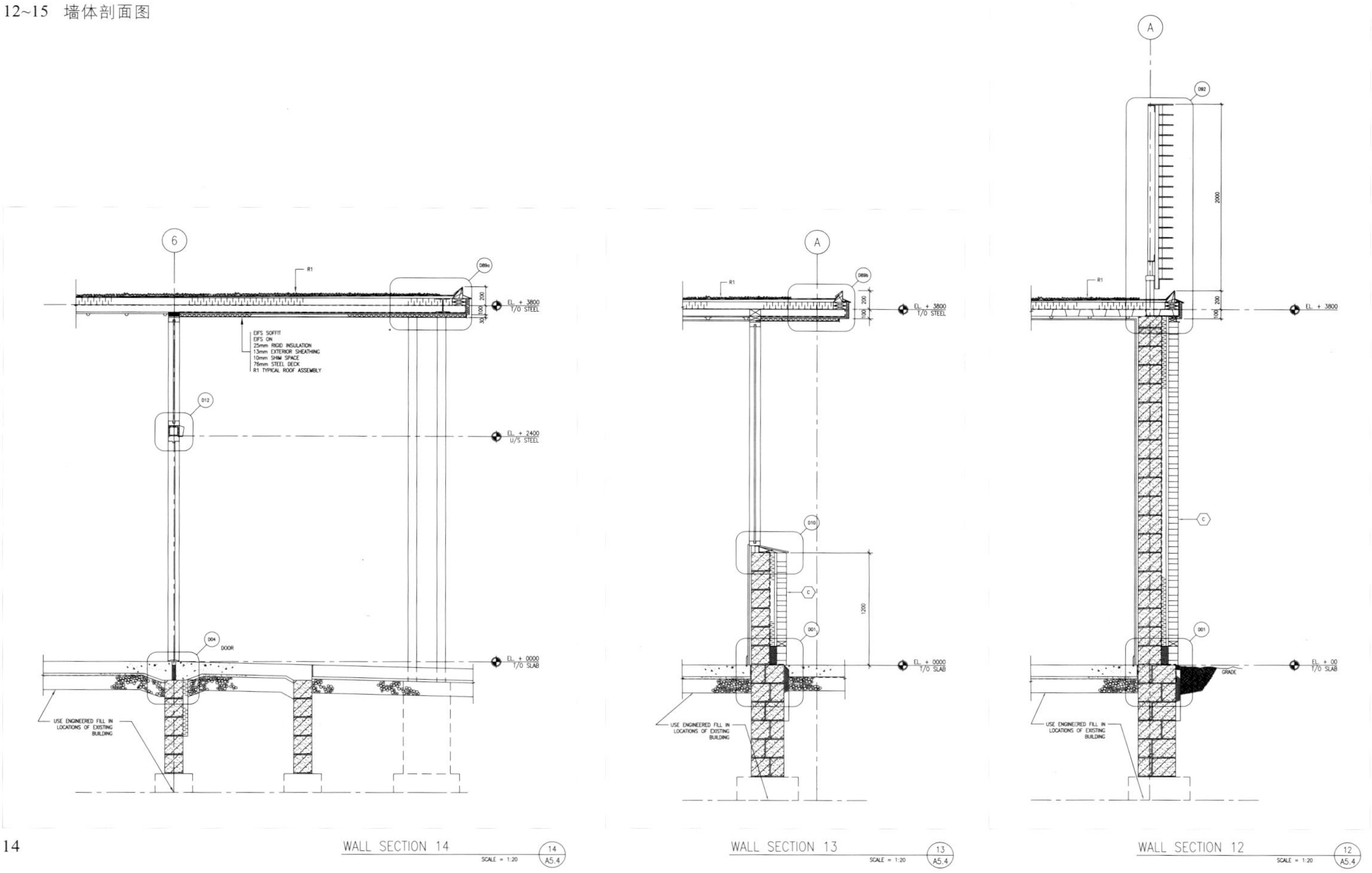

12~15 墙体剖面图

14

WALL SECTION 14

WALL SECTION 13

WALL SECTION 12

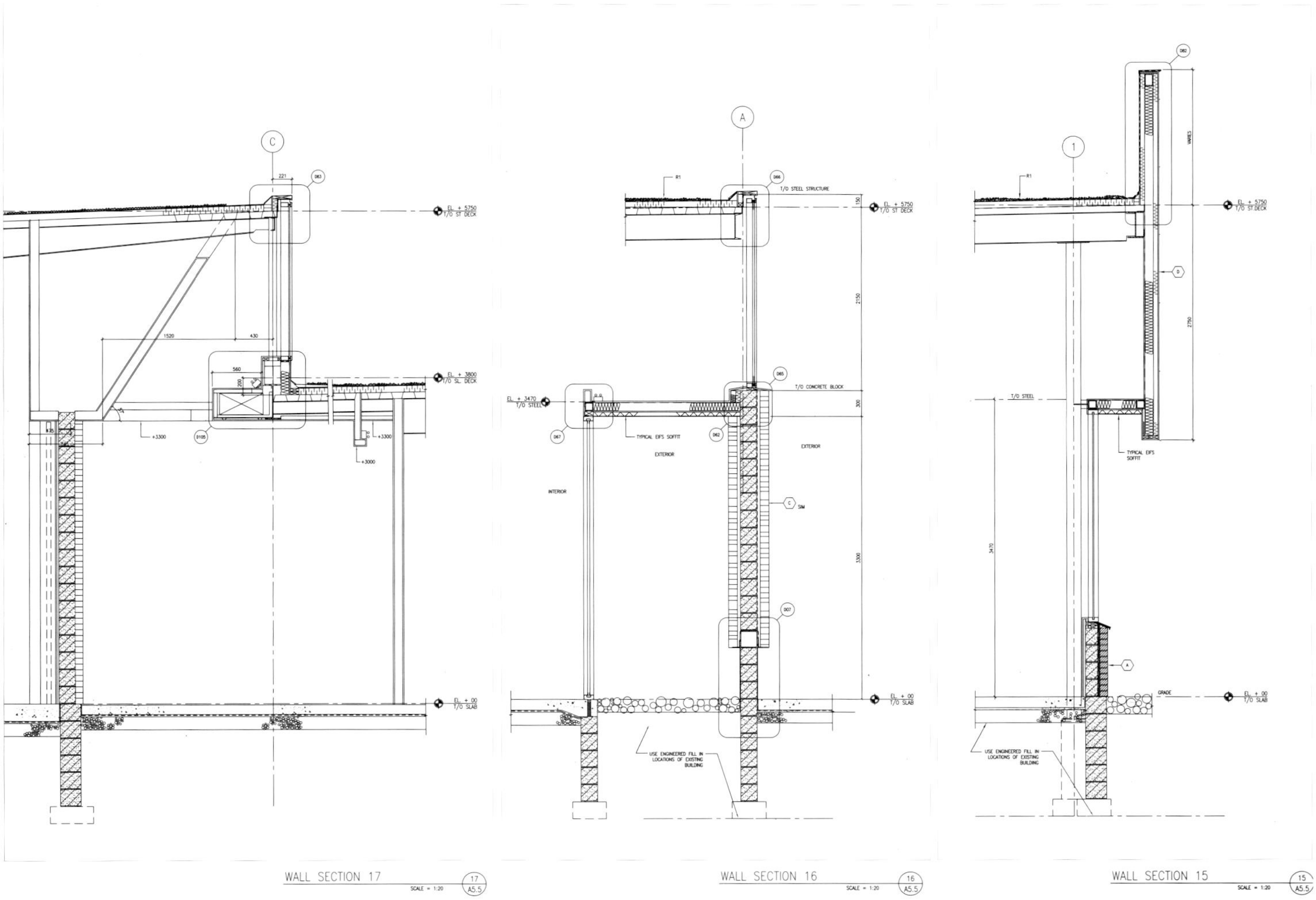

15

WALL SECTION 17

WALL SECTION 16

WALL SECTION 15

Early Learning Centre
早期教育中心

业　　主：多伦多大学
地　　点：多伦多市
规　　模：约 1 160m²
完成时间：2003 年 9 月

1　室内
2　早期教育中心入口
3　早期教育中心
4　立面详图

This project for 102 children amalgamates several existing campus facilities under one roof. This two and one half storey building incorporates shaded outdoor play space at every level, creating a variety of play environments. A strong interaction between the natural environment is reinforced with the use of a palate of natural materials in a treehouse-like setting, enclosed amongst mature trees to the south.

Playrooms are organized in pods, sharing service cores and a large multi-purpose room at each level. A play ramp and a series of lightwells connect these spaces and offer multilevel views to adjacent interior and exterior environments.

可容纳 102 个儿童的早期教育中心是重新组合已有的几个学校设施构成的。2 层半的建筑室内与有上盖的室外游戏空间相互结合，创造出不同的游戏空间。环绕在南方树丛中、树屋形的构筑物使用的天然材料更加强了建筑和自然环境之间的相互吸引力。

游戏室由位于每一层的小间、多功能服务中心和宽敞的多功能空间组成。一个斜坡和一系列天井连接这些空间，并提供多层次的室内外景观。

1

2

3

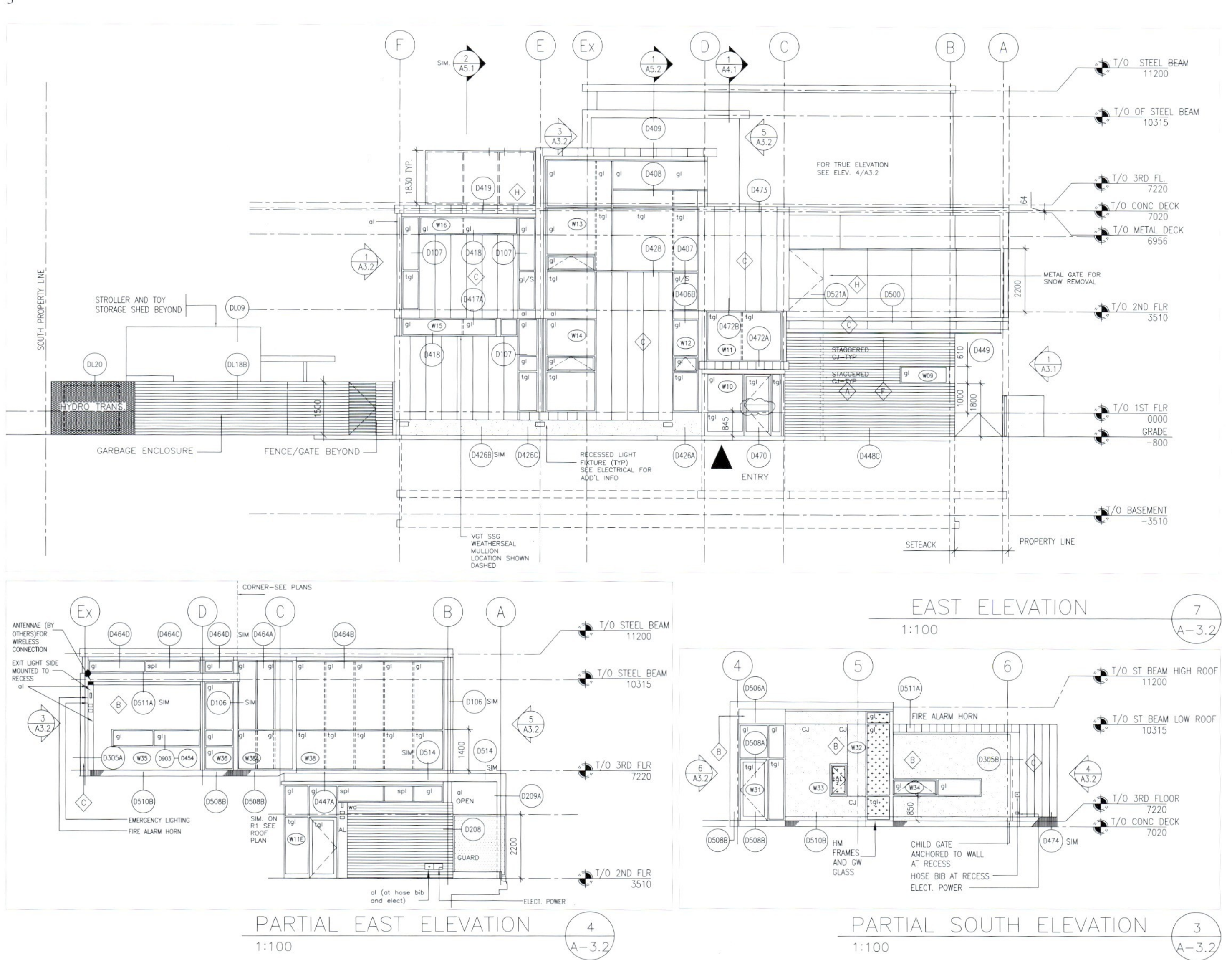
EAST ELEVATION
1:100
7
A-3.2
T/O STEEL BEAM 11200
T/O OF STEEL BEAM 10315
T/O 3RD FL. 7220
T/O CONC DECK 7020
T/O METAL DECK 6956
T/O 2ND FLR 3510
T/O 1ST FLR 0000
GRADE -800
T/O BASEMENT -3510
FOR TRUE ELEVATION SEE ELEV. 4/A3.2
METAL GATE FOR SNOW REMOVAL
STROLLER AND TOY STORAGE SHED BEYOND
SOUTH PROPERTY LINE
HYDRO TRANS.
GARBAGE ENCLOSURE
FENCE/GATE BEYOND
RECESSED LIGHT FIXTURE (TYP) SEE ELECTRICAL FOR ADD'L INFO
ENTRY
VGT SSG WEATHERSEAL MULLION LOCATION SHOWN DASHED
SETBACK
PROPERTY LINE
STAGGERED CJ-TYP
PARTIAL EAST ELEVATION
1:100
4
A-3.2
CORNER-SEE PLANS
ANTENNAE (BY OTHERS)FOR WIRELESS CONNECTION
EXIT LIGHT SIDE MOUNTED TO RECESS
EMERGENCY LIGHTING
FIRE ALARM HORN
SIM. ON R1 SEE ROOF PLAN
GUARD
al (at hose bib and elect)
ELECT. POWER
T/O STEEL BEAM 11200
T/O STEEL BEAM 10315
T/O 3RD FLR 7220
T/O 2ND FLR 3510
PARTIAL SOUTH ELEVATION
1:100
3
A-3.2
T/O ST BEAM HIGH ROOF 11200
T/O ST BEAM LOW ROOF 10315
T/O 3RD FLOOR 7220
T/O CONC DECK 7020
FIRE ALARM HORN
HM FRAMES AND GW GLASS
CHILD GATE ANCHORED TO WALL A" RECESS
HOSE BIB AT RECESS
ELECT. POWER

4

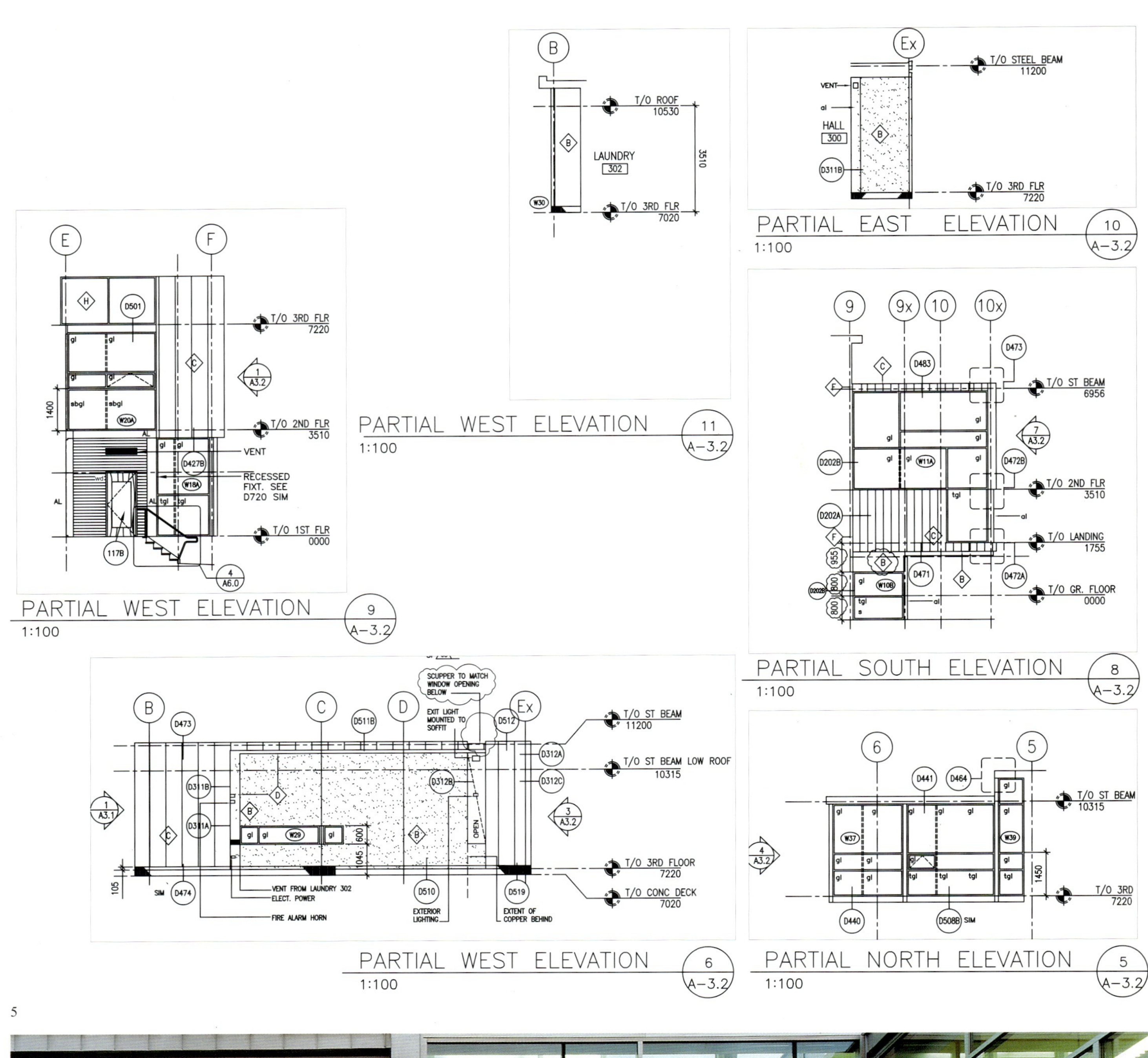
PARTIAL WEST ELEVATION
1:100
11
A-3.2
T/O ROOF
10530
LAUNDRY
302
T/O 3RD FLR
7020
3510
PARTIAL EAST ELEVATION
1:100
10
A-3.2
T/O STEEL BEAM
11200
HALL
300
T/O 3RD FLR
7220
PARTIAL WEST ELEVATION
1:100
9
A-3.2
T/O 3RD FLR
7220
T/O 2ND FLR
3510
T/O 1ST FLR
0000
VENT
RECESSED FIXT. SEE D720 SIM
PARTIAL SOUTH ELEVATION
1:100
8
A-3.2
T/O ST BEAM
6956
T/O 2ND FLR
3510
T/O LANDING
1755
T/O GR. FLOOR
0000
PARTIAL WEST ELEVATION
1:100
6
A-3.2
SCUPPER TO MATCH WINDOW OPENING BELOW
EXIT LIGHT MOUNTED TO SOFFIT
T/O ST BEAM
11200
T/O ST BEAM LOW ROOF
10315
T/O 3RD FLOOR
7220
T/O CONC DECK
7020
VENT FROM LAUNDRY 302
ELECT. POWER
FIRE ALARM HORN
EXTERIOR LIGHTING
EXTENT OF COPPER BEHIND
PARTIAL NORTH ELEVATION
1:100
5
A-3.2
T/O ST BEAM
10315
T/O 3RD
7220

5

6

5 立面详图
6 墙体细部
7 立面详图
8 坡道
9 立面详图

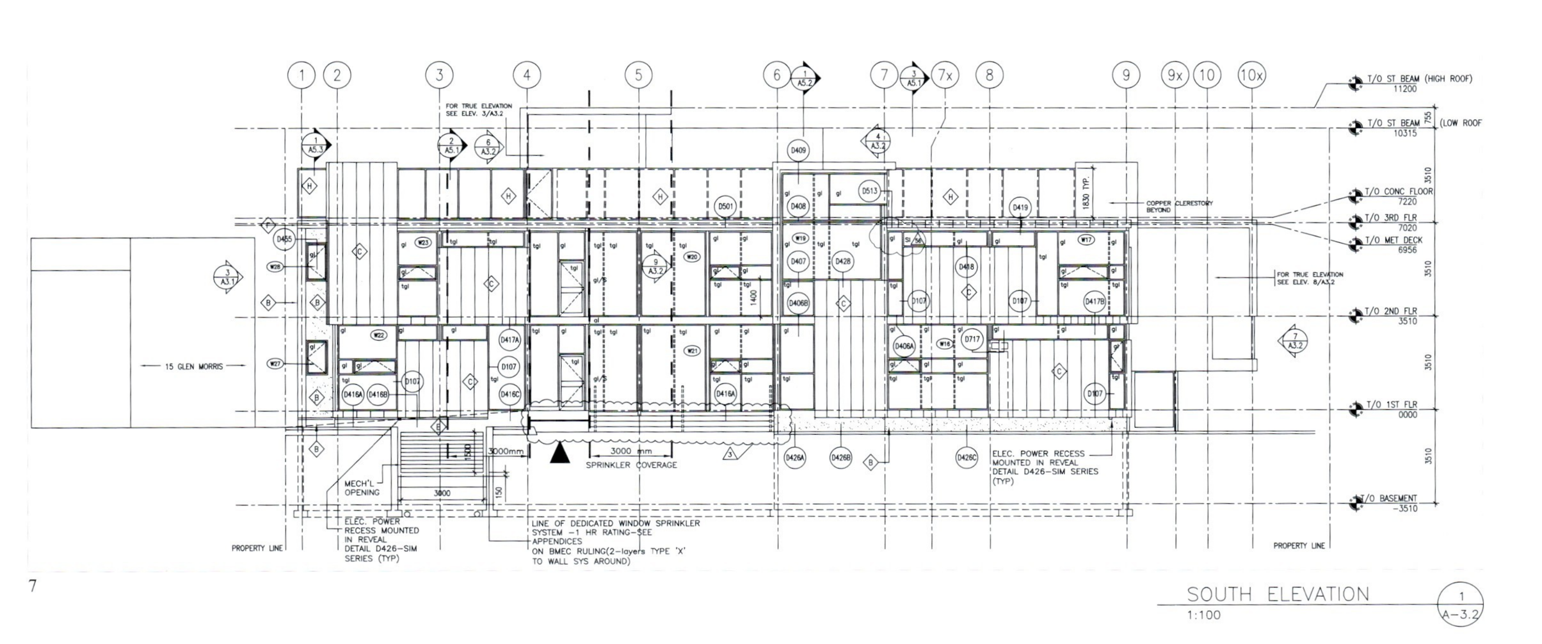

7

8

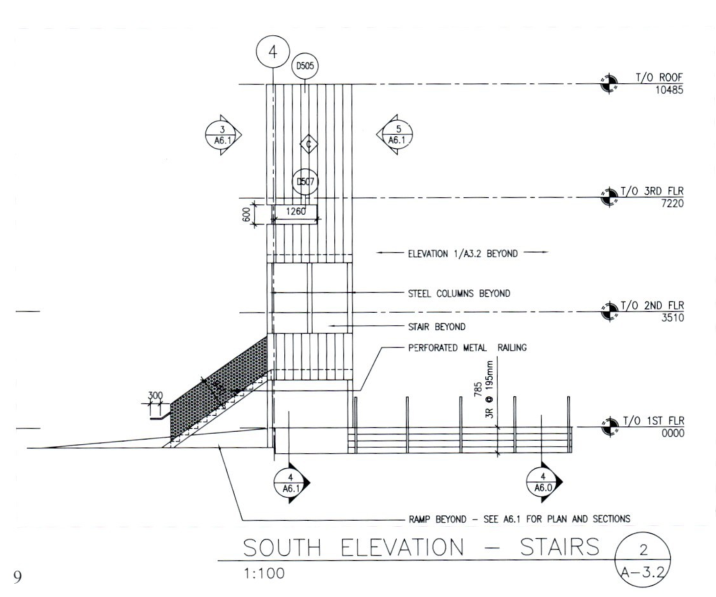

9

10 室内
11 剖面详图
12 室内

10

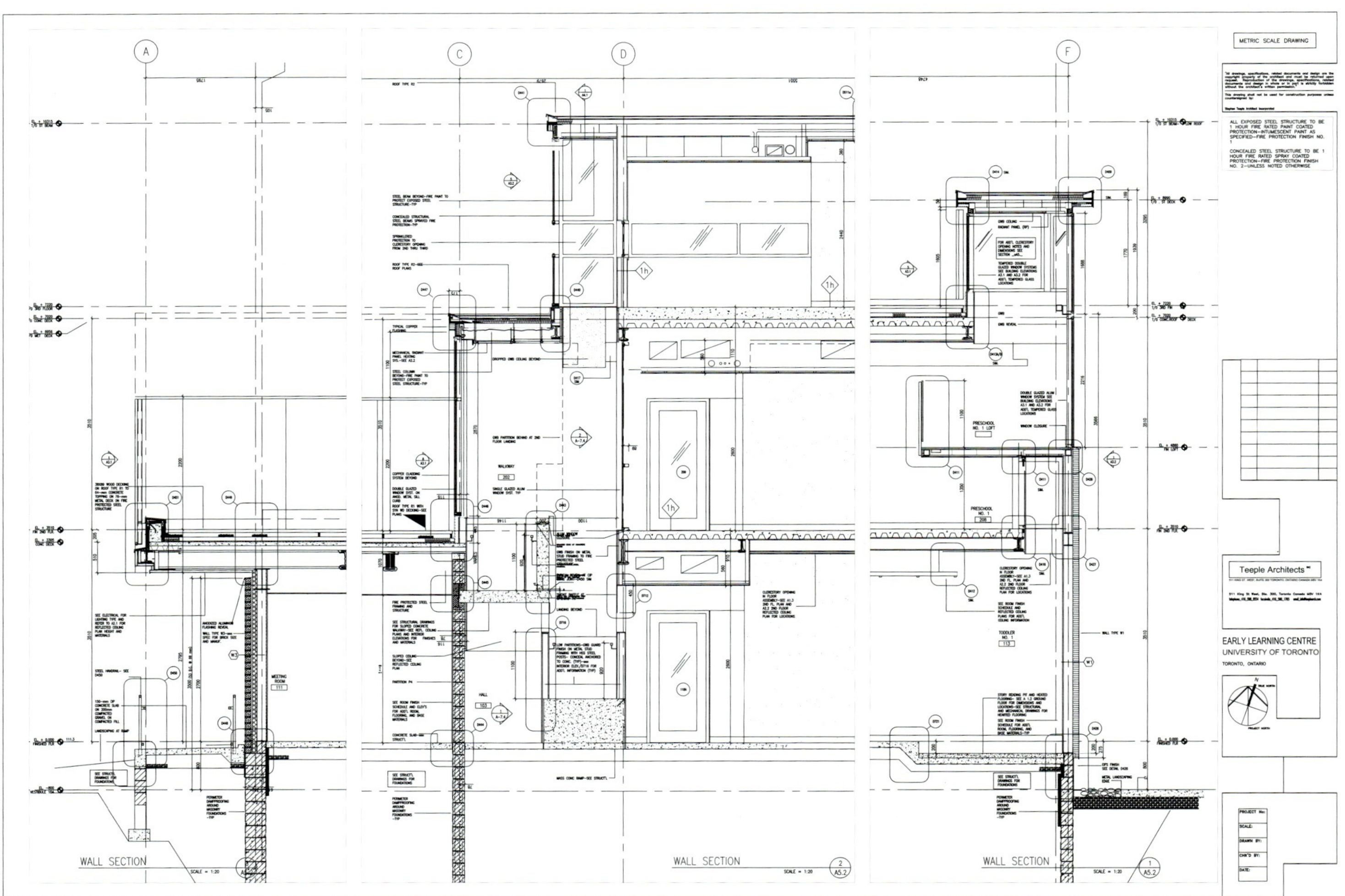

11

12

Academic Science Centre
科学技术中心

业　　主：Trent 大学
地　　点：彼得伯勒市
规　　模：3 600m²
完成时间：2003 年 1 月

This Academic Science Centre houses new state-of-the-art including teaching and computer laboratories, and two wings of high level, flexible research laboratories. The Centre includes four large chemistry-teaching labs, containing piped services and twenty-four fume hoods, as well as twelve bays of 600 sq.ft. Research labs, custom designed to each researcher, with piped natural gas, argon, nitrogen and reverse-osmosis water and seventeen fume hoods. A 3,000 sq.ft. state-of-the-art water quality laboratory, a central feature of the design, has become the prime facility of its type in Canada. Dry teaching labs and administrative offices are also provided.

In a campus that has strong ties with the landscape of the Ottanabee River, this single storey laboratory building becomes a landscape itself - the landscape extends over the roofs of the building, as a living roof, integrating the building with its beautiful site. Further sustainable design features include low-flow fume hoods that minimize energy consumption and heat recovery systems, and metal rain screen cladding systems provide longevity and exceptional energy efficiency.

(In association with Shore Tilbe Irwin & Partners, Toronto)

Trent 大学科学技术中心设有教学和计算机实验室。高空间的多用途的研究实验室位于主体建筑的两翼。四个大化学教学实验室装配有上下水设施和20个烟罩。设有12个隔间的600平方英尺的研究实验室是根据研究者的需求设计的，并装配有天然气管、氩气管、氮气管、水管和17个烟罩。一个3000平方英尺的最先进的水质实验室是科学技术中心的最重要的设施，也是加拿大同类型实验室中最先进的。科学技术中心内还设有干燥实验室和行政办公室。

Trent 大学的校园建筑与Ottanabee河的景观融为一体，科学中心的单层建筑本身就是大学的景观。园艺花园像一个生活屋顶延伸至整个屋顶,成为建筑景观的一部分。科学中心设有多个可持续化设计，包括低流量烟罩、最小化能量消耗和供热系统、金属雨刮系统可以延长使用寿命和提高能效。

（与Shore Tilbe Irwin & Partners建筑设计事务所共同完成。）

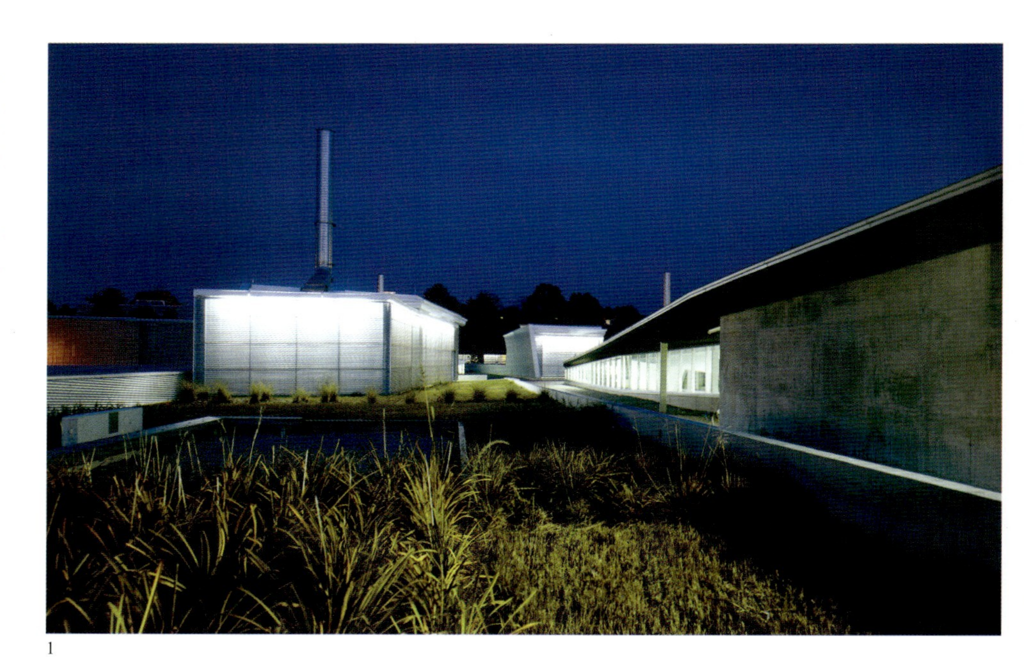

1

2

1 中心建筑夜景
2 中心屋顶景观
3 建筑电脑效果图
4 总平面图

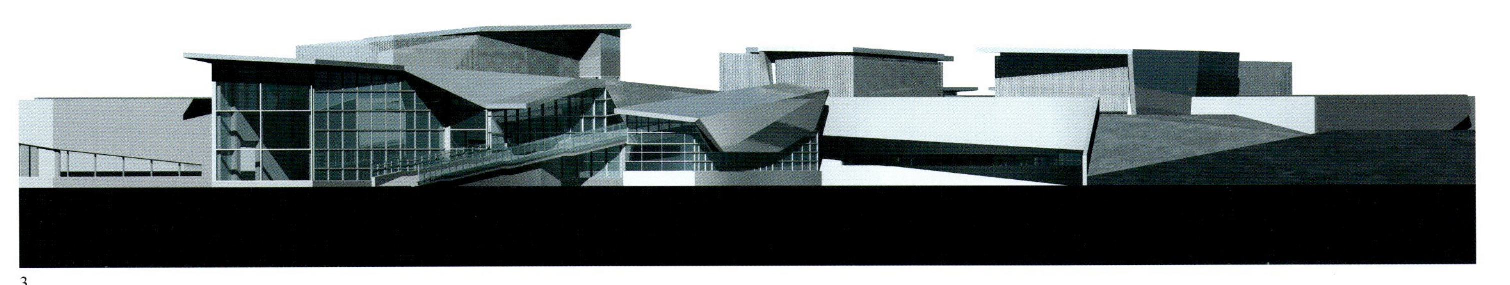

3

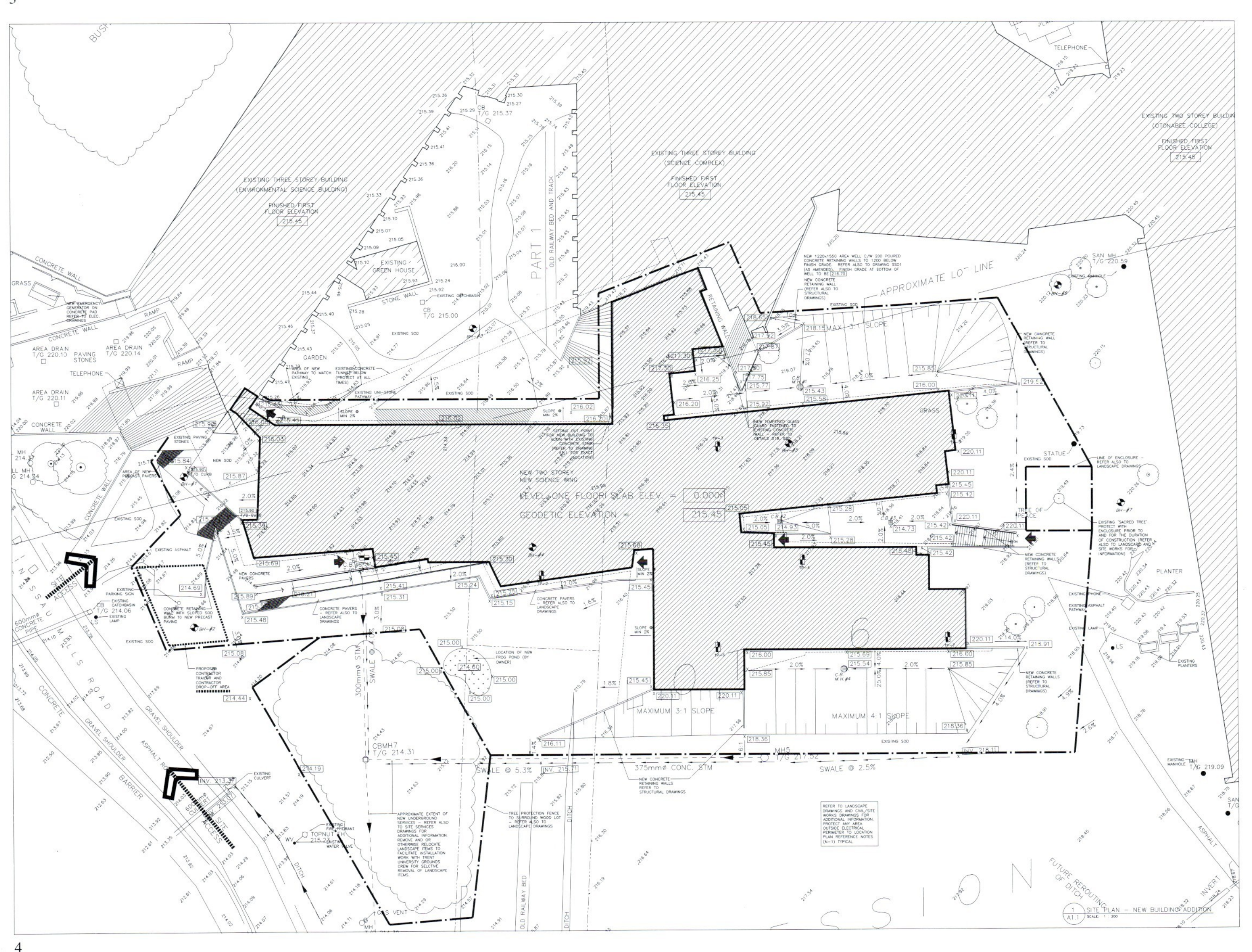

4

5

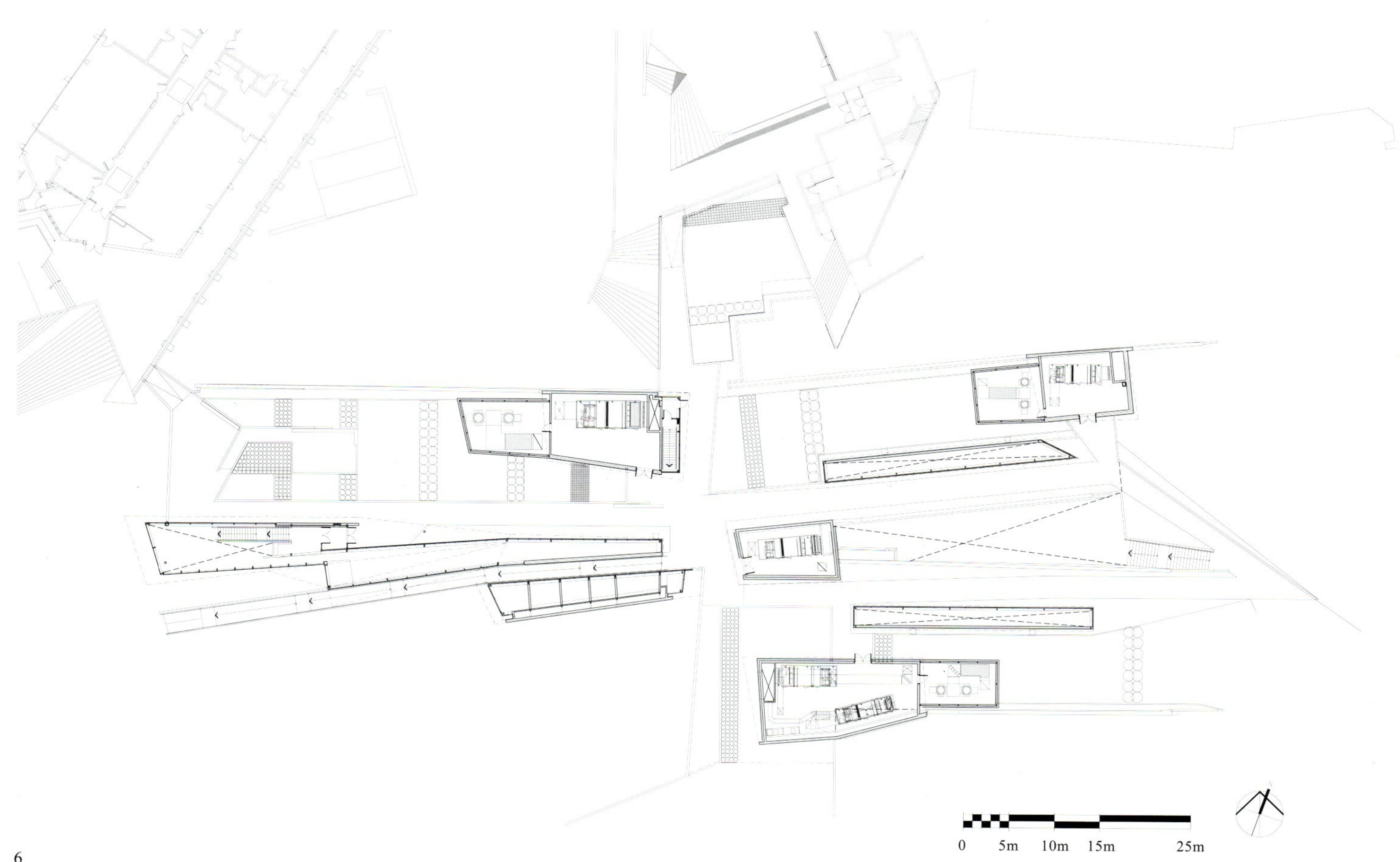

6

5　中心夜景
6　一层平面图
7　建筑电脑效果图
8　坡道电脑效果图
9　坡道电脑效果图
10~11　剖面图

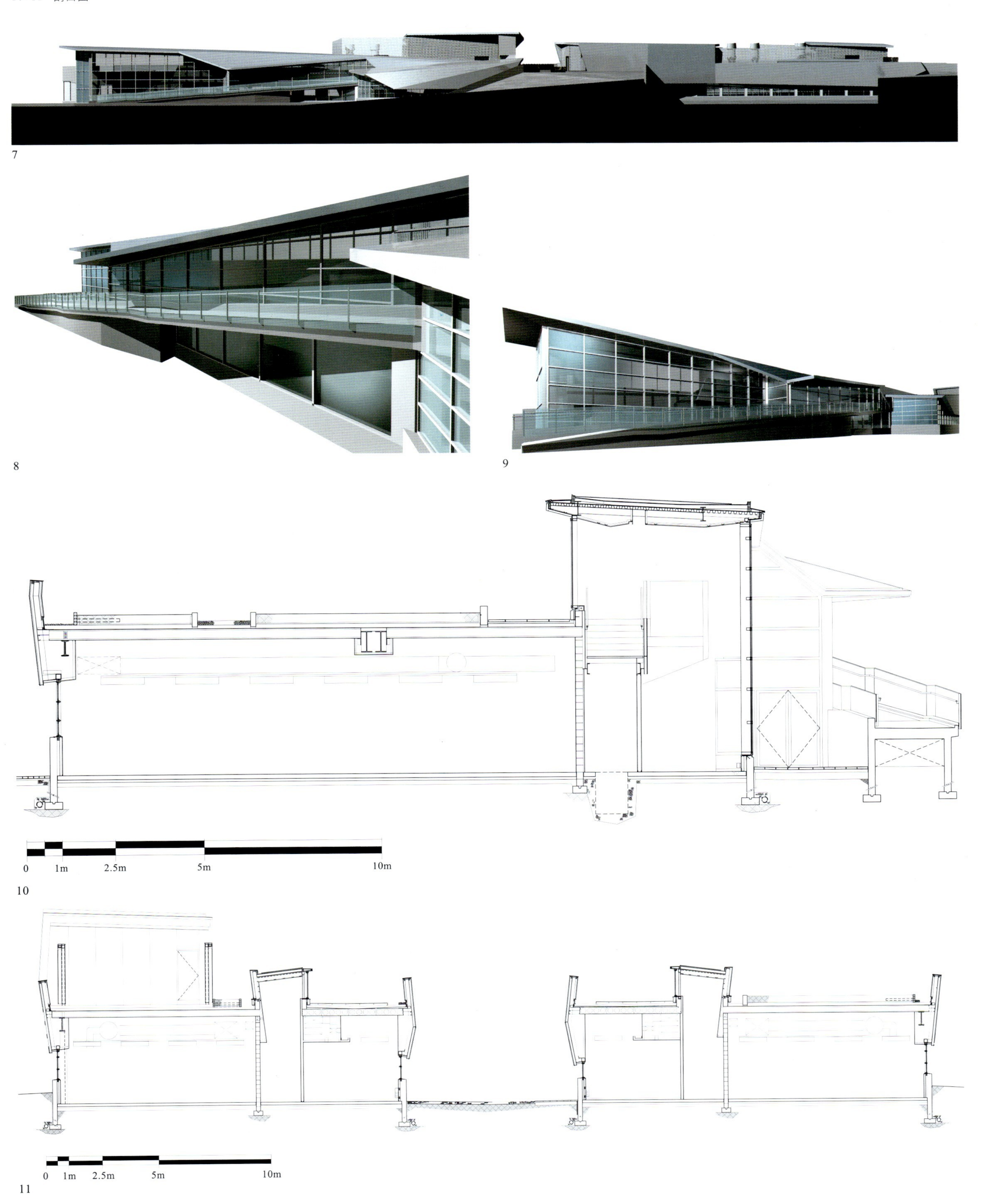

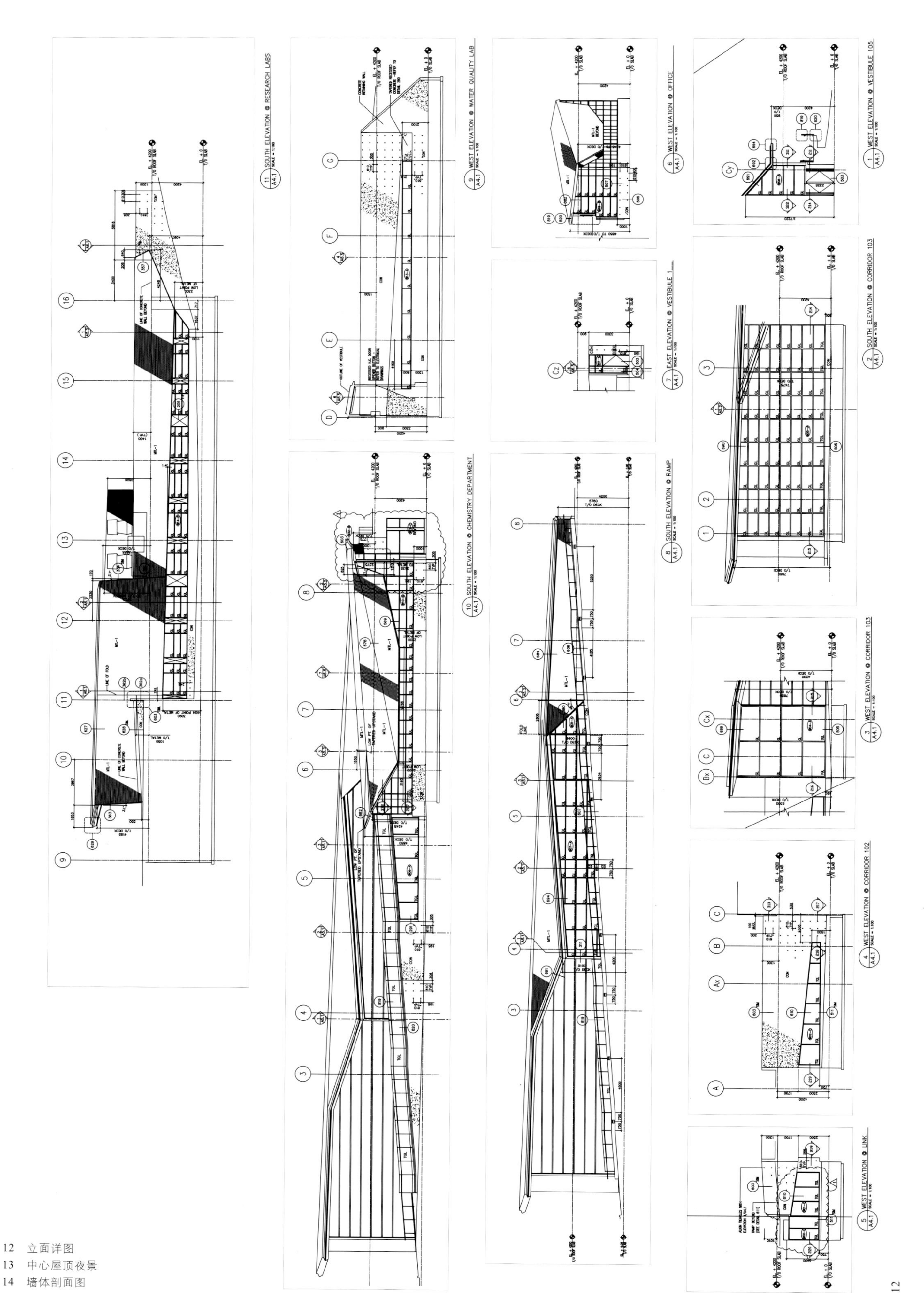

12 立面详图
13 中心屋顶夜景
14 墙体剖面图

13

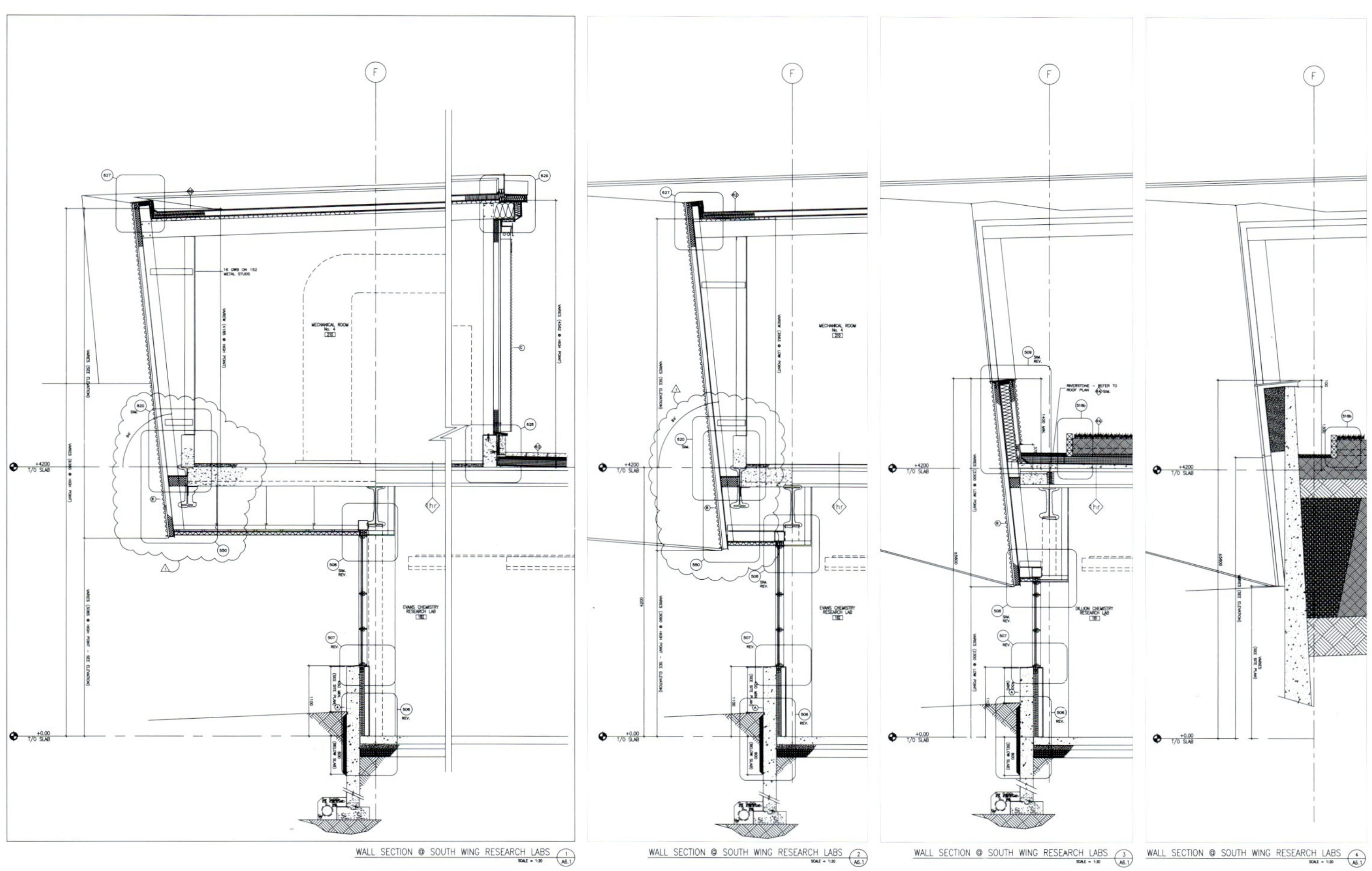
WALL SECTION @ SOUTH WING RESEARCH LABS
WALL SECTION @ SOUTH WING RESEARCH LABS
WALL SECTION @ SOUTH WING RESEARCH LABS
WALL SECTION @ SOUTH WING RESEARCH LABS

14

15 内院景观
16 中心建筑独有的视觉组织
17 墙体剖面图
18 中心建筑屋面景观
19 内院组织
20 墙体剖面图

15

16

17

18

19

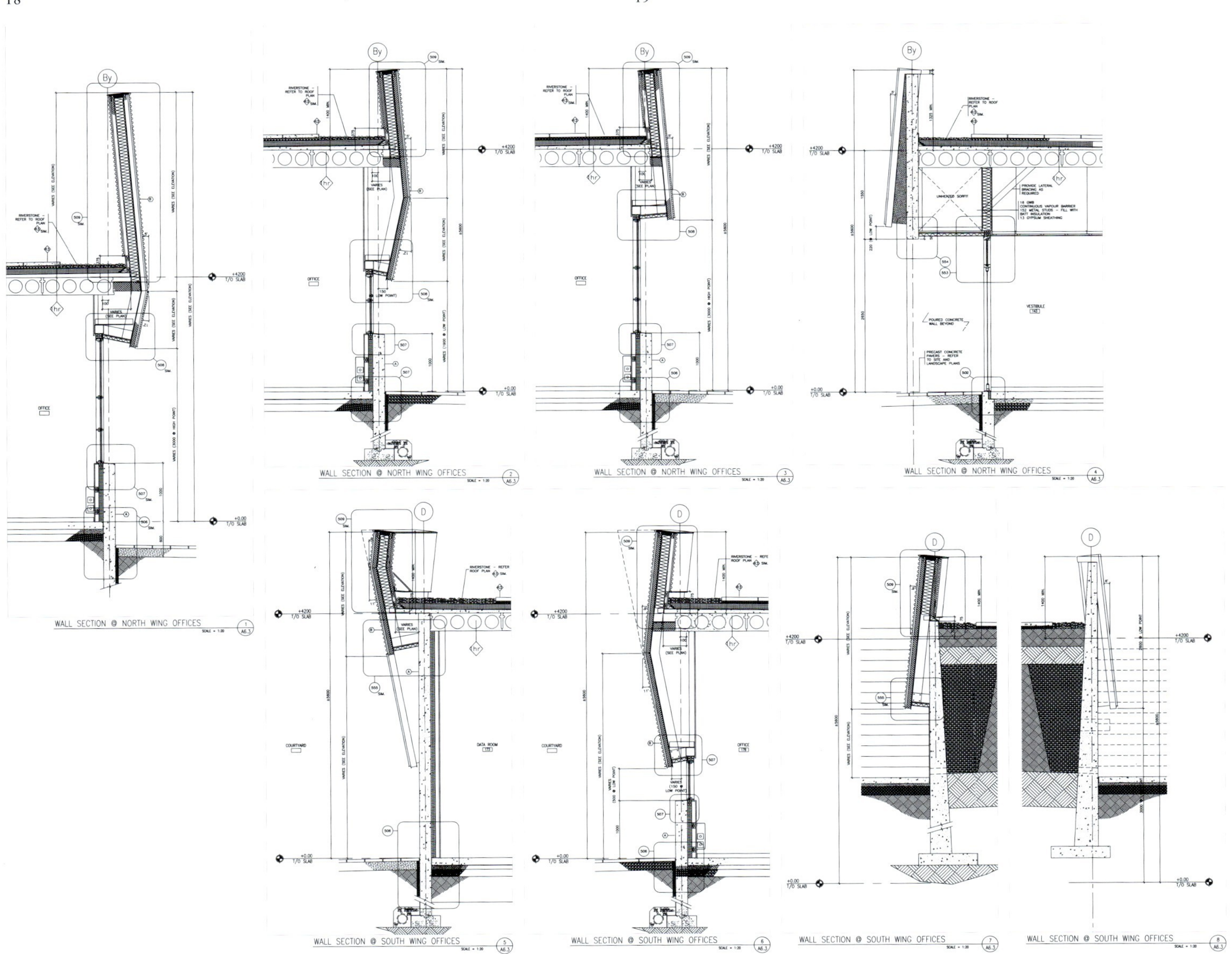

WALL SECTION @ NORTH WING OFFICES
WALL SECTION @ NORTH WING OFFICES
WALL SECTION @ NORTH WING OFFICES
WALL SECTION @ NORTH WING OFFICES
WALL SECTION @ SOUTH WING OFFICES
WALL SECTION @ SOUTH WING OFFICES
WALL SECTION @ SOUTH WING OFFICES
WALL SECTION @ SOUTH WING OFFICES

20

21

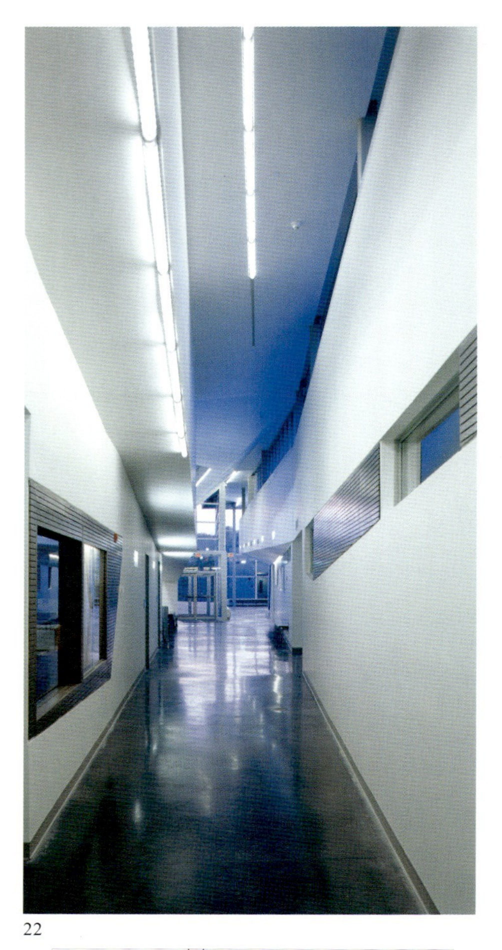

22

23

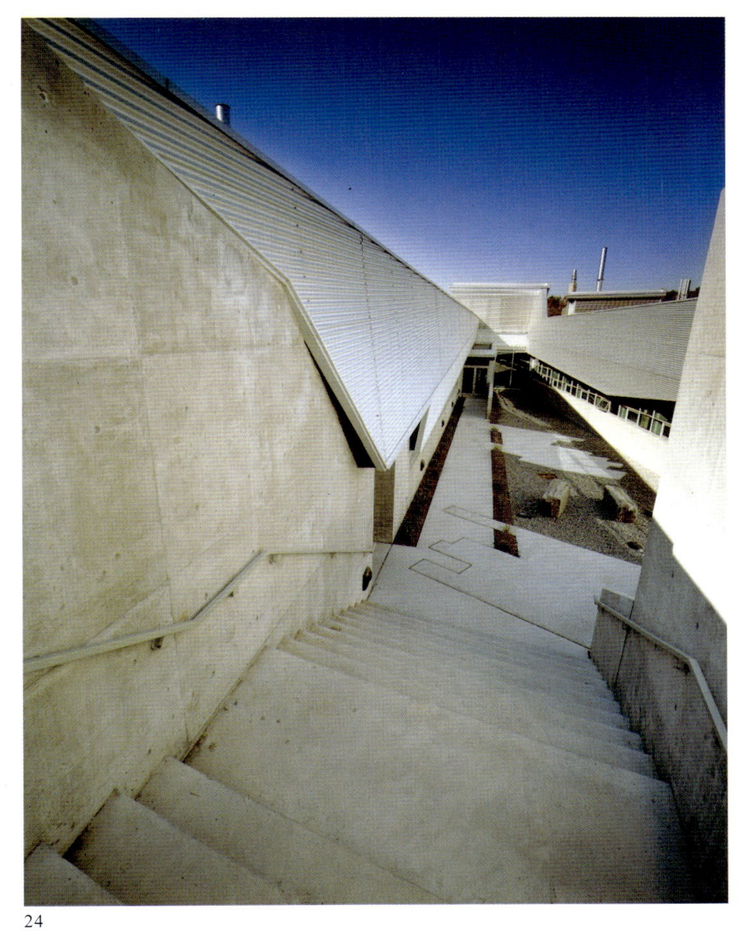

24

21 建筑室内
22 建筑室内
23 剖面图
24 内院景观
25 剖面图

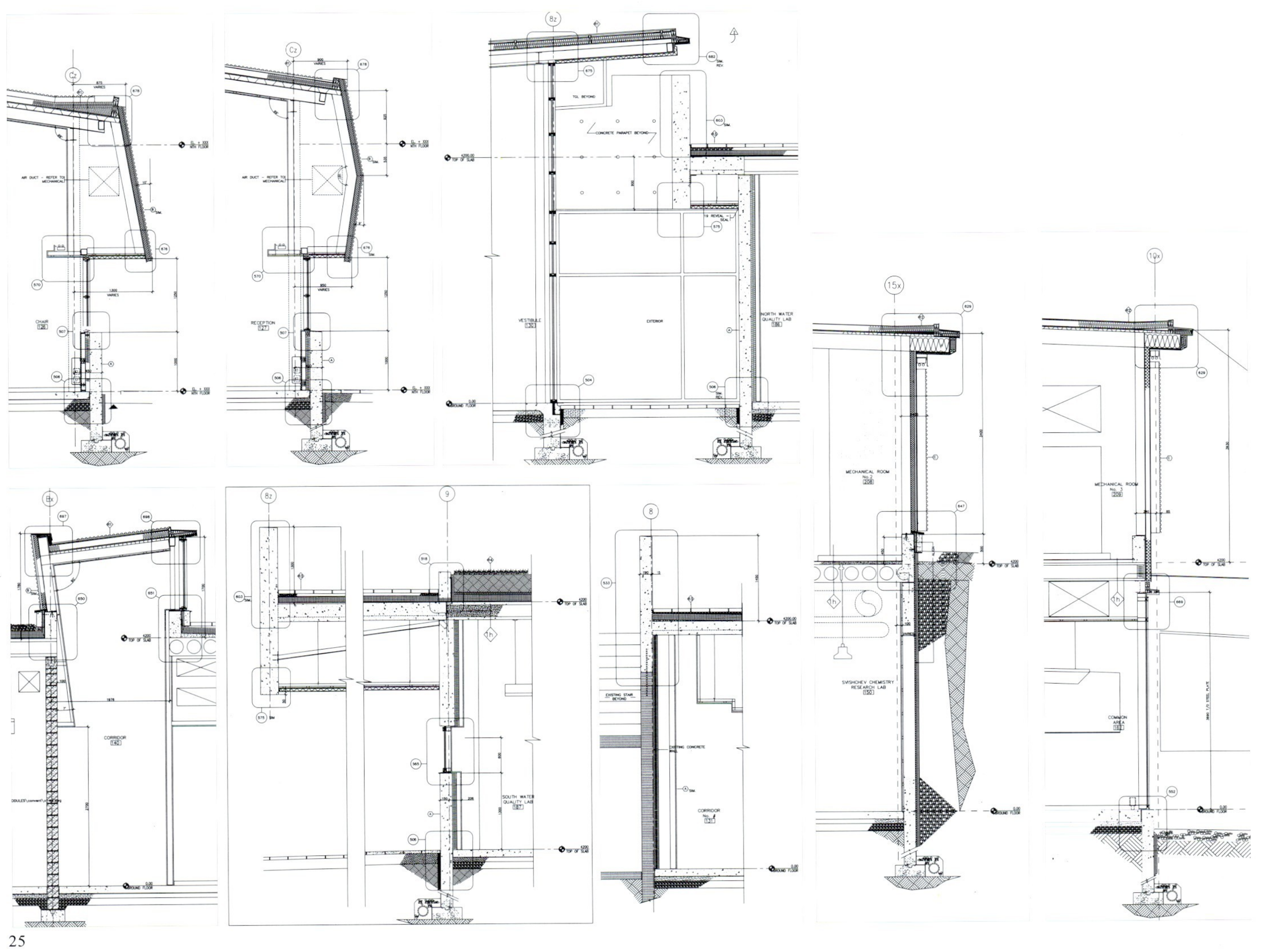

25

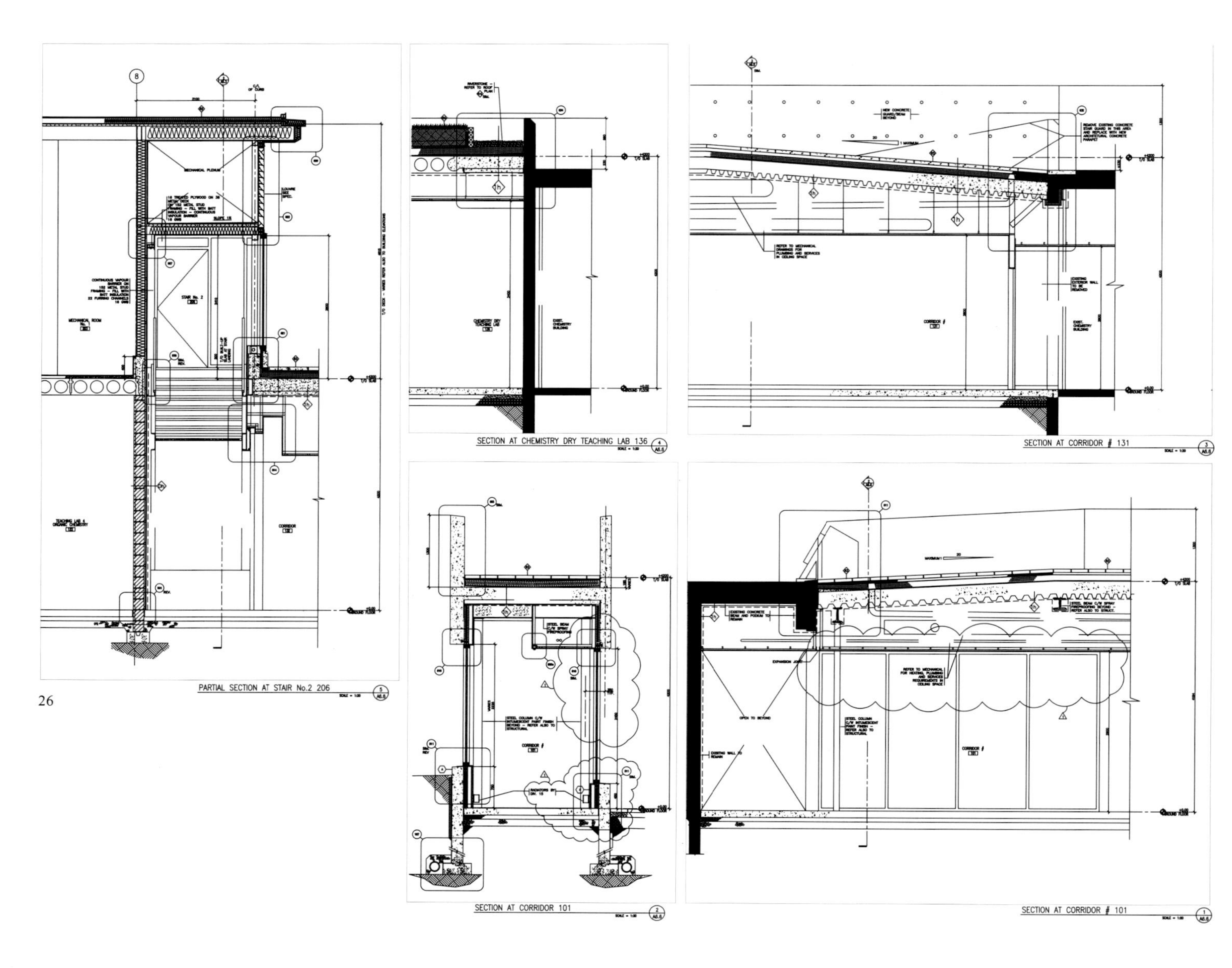
PARTIAL SECTION AT STAIR No.2 206
SECTION AT CHEMISTRY DRY TEACHING LAB 136
SECTION AT CORRIDOR # 131
SECTION AT CORRIDOR 101
SECTION AT CORRIDOR # 101

26

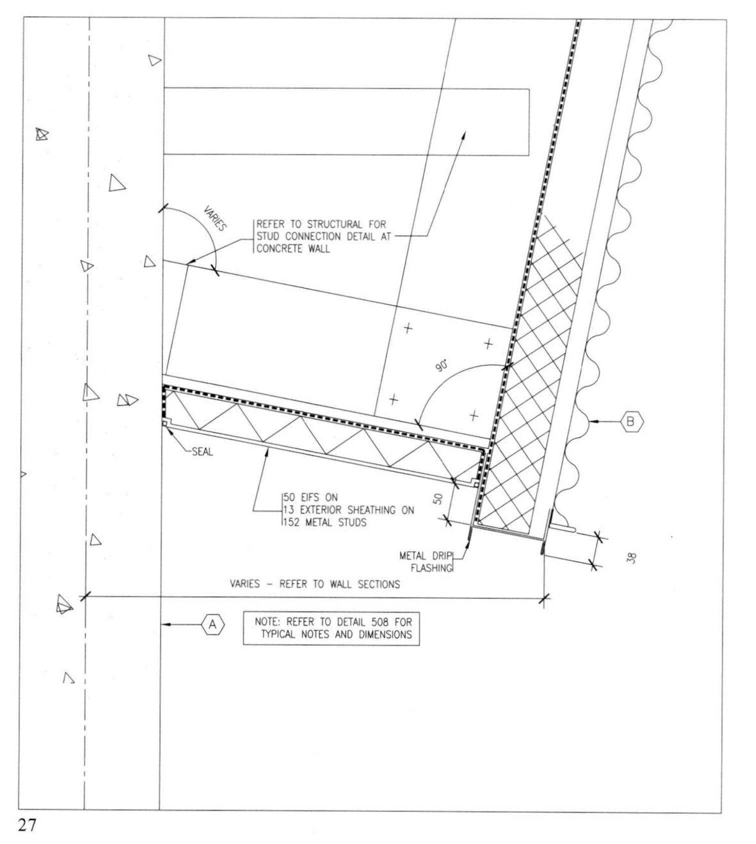
VARIES
REFER TO STRUCTURAL FOR
STUD CONNECTION DETAIL AT
CONCRETE WALL
90°
B
SEAL
50 EIFS ON
13 EXTERIOR SHEATHING ON
152 METAL STUDS
50
METAL DRIP
FLASHING
38
VARIES – REFER TO WALL SECTIONS
A
NOTE: REFER TO DETAIL 508 FOR
TYPICAL NOTES AND DIMENSIONS

27

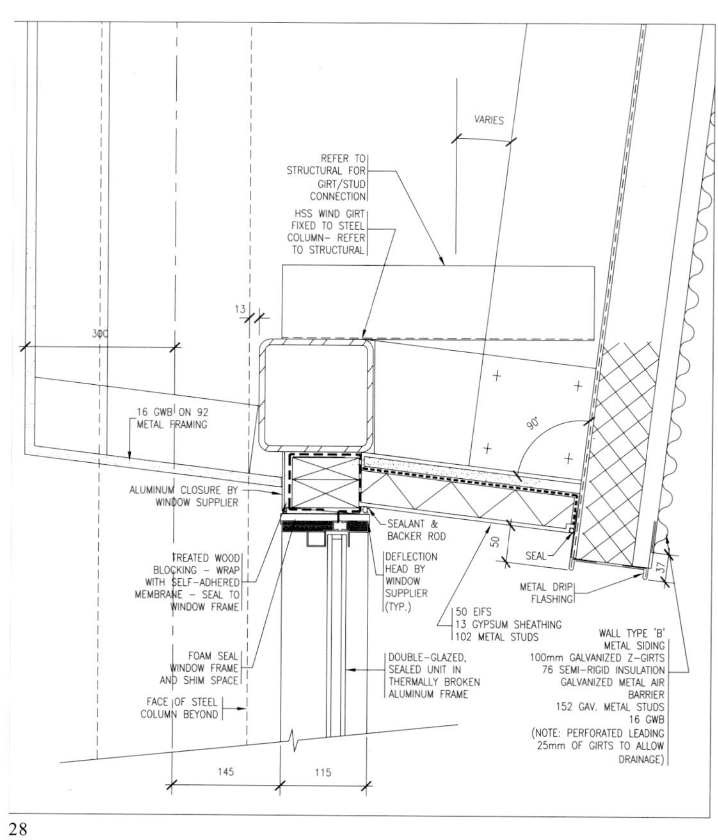
VARIES
REFER TO
STRUCTURAL FOR
GIRT/STUD
CONNECTION
HSS WIND GIRT
FIXED TO STEEL
COLUMN– REFER
TO STRUCTURAL
13
300
90°
16 GWB ON 92
METAL FRAMING
ALUMINUM CLOSURE BY
WINDOW SUPPLIER
SEALANT &
BACKER ROD
TREATED WOOD
BLOCKING – WRAP
WITH SELF-ADHERED
MEMBRANE – SEAL TO
WINDOW FRAME
DEFLECTION
HEAD BY
WINDOW
SUPPLIER
(TYP.)
50
SEAL
METAL DRIP
FLASHING
37
50 EIFS
13 GYPSUM SHEATHING
102 METAL STUDS
WALL TYPE 'B'
METAL SIDING
100mm GALVANIZED Z-GIRTS
76 SEMI-RIGID INSULATION
GALVANIZED METAL AIR
BARRIER
152 GAV. METAL STUDS
16 GWB
(NOTE: PERFORATED LEADING
25mm OF GIRTS TO ALLOW
DRAINAGE)
FOAM SEAL
WINDOW FRAME
AND SHIM SPACE
DOUBLE-GLAZED,
SEALED UNIT IN
THERMALLY BROKEN
ALUMINUM FRAME
FACE OF STEEL
COLUMN BEYOND
145
115

28

26 剖面图
27~29 墙体节点详图
30 女儿墙节点详图
31~32 檐口节点详图

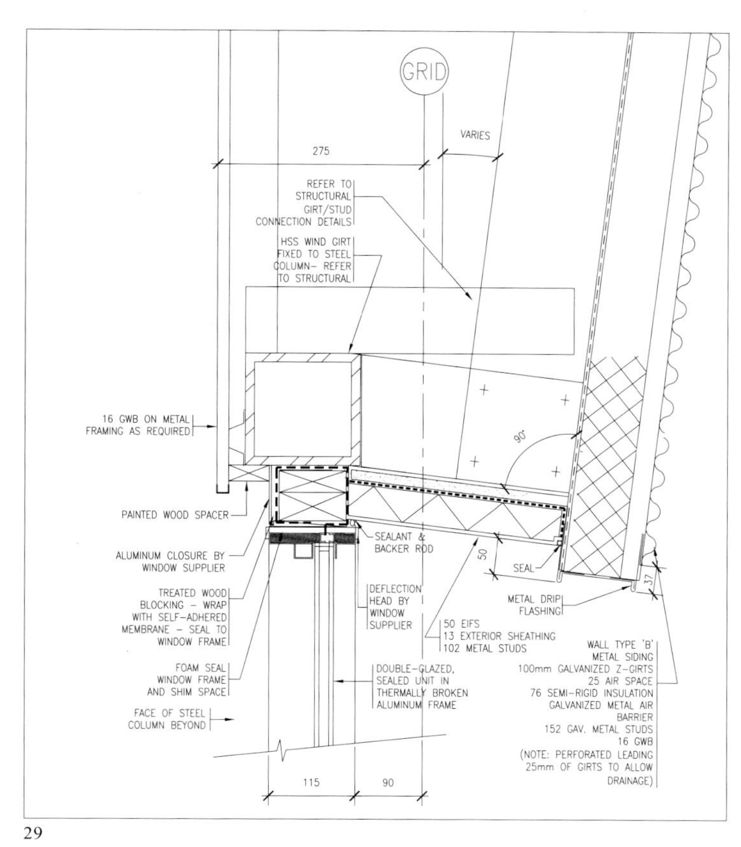

29

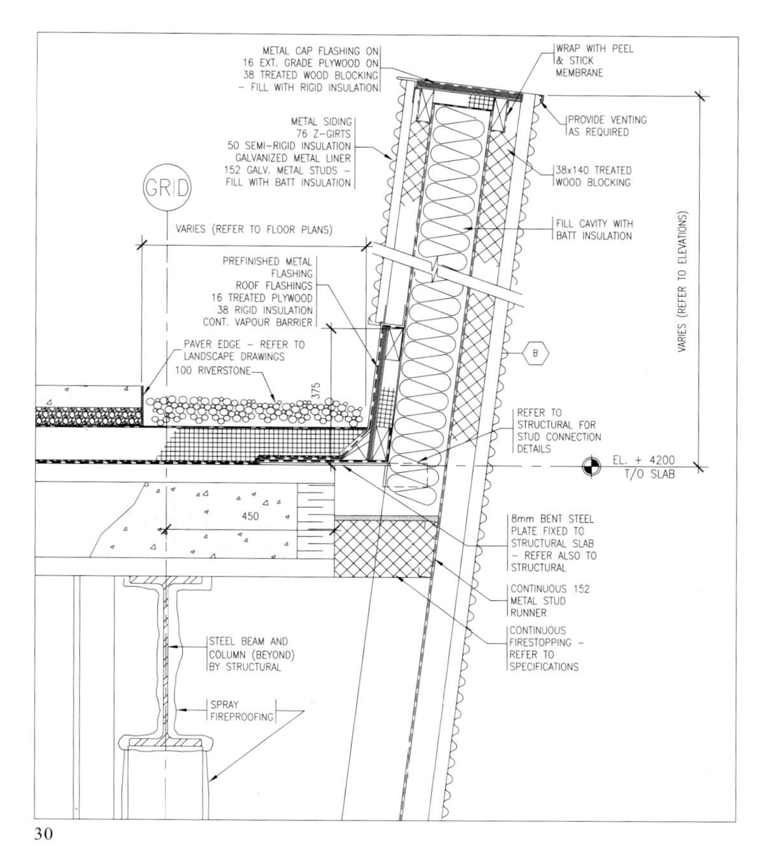

30

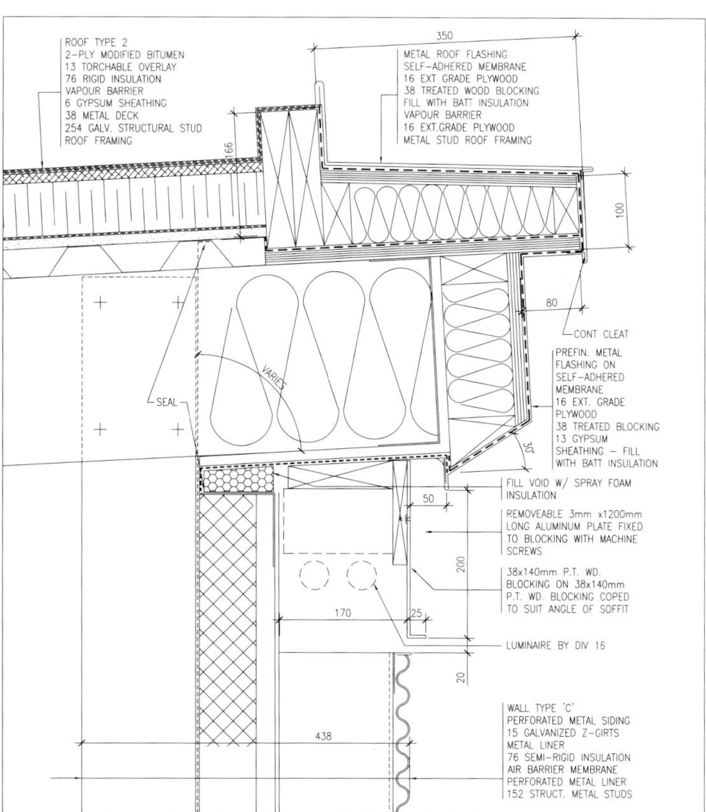

31

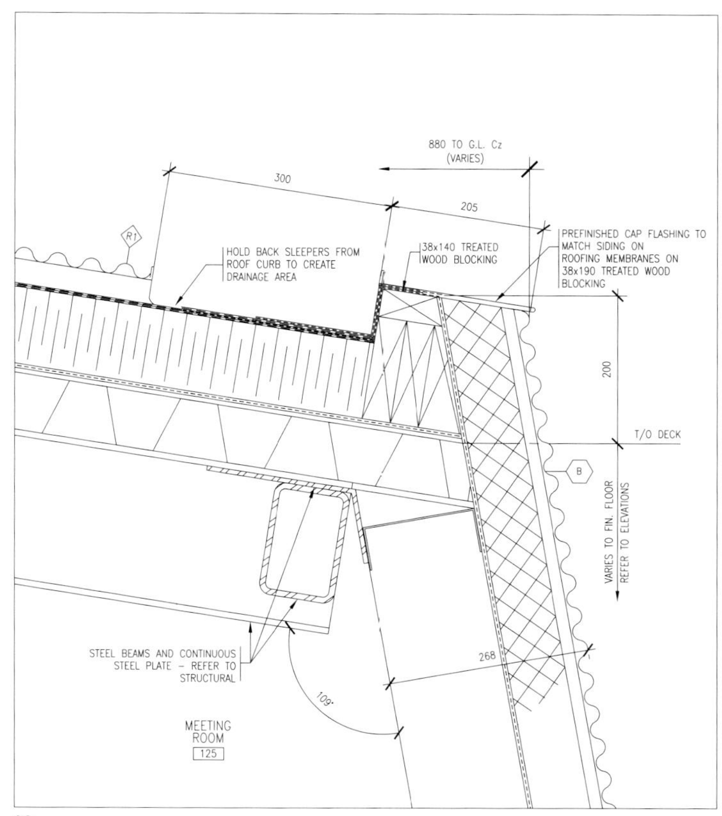

32

Quinte Enhanced Technology Learning Center
Quinte 高科技学习中心

业　　　主：Loyalist 学院
地　　　点：贝勒维尔市
规　　　模：约 4 800m²
完 成 时 间：2003 年 8 月

1　中心建筑的独特造型
2　总平面图
3　中心建筑外观
4　图书馆平面图

This high tech learning resource centre brings all the students together in a state-of-the-art facility that combines digital and print media design in a flexible environment. The learning resource centre consists of traditional print material with clustered computer stations, study spaces, student lounge, circulation desk/workroom and offices on the main floor. A mezzanine level, with barrier free access, includes two student seminar rooms, fully wired study carrels and general study spaces that overlook the library. This facility will also display a comprehensive collection of paintings from a famous Canadian artist. The Computer Commons consists of a multi-use 24 hour student access laboratory, and high-speed fibre optic cabling to over 180 computer stations on two levels. New classrooms have also been created to facilitate multi-task interactive teaching environments, as well as a fully wired 100-seat lecture theatre.

The complex creates strong links between the college and its rural landscape while presenting a new image for the college toward the entry of the campus. The new library is sited to take advantage of the natural landscape and vistas of the campus, and provide much needed natural light.

1

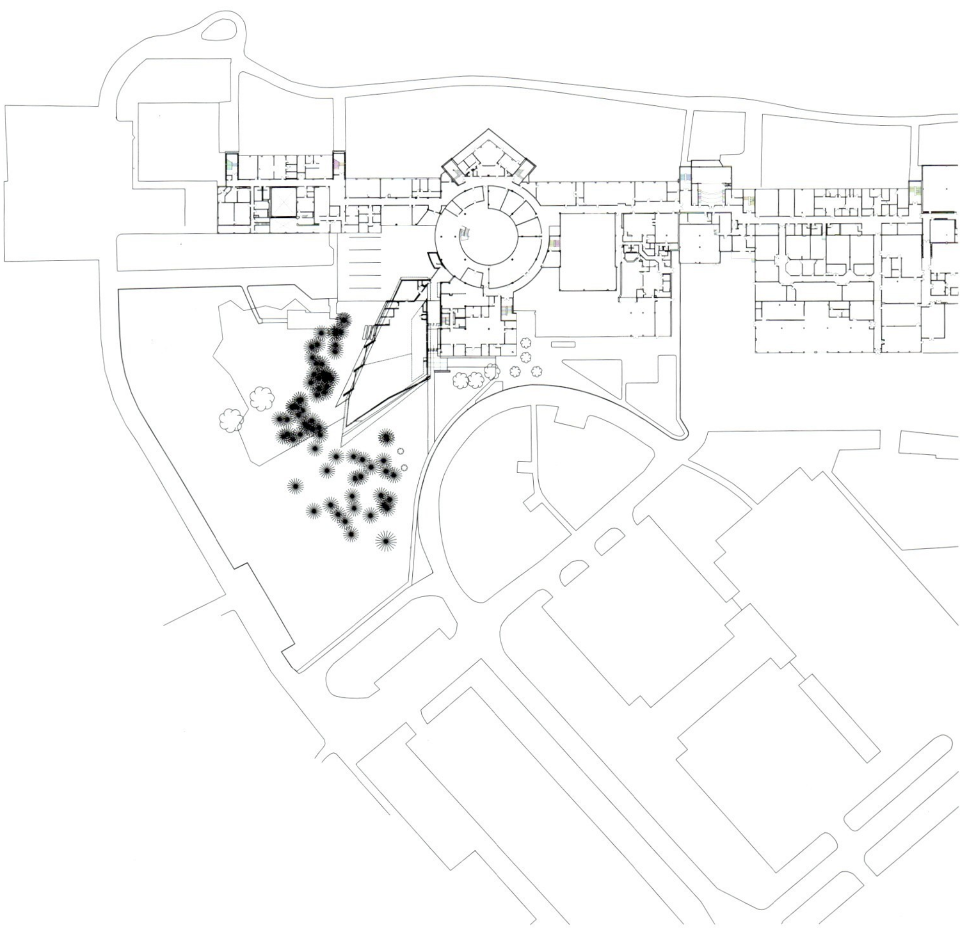

2

Quinte 高科技学习中心给学生提供可以集中学习最先进的包括数字化和印刷科技的教育设施，其空间可以灵活组合。中心包括传统的印刷资源：计算机中心、自习空间、学生休息处、工作室和首层办公室。位于夹层的两个无障碍教室，装备全网络阅读空间和俯视图书馆的学习空间。Quinte 高科技学习中心同样展示着加拿大当代艺术家的绘画作品。计算机房设有 24 小时对学生开放的多用途实验室和占两层多达 180 个高速光纤计算机工作站。新教室和一个全装备的 100 座教室同样也为中心提供了多功能的教学空间。

新的Quinte 高科技学习中心的建成改变了 Loyalist 学院的入口景观，同时也加强了学院建筑和当地乡村环境景观之间的交流。新建筑成为当地美丽自然景观和学院林荫道的一部分。Quinte 高科技学习中心同样大量地利用了自然光线。

3

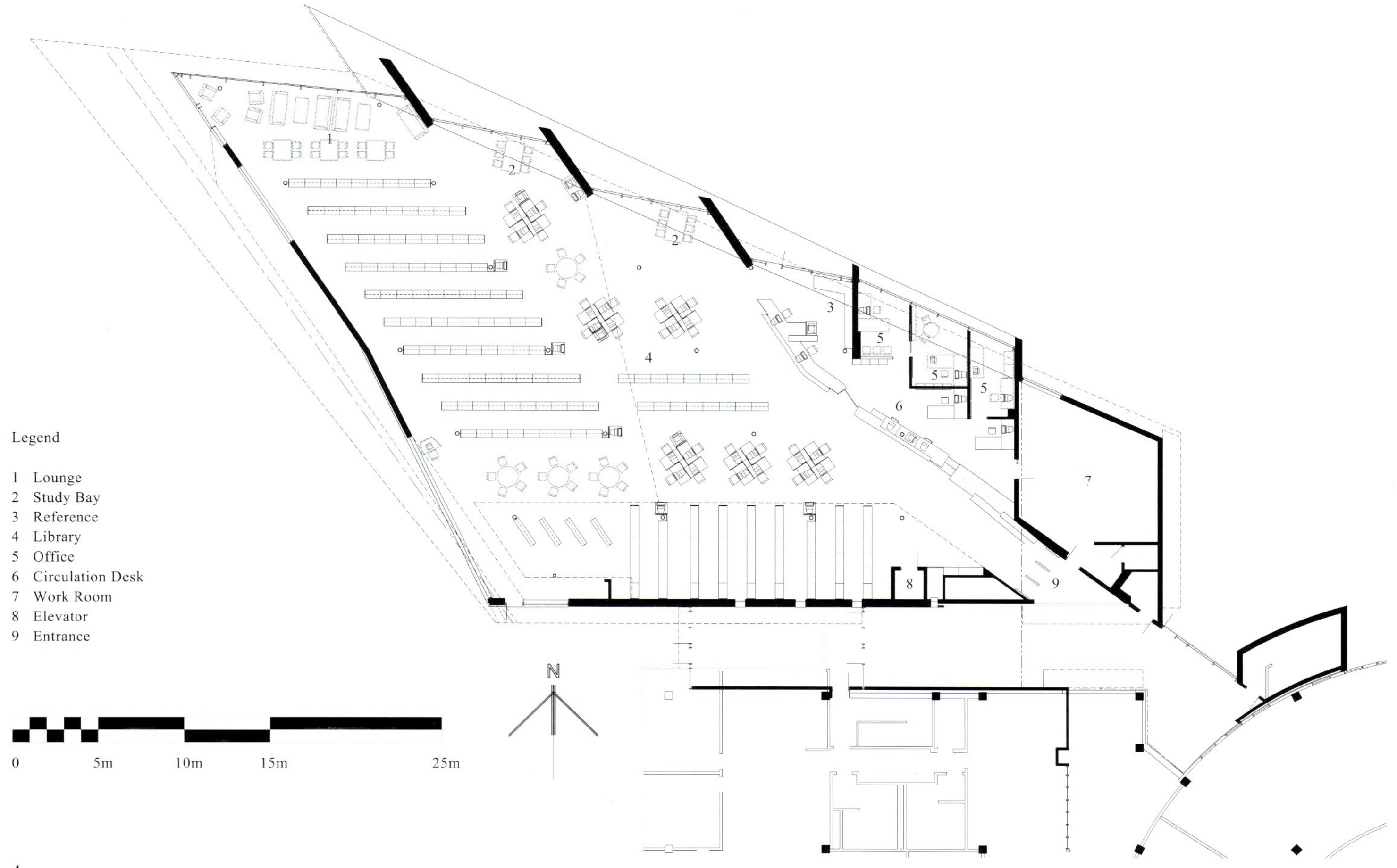
Legend
1 Lounge
2 Study Bay
3 Reference
4 Library
5 Office
6 Circulation Desk
7 Work Room
8 Elevator
9 Entrance
0
5m
10m
15m
25m
N
1
2
2
3
4
5
5
5
6
7
8
9

4

5

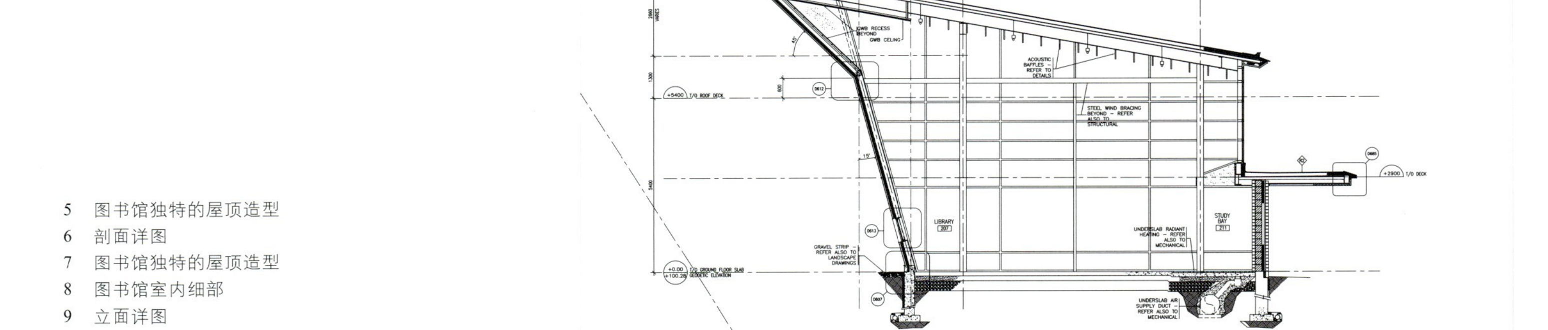

5　图书馆独特的屋顶造型
6　剖面详图
7　图书馆独特的屋顶造型
8　图书馆室内细部
9　立面详图

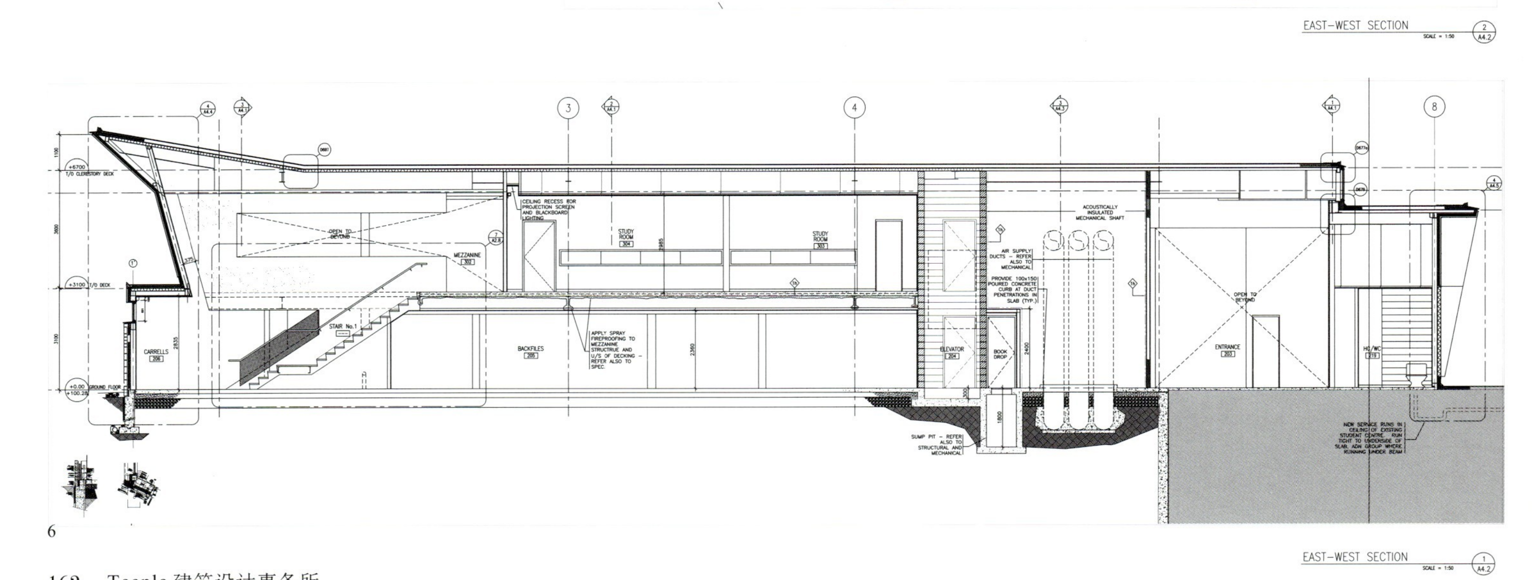

6

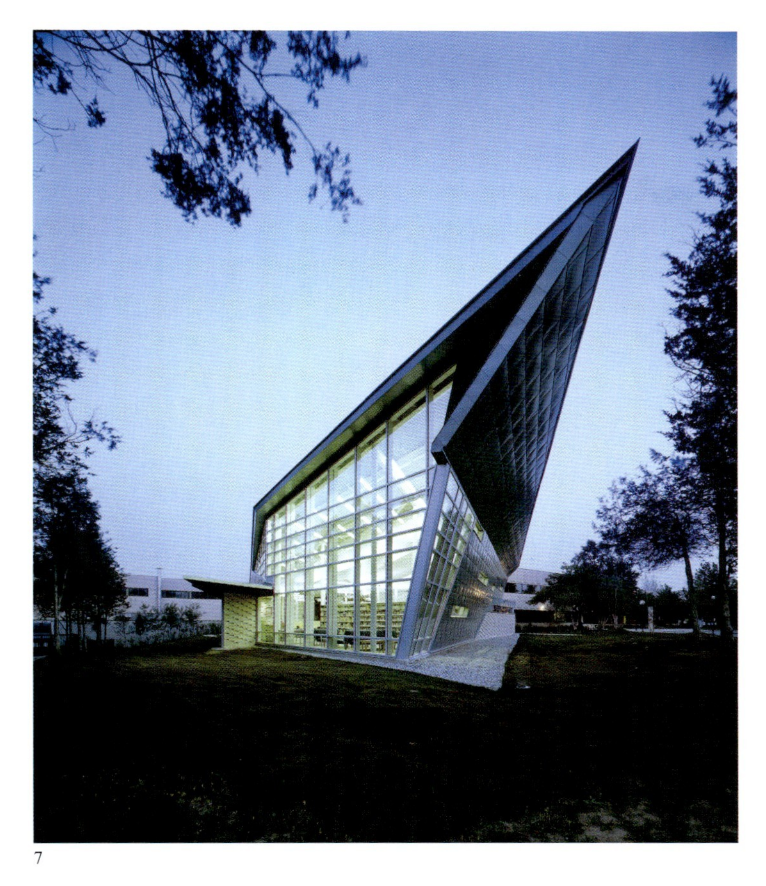

7

8

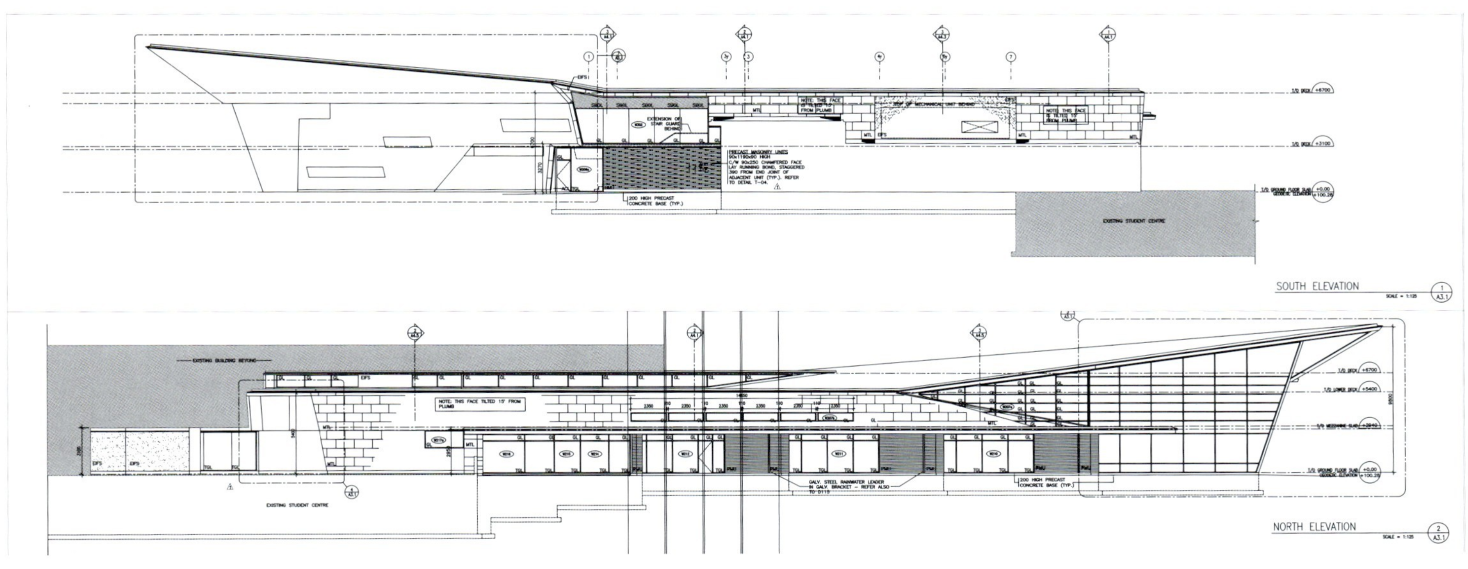
EXISTING STUDENT CENTRE
SOUTH ELEVATION
EXISTING BUILDING BEYOND
EXISTING STUDENT CENTRE
NORTH ELEVATION

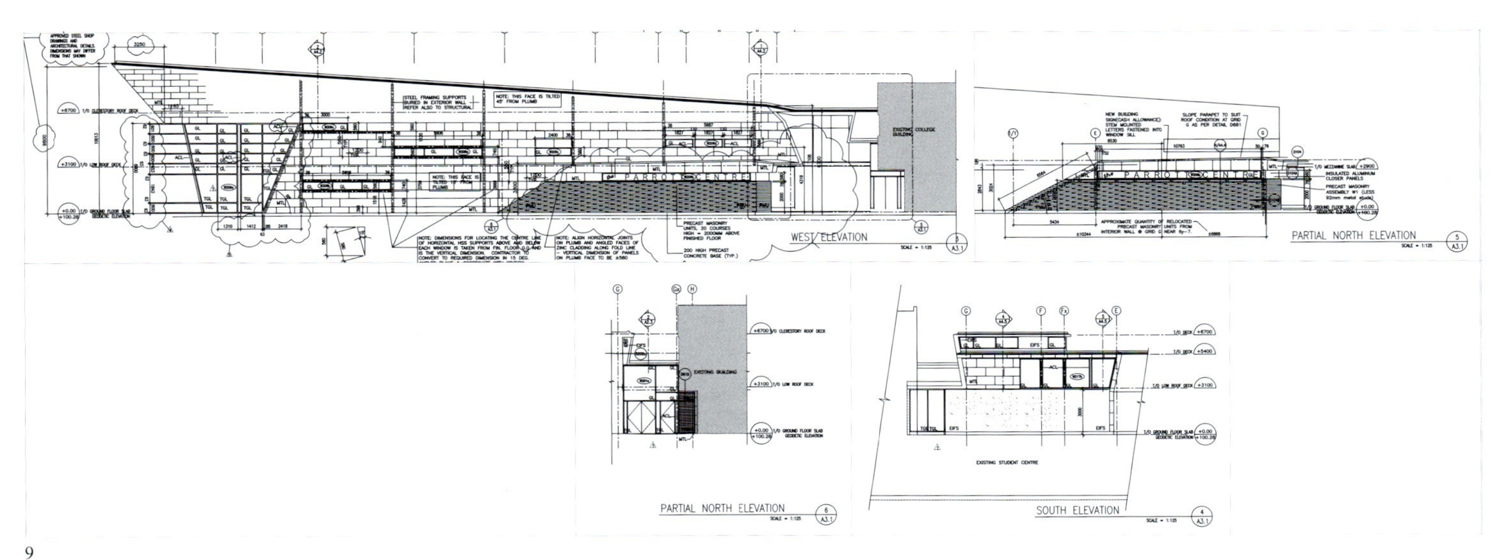
WEST ELEVATION
PARTIAL NORTH ELEVATION
PARTIAL NORTH ELEVATION
SOUTH ELEVATION

9

10

11

12

13

10　室内细部
11　简洁明快的室内设计
12　墙体剖面详图
13　建筑入口
14　墙体剖面详图

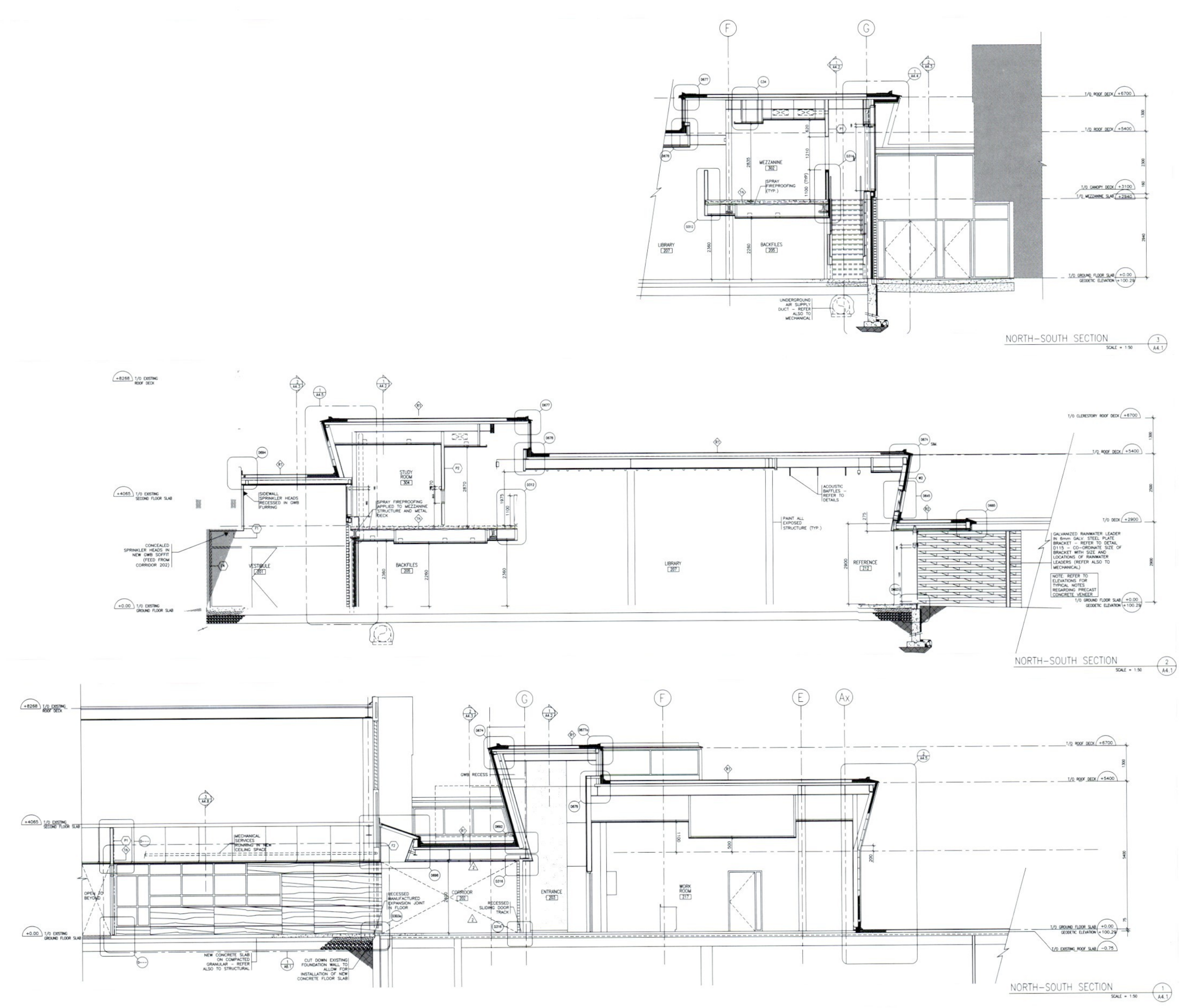

14

15

15　墙体细部
16　墙体剖面详图

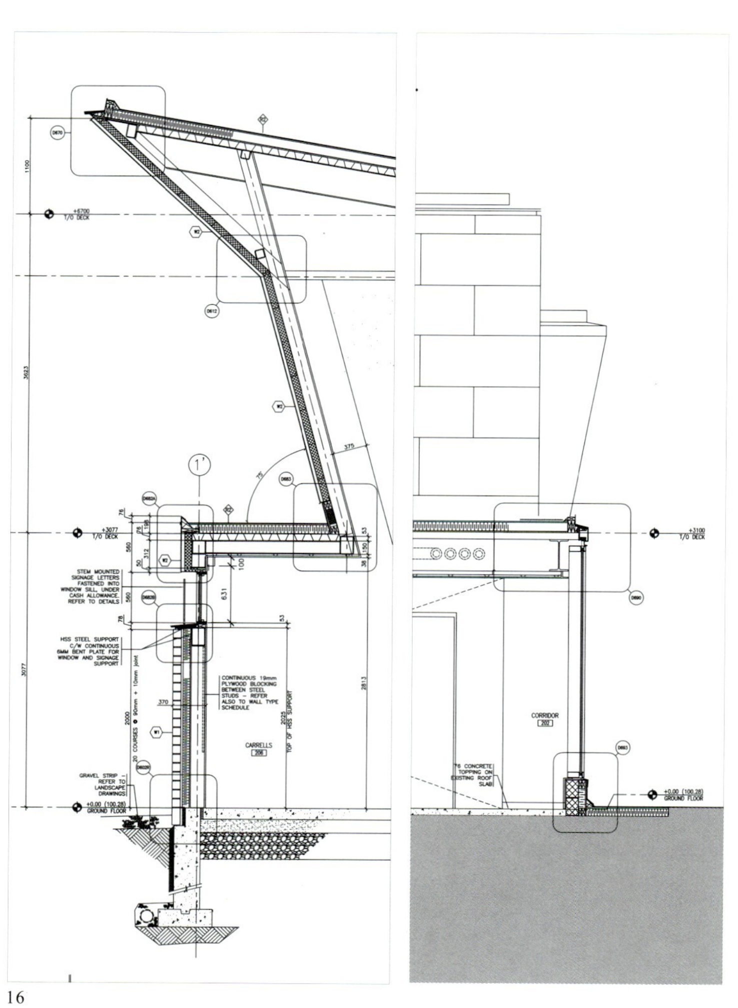

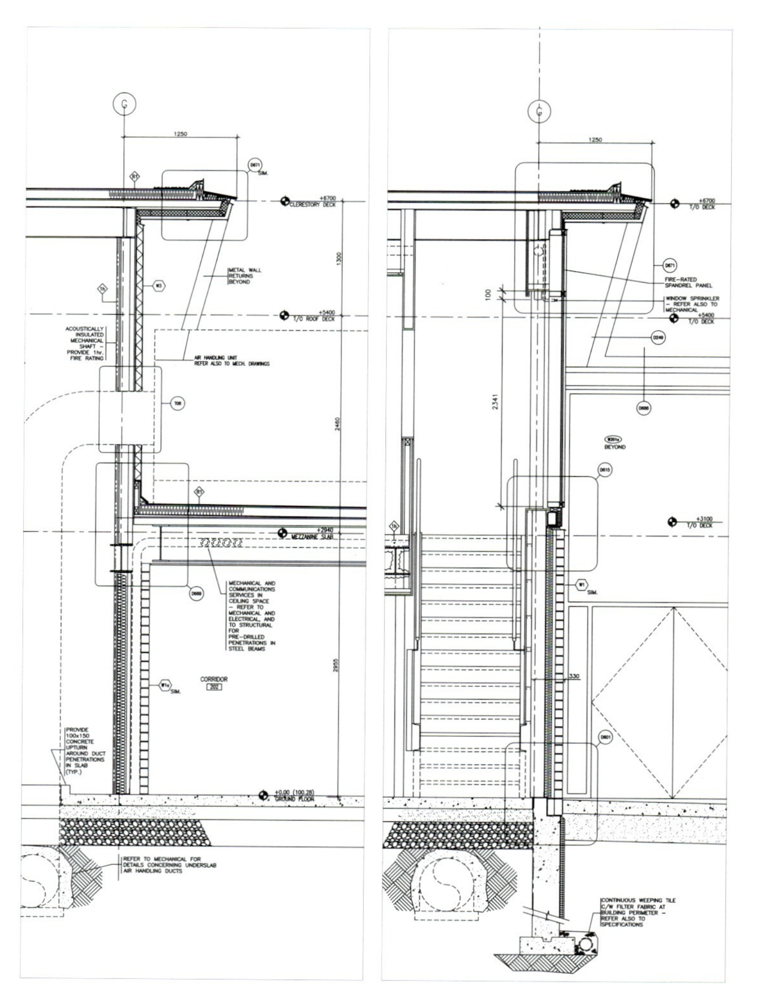

16

Downs/Archambault & Partners

Downs/Archambault 及合伙人建筑设计事务所

Downs/Archambault & Partners has been in continuous practice in Vancouver for 34 years and presently operates with 3 partners and a staff compliment of 19. The Three Partners are David Galpin MAIBC, Ron Beaton MAIBC and Mark Ehman MAIBC.

The firm's work is primarily focused in the lower mainland, with significant Master Planning work and large projects completed in Whistler, Victoria, Seattle and Beijing.

Downs/Archambault & Partners has successfully completed over 400 projects within British Columbia. The firm is proud of its reputation as a creative, resourceful, and congenial firm of dedicated architects and technicians. The partnership is based on the following convictions:

• A close collaborative and interactive relationship with the client is vital to ensure that the client's objectives are fully realized.

• Architecture must achieve a social objective by satisfying the physical and emotional needs and aspirations of stakeholders and users.

• The process of master planning and building design must respond to the site and its context, minimize the environmental impact, and be expressive of its logic and purpose.

• That a multi disciplinary design team must be comprise of the best-qualified members available to provide the highest level of ability, commitment and service throughout the project.

To this end the policy of the firm is that each project receives the direct and continuing participation of senior partner and associates in every phase of the work. The principals are practicing architects and planners directly involved in project programming, site planning, design development, production of the drawings and contract administration.

The firm has a proven ability to communicate honestly and directly with client... a collective "personality"that has allowed them to work in partnership with their clients on demanding projects.

Downs/Archambault 及合伙人建筑设计事务所有34年的执业历史。现今拥有19个成员和三个合伙人：大卫 · 加尔平（David Galpin）、罗恩 · 比顿（Ron Beaton）和马克 · 埃曼（Mark Ehman）。

Downs/Archambault 及合伙人建筑设计事务所的业务主要集中在温哥华低陆平原。他们的城市规划和大型建筑项目位于惠斯勒、维多利亚、西雅图和北京。

Downs/Archambault 及合伙人建筑设计事务所在不列颠哥伦比亚省完成了400多个工程项目，并以创新的、富于想像的设计和团结的事务所而著称。其设计目标有：

• 与业主之间密切协作，全面的理解业主的想法。

• 建筑师通过满足股东和使用者的愿望和物质上与精神上的需求来实现建筑物的社会功能。

• 总体规划和建筑设计必须对场地及其历史文脉作出回应，使新的设计对原有环境的影响最小化，并且对设计的合理性和目的有很好的表达。

• 复合型的、训练有素的设计团队拥有最富经验的设计师，他们能够确保提供最高水平的服务。

为达到这些目标，在每一个项目中，Downs/Archambault 及合伙人建筑设计事务所都与他们的合作伙伴和其他设计师共同参与设计的每一个细节。总负责人是执业建筑师和规划师，他们直接参与项目的立项、规划、初步设计、施工图设计和施工服务。

他们被证实具有与业主合作顺畅、诚实的沟通的能力。这些特质使他们能够在一些困难重重的项目中与业主有良好的合作。

Tower G- 1650 Bayshore Drive
海湾大道1650号G座大厦

地　　点：温哥华
规　　模：11 000m²
完成时间：2003年

As one of the final residential towers to be built in the Bayshore Gardens neighborhood, Tower G's inevitable waterfront location demanded a unique architectural response. The building's fan shaped plan provides all units with unobstructed views of CoalHarbour, Stanley Park and the North Shore mountains beyond. The three-story townhouses carefully relate in scale to the Westin Bayshore Hotel complex across the street. An extensive water garden with lush landscaping further enhances this prestigious address.

The unique entry canopy and "bridge" extends across an extensive reflecting pool, providing transition and separation from the street. The cascading water garden above the 5 level parking structure provides a visual and auditory amenity to both the public and the residents. The accompanying drawings illustrate details of both the entry canopy and the water garden.

作为海湾大道花园社区的最后一栋建筑，海湾大道1650号G座大厦位于海滨独特的地理位置，其设计要求十分苛刻。风车状的平面设计使每个居住单位都有Coal海湾、史坦利公园和北岸山脉的良好景色。三层高的城市别墅的尺度与街对面的Westin海湾酒店综合楼的尺度相配合。一个以水为主题、植被茂盛的花园进一步突出了G座大厦地理位置优越的显著特征。

一个独特的入口雨篷和"桥"设置在水池之上，连接出入口和街道，同时也提供了与街道的一个屏障。跌落的水池花园位于5层的停车场屋顶上，是公众和住户视觉和声觉上的一个绿洲。后面所附的图纸是大厦入口雨篷和水池花园的细部大样。

1

1　沿城市立面
2　总平面图

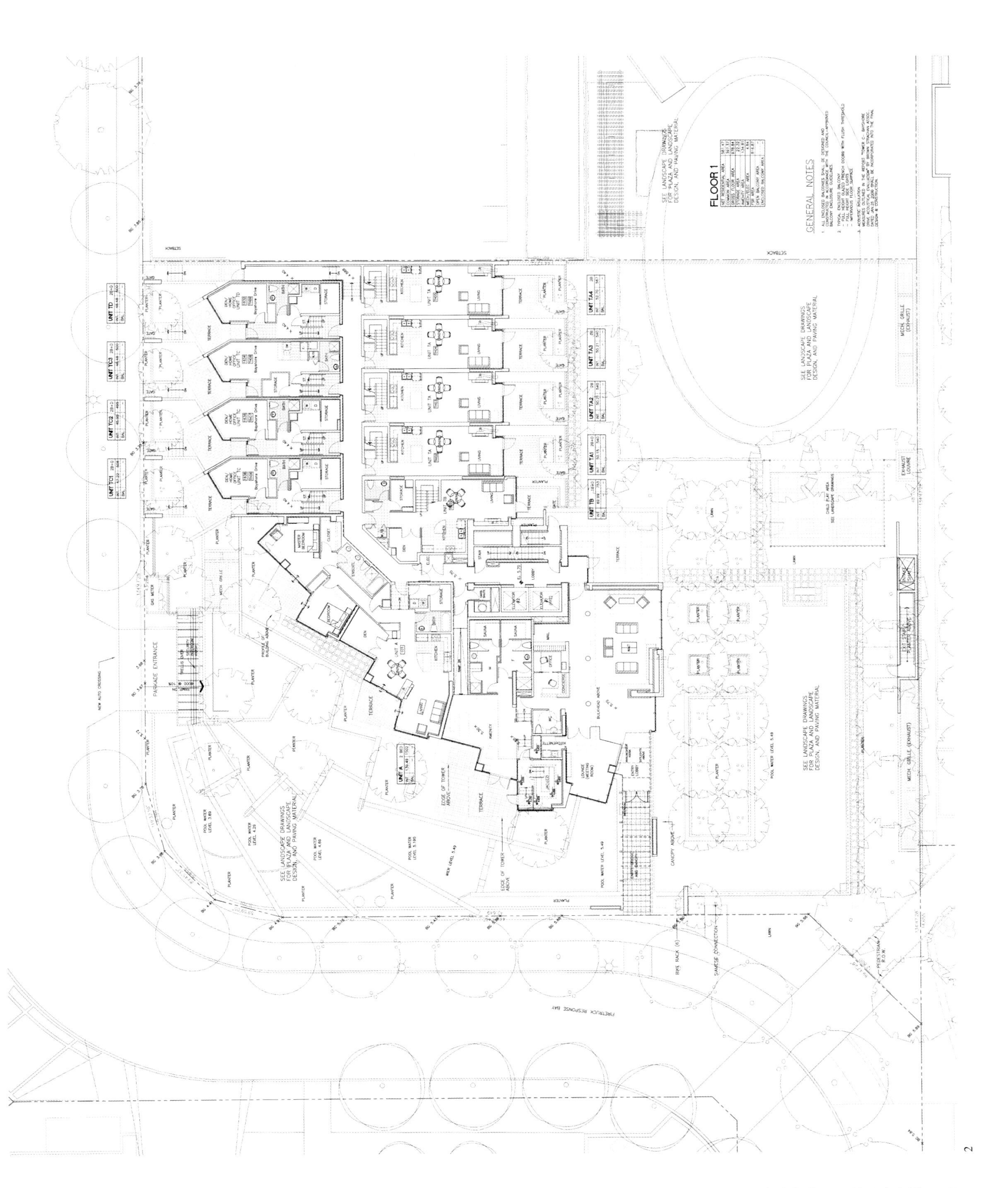

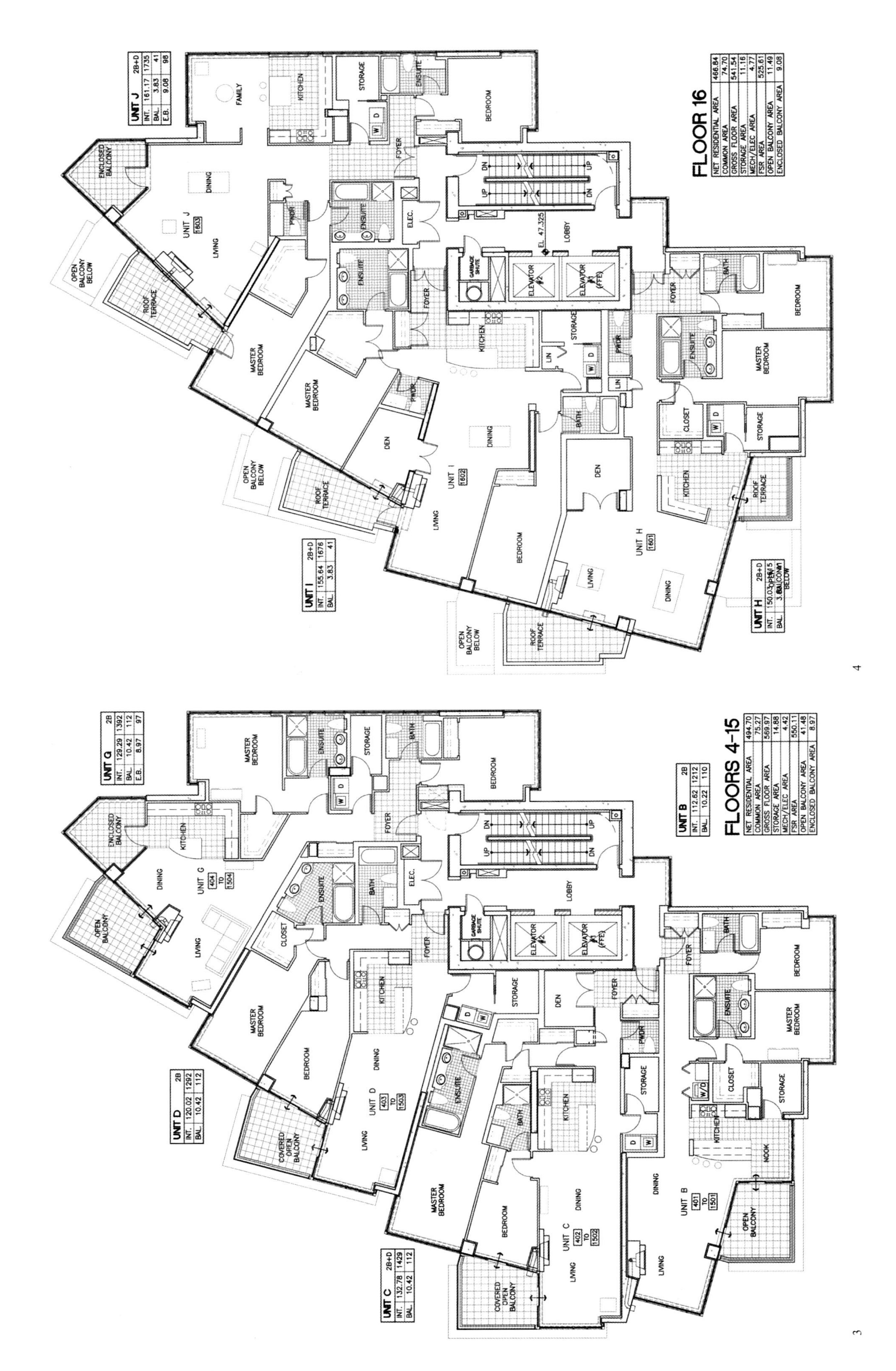
FLOOR 16
NET RESIDENTIAL AREA 466.84
COMMON AREA 74.70
GROSS FLOOR AREA 541.54
STORAGE AREA 11.16
MECH/ELEC AREA 4.77
FSR AREA 525.61
OPEN BALCONY AREA 11.49
ENCLOSED BALCONY AREA 9.08
UNIT J 2B+D
INT. 161.17 1735
BAL. 3.83 41
E.B. 9.08 98
UNIT I 2B+D
INT. 155.64 1676
BAL. 3.83 41
4
FLOORS 4–15
NET RESIDENTIAL AREA 494.70
COMMON AREA 75.27
GROSS FLOOR AREA 569.97
STORAGE AREA 14.88
MECH/ELEC AREA 4.42
FSR AREA 550.11
OPEN BALCONY AREA 41.48
ENCLOSED BALCONY AREA 8.97
UNIT Q 2B
INT. 129.29 1392
BAL. 10.42 112
E.B. 8.97 97
UNIT B 2B
INT. 112.62 1212
BAL. 10.22 110
UNIT D 2B
INT. 120.02 1292
BAL. 10.42 112
UNIT C 2B+D
INT. 132.78 1429
BAL. 10.42 112
3

3　4~15层平面图
4　16~18层平面图
5　花园水池
6　花园水池
7　花园水池详图
8　花园水池详图

5

6

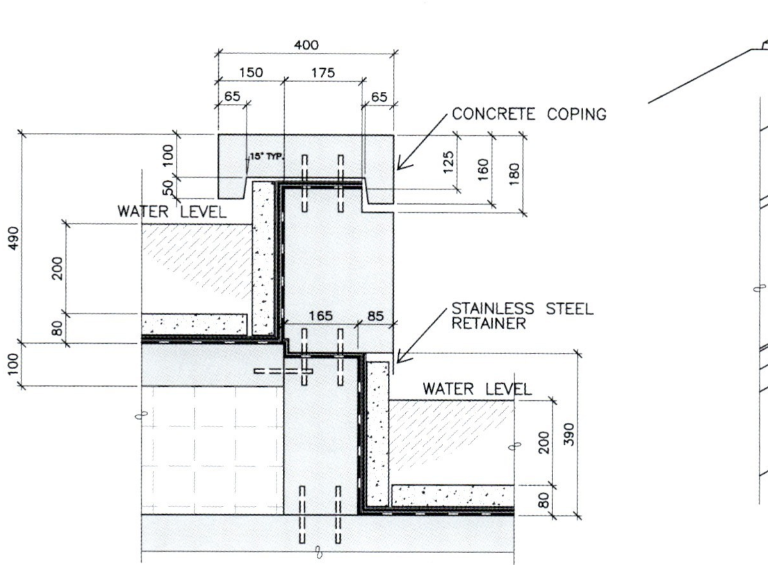

7

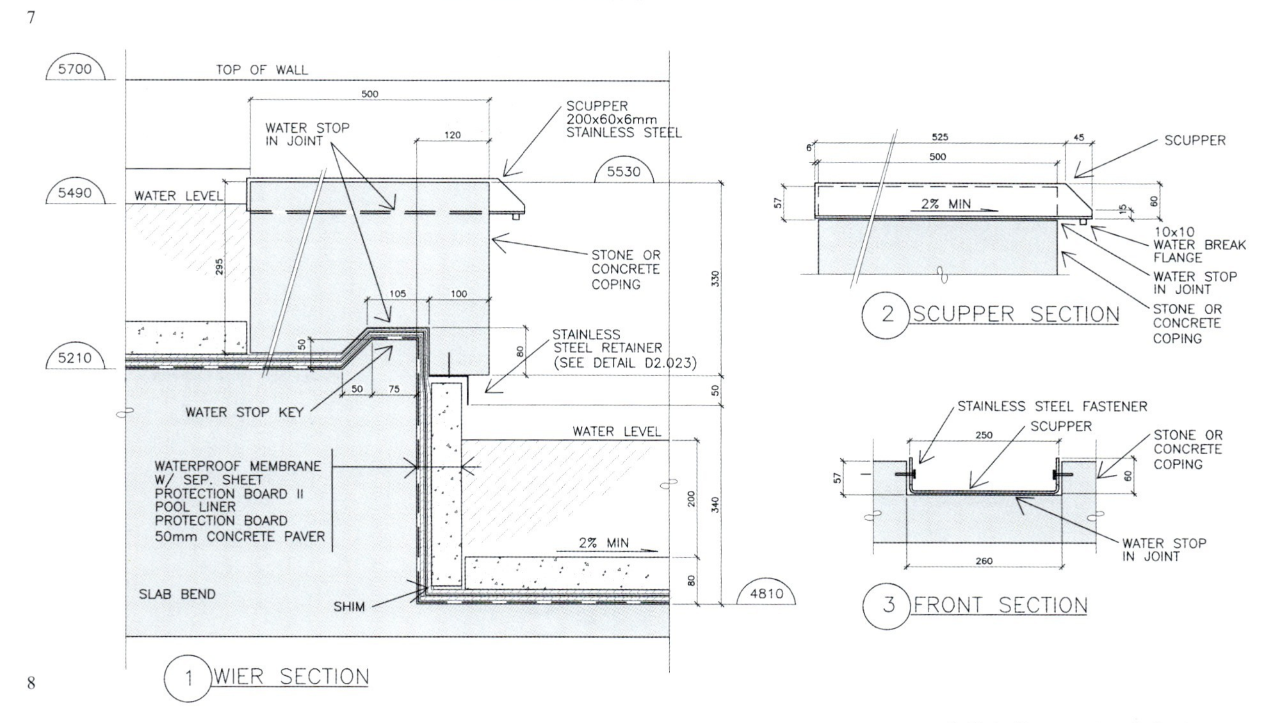

8

9

10

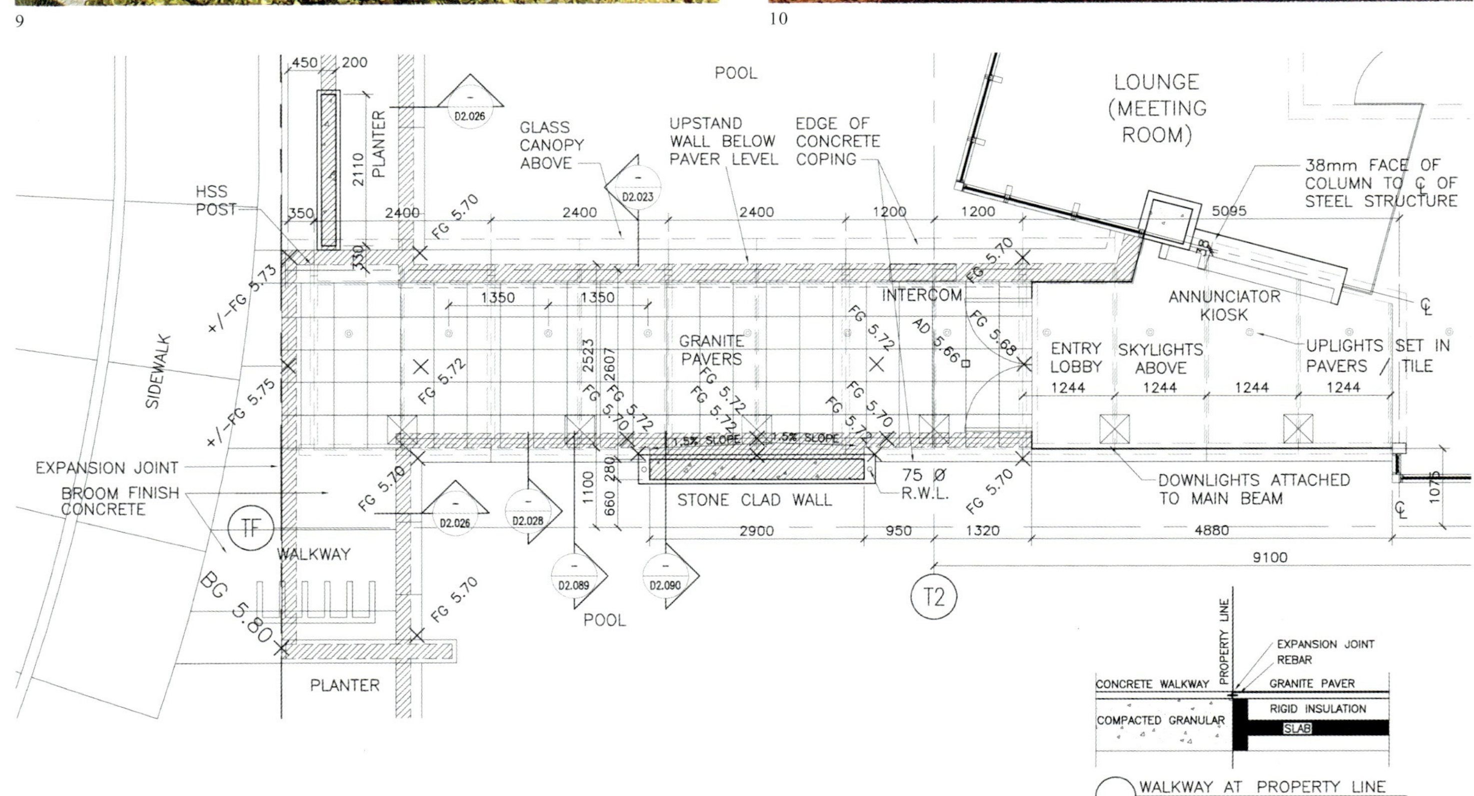
POOL
LOUNGE
(MEETING
ROOM)
PLANTER
GLASS
CANOPY
ABOVE
UPSTAND
WALL BELOW
PAVER LEVEL
EDGE OF
CONCRETE
COPING
38mm FACE OF
COLUMN TO ℄ OF
STEEL STRUCTURE
HSS
POST
SIDEWALK
INTERCOM
GRANITE
PAVERS
ANNUNCIATOR
KIOSK
ENTRY
LOBBY
SKYLIGHTS
ABOVE
UPLIGHTS SET IN
PAVERS / TILE
EXPANSION JOINT
BROOM FINISH
CONCRETE
STONE CLAD WALL
75 Ø
R.W.L.
DOWNLIGHTS ATTACHED
TO MAIN BEAM
WALKWAY
BG 5.80
POOL
PLANTER
PROPERTY LINE
EXPANSION JOINT
REBAR
CONCRETE WALKWAY
GRANITE PAVER
COMPACTED GRANULAR
RIGID INSULATION
SLAB
WALKWAY AT PROPERTY LINE

11

9　沿海滨立面
10　入口雨篷和"桥"
11　入口雨篷和"桥"平面图及节点详图
12　雨篷详图
13　雨篷节点详图
14　雨篷详图

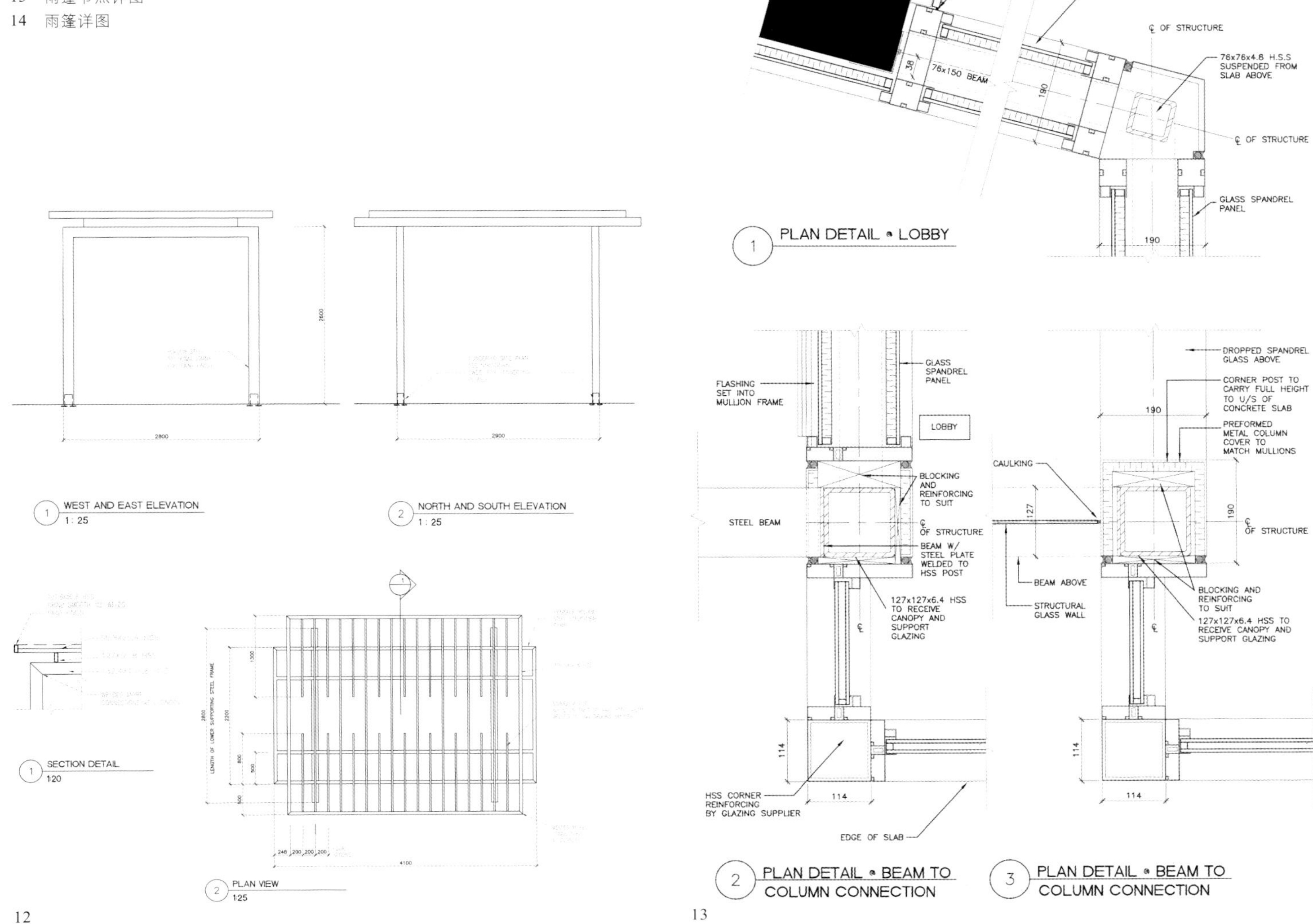

12

13

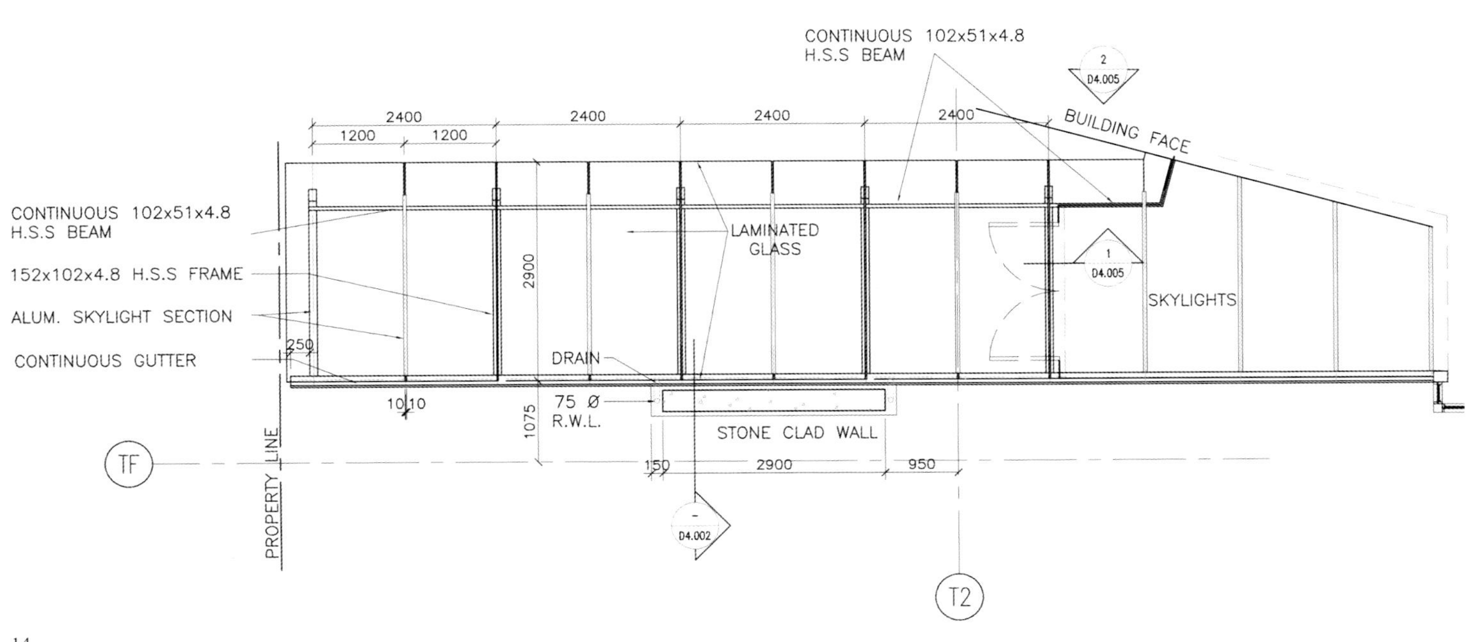

14

Westin Bayshore Hotel

Westin 海湾酒店

地　　点：温哥华
完成时间：2000 年

The Westin Bayshore has been one of the preeminent Hotels in Vancouver for more than 30 years.

Downs/Archambault & Partners was commissioned to oversee the revitalization of the Hotel in 1998, with the new facilities consisting of an entirely new conference facility and a complete reconstruction of the hotel facilities with seismic upgrading and a total project upgrade, including a new porte cochere that provides a welcoming arrival for guests.

The porte cochere is a simple barrel vault with its end radiuses in plan. The compound radius produces a "wave" form that expresses the water and harbor setting and meets the size requirements of the drive area while maintaining a welcoming pedestrian scale.

The linear geometry of the associated canopied walkway complements the porte cochere

Westin海湾酒店作为温哥华的豪华酒店已经有超过30年的历史了。Downs/Archambault及合伙人建筑设计事务所于1998年承接了整体改造Westin海湾酒店的工程，其任务包括：全新的会议设施，提高酒店的抗震设施和整体全面的改建。作为工程的一部分，一个独特的入口雨篷可以给人热情温暖的感觉。

入口雨篷是一个简单的筒拱形屋顶，平面为弧线形。这个复合弧线形的雨篷既从形式上呼应了海水和海港主题，又满足了来宾车辆到达入口所需要的雨篷尺度。

直线几何形的连廊雨篷完善了入口雨篷的形式。

1

1　Westin 海湾酒店鸟瞰
2　酒店入口夜景
3　改建立面图

WESTIN

2

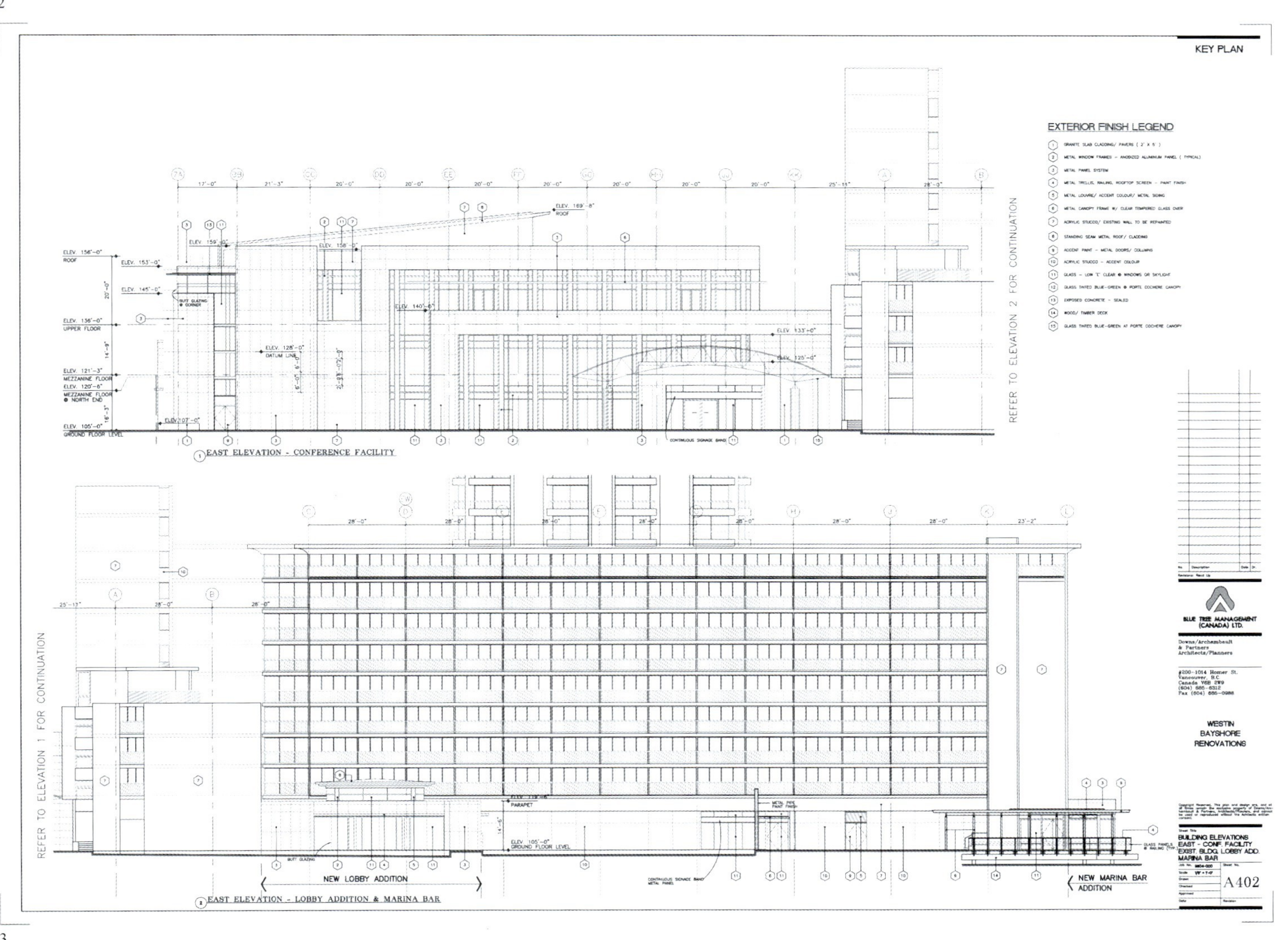
KEY PLAN
EXTERIOR FINISH LEGEND
EAST ELEVATION - CONFERENCE FACILITY
REFER TO ELEVATION 2 FOR CONTINUATION
REFER TO ELEVATION 1 FOR CONTINUATION
NEW LOBBY ADDITION
NEW MARINA BAR ADDITION
EAST ELEVATION - LOBBY ADDITION & MARINA BAR
BLUE TREE MANAGEMENT (CANADA) LTD.
WESTIN BAYSHORE RENOVATIONS
A402

3

4

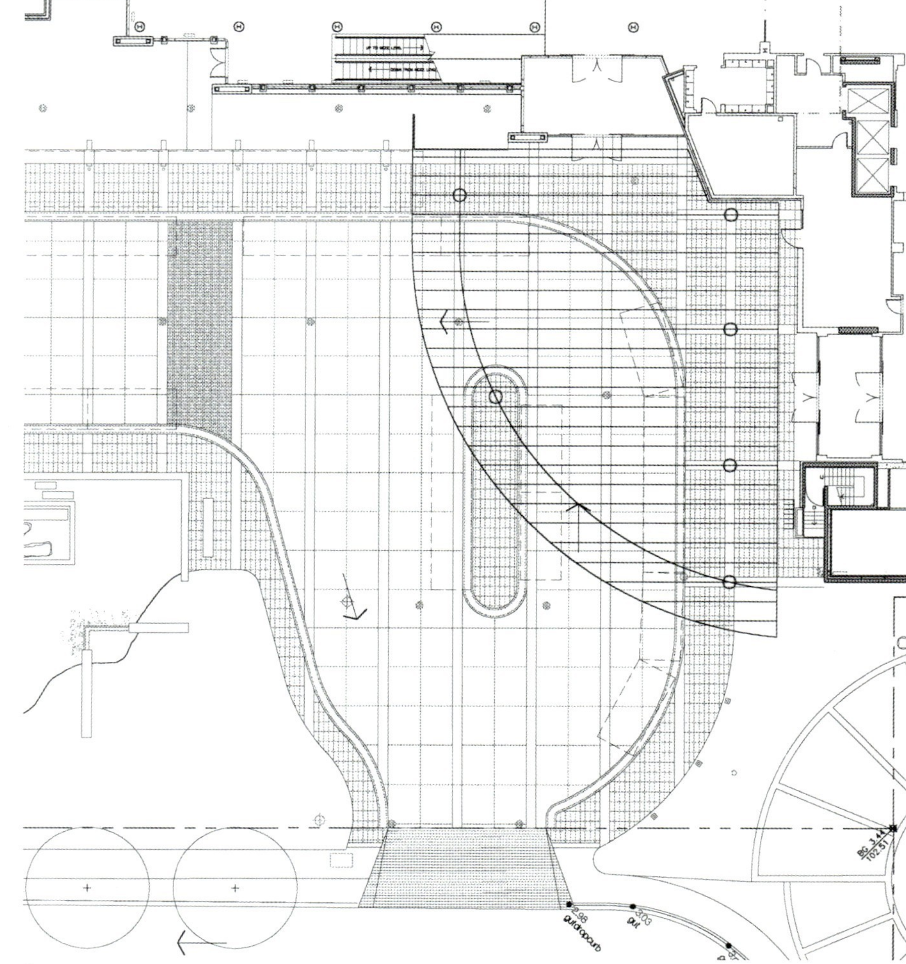

5

4　酒店入口
5　入口上盖平面图
6　入口上盖立面图
7　入口上盖详图
8　入口上盖柱与屋面节点详图

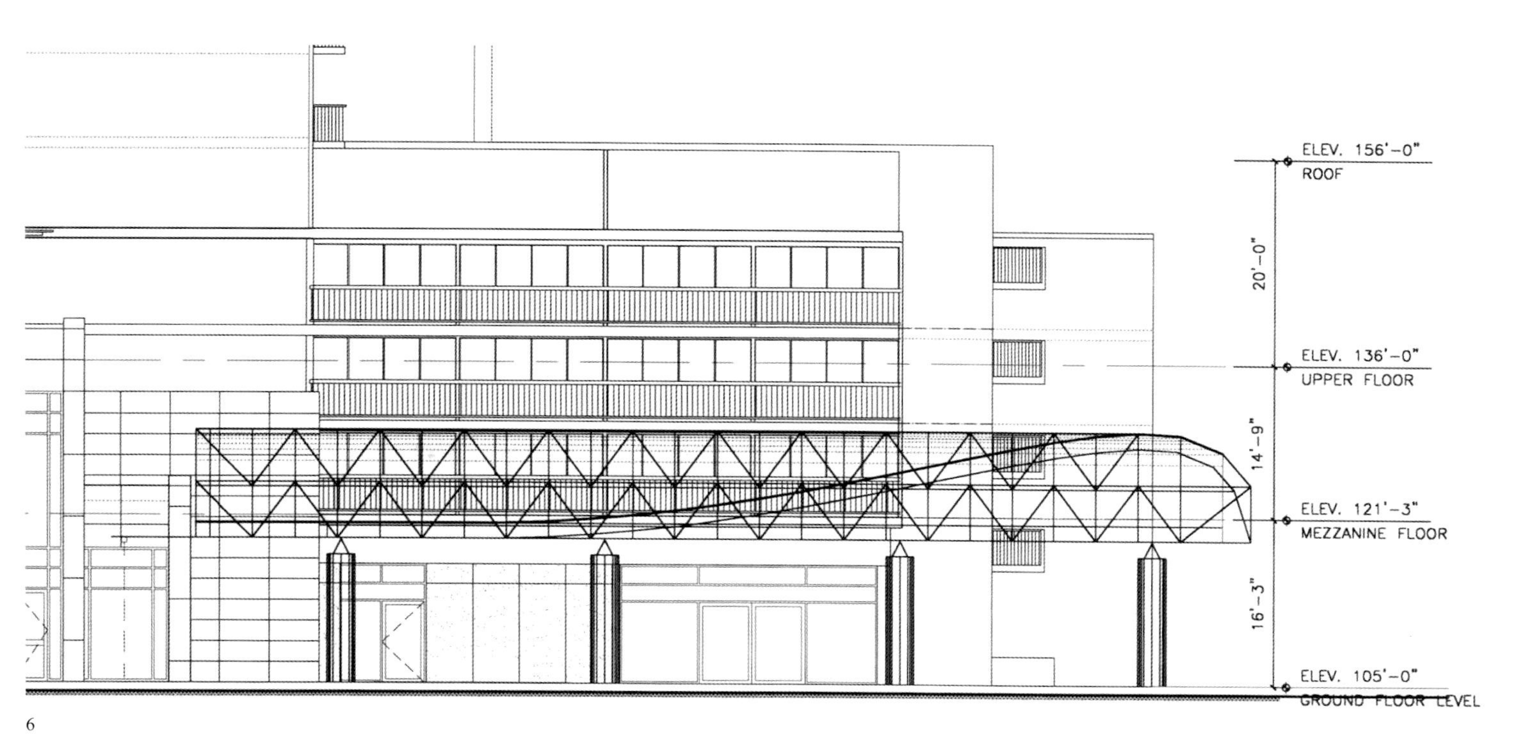
ELEV. 156'-0"
ROOF
20'-0"
ELEV. 136'-0"
UPPER FLOOR
14'-9"
ELEV. 121'-3"
MEZZANINE FLOOR
16'-3"
ELEV. 105'-0"
GROUND FLOOR LEVEL

6

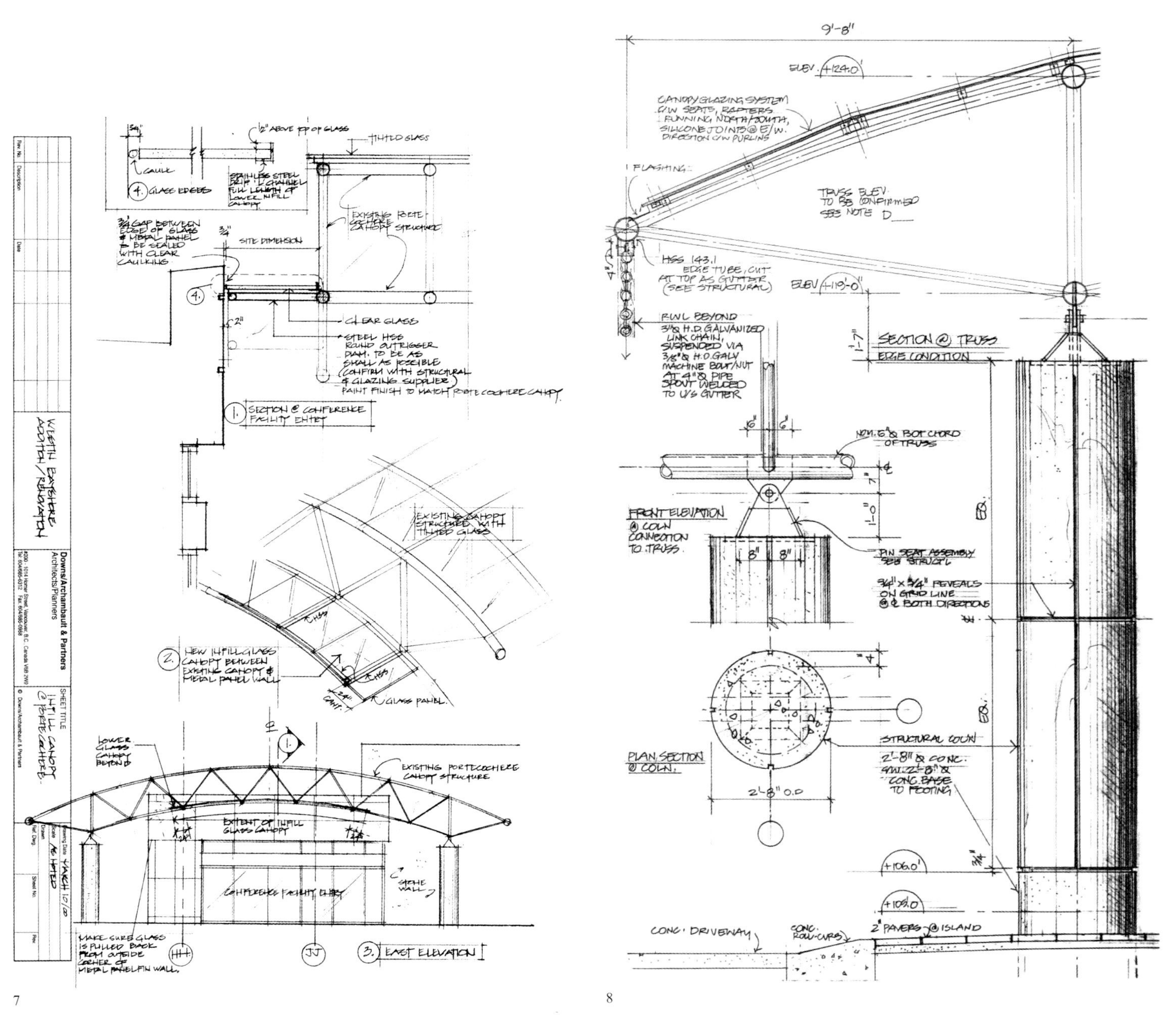
Downs/Archambault & Partners
Architects/Planners
WESTIN BAYSHORE ADDITION / RENOVATION
SHEET TITLE
4. GLASS EDGES
CLEAR GLASS
1. SECTION @ CONFERENCE FACILITY ENTRY
2. NEW INFILL GLASS CANOPY BETWEEN EXISTING CANOPY & METAL PANEL WALL
GLASS PANEL
EXISTING PORTECOCHERE CANOPY STRUCTURE
EXTENT OF INFILL GLASS CANOPY
CONFERENCE FACILITY ENTRY
3. EAST ELEVATION
9'-8"
ELEV. +124.0'
FLASHING
TRUSS ELEV. TO BE CONFIRMED SEE NOTE D
ELEV. +119.0'
SECTION @ TRUSS
EDGE CONDITION
FRONT ELEVATION @ COLN CONNECTION TO TRUSS
PIN SEAT ASSEMBLY SEE STRUCT'L
PLAN SECTION @ COLN.
2'-8" O.D
STRUCTURAL COLN
+106.0'
+105.0'
CONC. DRIVEWAY
2" PAVERS @ ISLAND

7

8

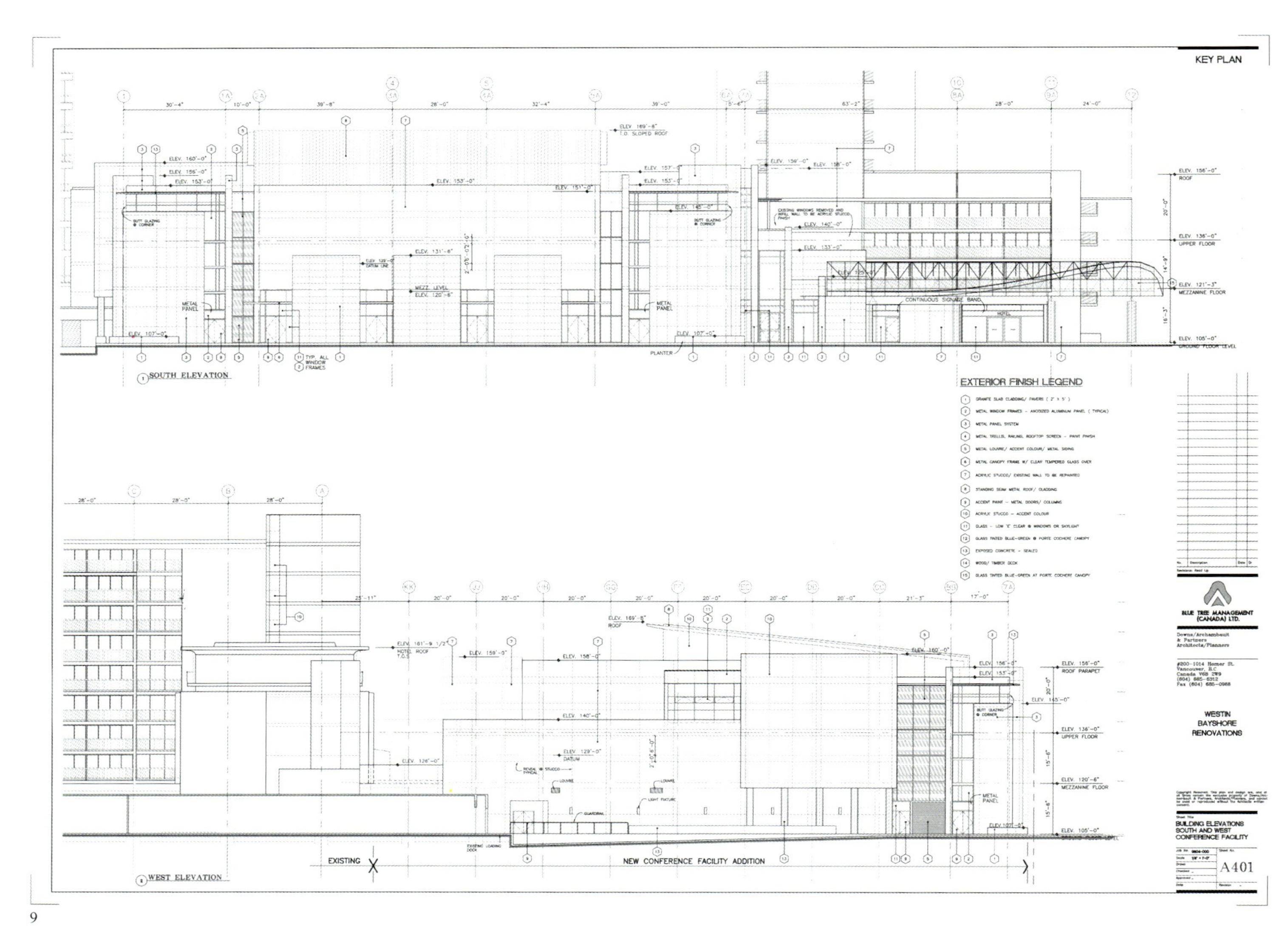

9

10

9 改建立面图
10 直线几何形的走廊雨篷
11~12 直线几何形的走廊雨篷详图

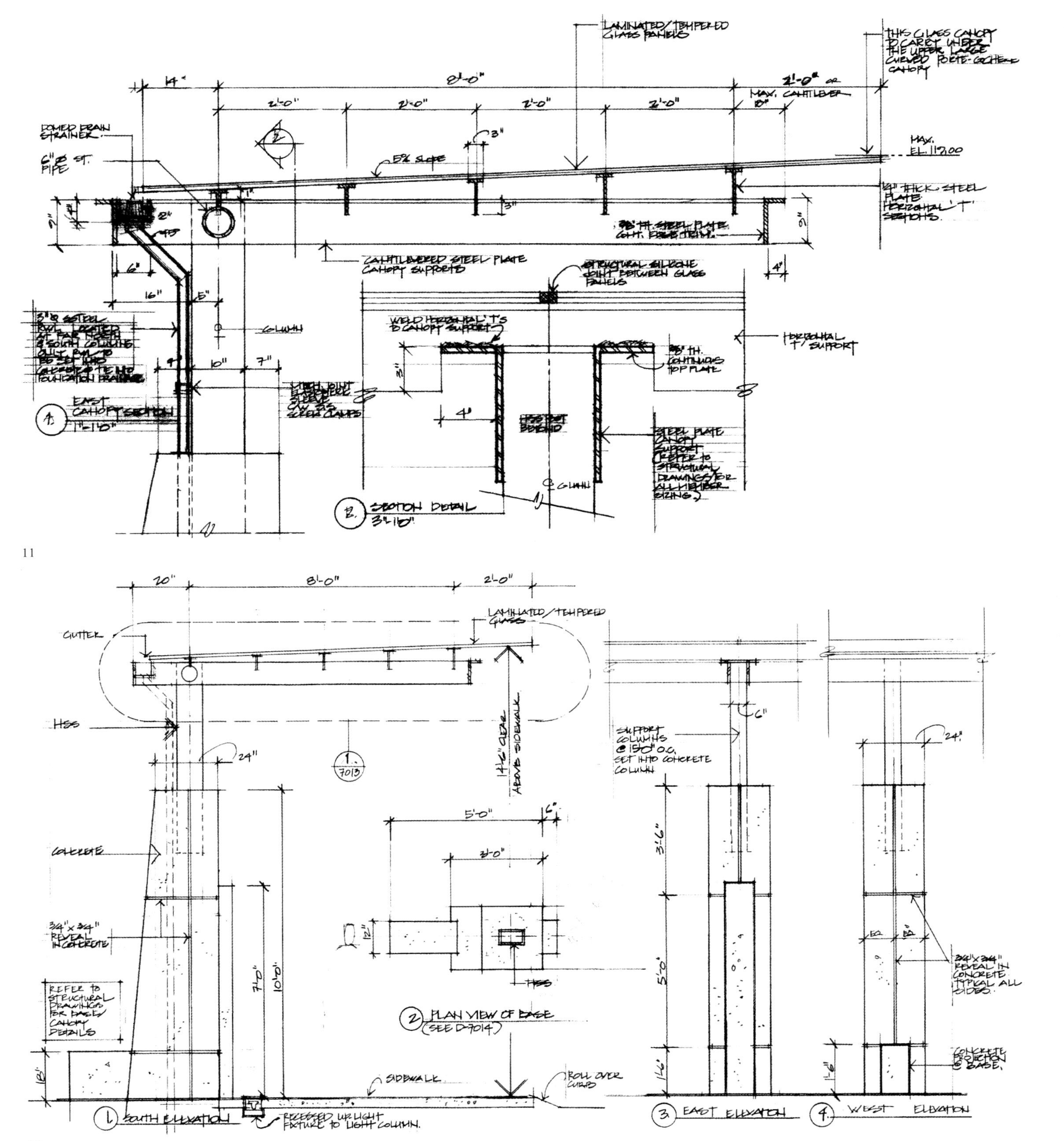

LAMINATED/TEMPERED GLASS PANELS
THIS GLASS CANOPY TO CARRY UNDER THE UPPER LARGE CURVED PORTE-COCHERE CANOPY
14"
8'-0"
2'-0"
MAX. CANTILEVER
DOMED DRAIN STRAINER
5% SLOPE
CANTILEVERED STEEL PLATE CANOPY SUPPORTS
STRUCTURAL SILICONE JOINT BETWEEN GLASS PANELS
WELD HORIZONTAL 'T'S TO CANOPY SUPPORT
HORIZONTAL T' SUPPORT
COLUMN
EAST CANOPY SECTION
SECTION DETAIL
STEEL PLATE CANOPY SUPPORT REFER TO STRUCTURAL DRAWINGS FOR ALL MEMBER SIZING
20"
GUTTER
LAMINATED/TEMPERED GLASS
HSS
24"
14'-6" CLEAR ABOVE SIDEWALK
CONCRETE
3/4" x 3/4" REVEAL IN CONCRETE
REFER TO STRUCTURAL DRAWINGS FOR BASE/CANOPY DETAILS
5'-0"
3'-0"
PLAN VIEW OF BASE (SEE D-7014)
SIDEWALK
ROLL OVER CURB
SOUTH ELEVATION
RECESSED UPLIGHT FIXTURE TO LIGHT COLUMN.
SUPPORT COLUMNS @ 15'-0" O.C. SET INTO CONCRETE COLUMN
3/4"x3/4" REVEAL IN CONCRETE TYPICAL ALL SIDES.
CONCRETE PROJECTION @ BASE.
EAST ELEVATION
WEST ELEVATION

11

12

VIA ARCHITECTURE

VIA 建筑设计事务所

VIA ARCHITECTURE is "community-building". We create environments that people love, from gallery and museum to home and cafe, including the bench outside the railway station. From this attention to delight in the small, leads the path to true richness at the scale of city building.

With offices in Vancouver, Seattle and Shanghai, founded in 1984 in Vancouver, VIA Architecture has travelled the globe. Our projects span Canada, China, Malaysia, Mexico, Philippines, Russia, and the USA; ranging in scope and complexity from retail kiosks to billion dollar rapid transit systems.

VIA 建筑设计事务所的宗旨是建筑与人共存。不论是展览馆、博物馆、教堂、家庭、咖啡屋还是公共车站外的坐椅，都力图创造公众喜爱的空间。从每一个细节入手，VIA建筑在城市中创造出浓墨重彩的一笔。

VIA 建筑设计事务所有温哥华、西雅图和上海分公司，创建于 1984 年。从加拿大、美国、俄罗斯、中国、马来西亚，到墨西哥，工作遍及世界各地。从零售亭到造价高昂的城市快速交通系统，VIA 建筑的工作涉及广泛的领域。

VIA 建筑涉及的领域包括：

- 多功能建筑发展计划
- 公共住宅区
- 娱乐休闲建筑（宾馆，餐馆）
- 城市规划及设计
- 大规模公共交通系统工程
- 旧建筑改建，扩建

Commercial SkyTrain Station
商业街轨道交通车站

地　　点：温哥华
完成时间：2002年

At the interchange station for the Expo and Millennium Lines, Commercial Station rises 20 meters from its foundations in the railway cut and bridges the cut to connect to the existing Broadway Station. Great care was required to avoid disruption of the adjacent neighborhood above and the Burlington Northern / Santa Fe and Amtrak elevations below within this deep, narrow, operating heavy rail corridor that runs through this residential Vancouver community.

Within demanding constraints and requirements, steel was chosen for the station superstructure because of its ability to be prefabricated off site and inserted into the project without disruption to the existing rail operations. Combining the steel superstructure with glulam beams, roof panels, and skylights results in a design that fits into the community as if it has always been there.

作为温哥华轨道交通博览线和新千年线的中转车站，商业街轨道交通车站位于铁轨隧道基础20m之上，于现有的百老汇街车站有步行天桥连接。车站的设计需要避免与相邻的街区的干扰，和位于相邻深而狭窄的铁轨隧道内的Burlington Northern / Santa Fe和Amtrak铁路公司的轨道之间的干扰。

因条件限制的需要，商业街轨道交通车站的主要建筑材料为钢。钢可以预制，而且不影响现有的轨道的使用。与钢一起使用的是木架，屋顶板和天窗。通过组合，这些材料成为社区的一部分——存在了很久的构筑物。

1

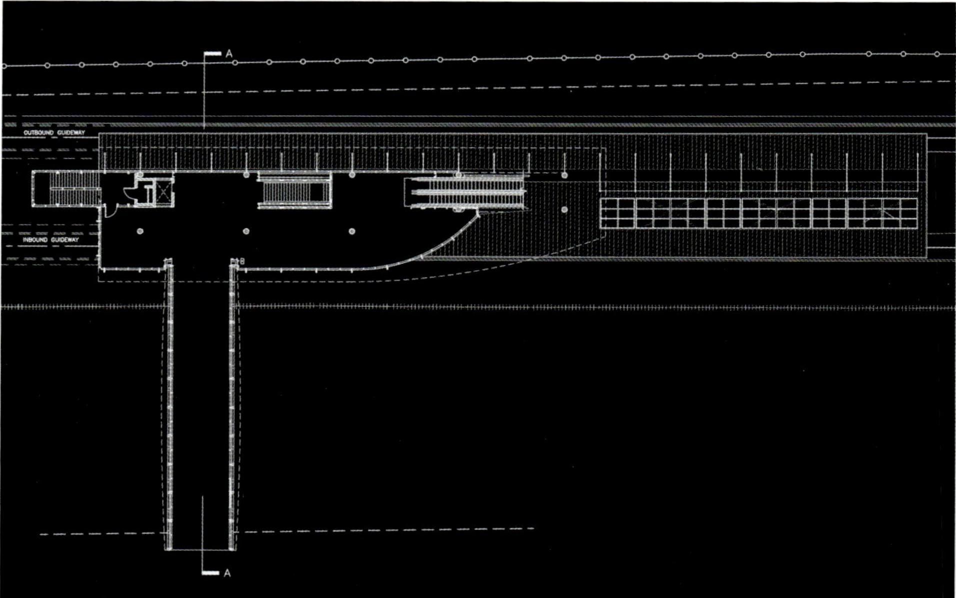

2

3

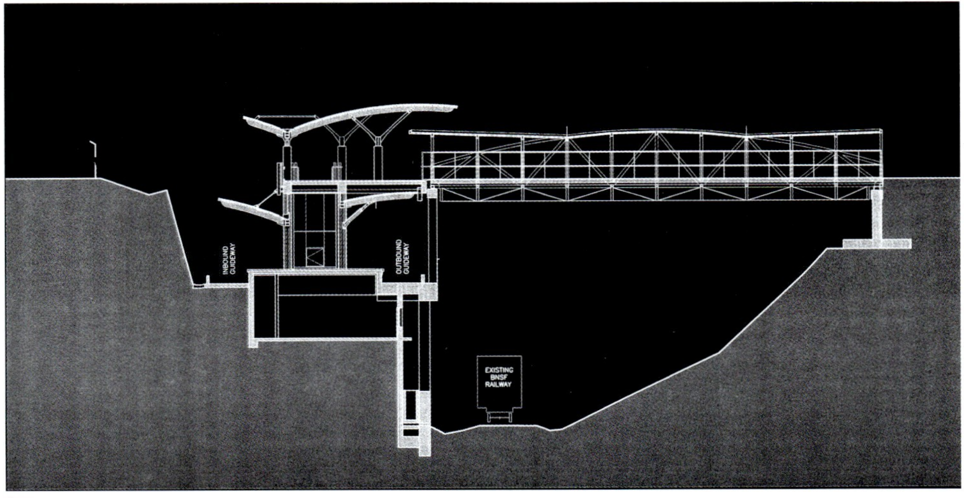

4

1　俯视车站
2　车站平面图
3　车站绿化挡土墙
4　剖面图

5

6

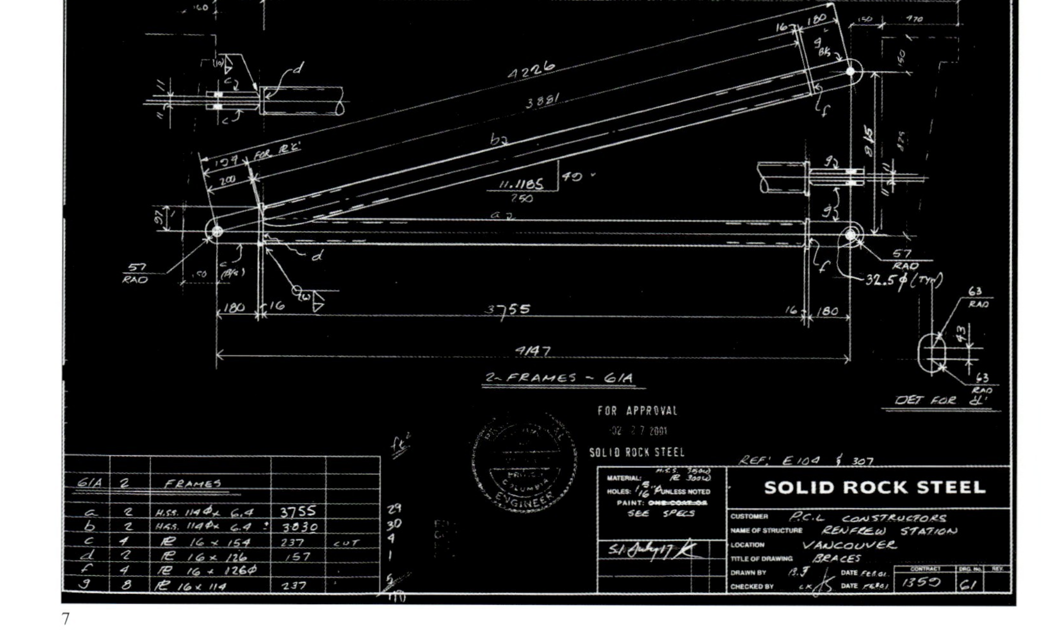

7

5　车站月台俯视街道
6　屋顶
7　屋顶钢构件详图
8　屋顶构件详图
9~10　屋顶详图
11~12　屋顶细部
13　钢肋构件详图

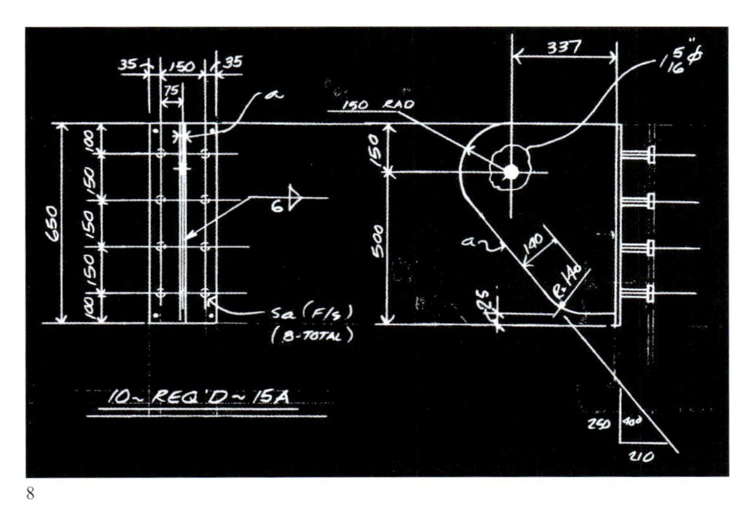

337
150 RAD
10~ REQ'D ~ 15A
5a (F/s)
(8-TOTAL)

8

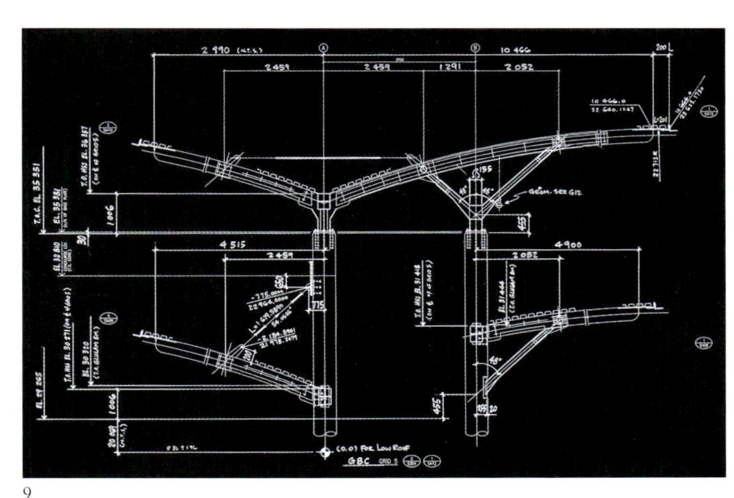

(0.0) FOR LOW ROOF
GBC

9

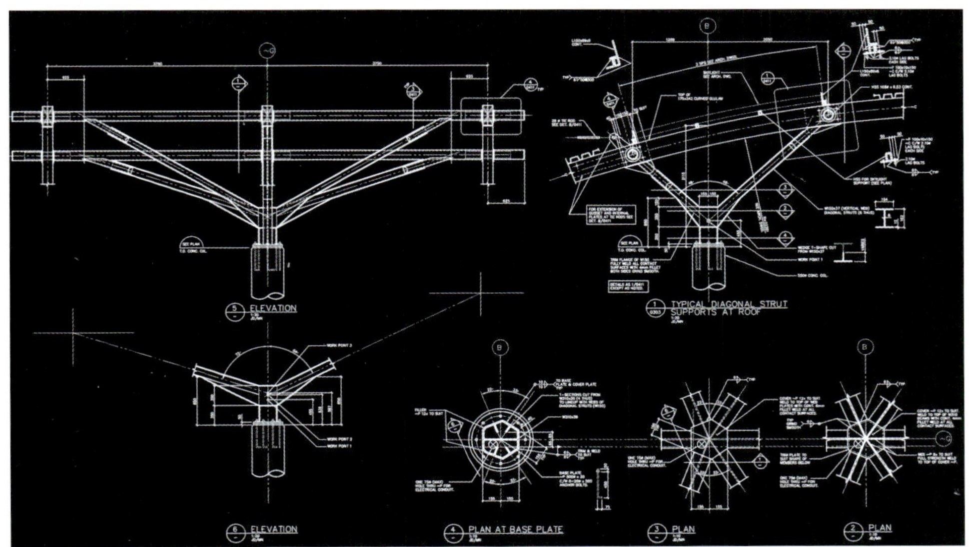

ELEVATION
TYPICAL DIAGONAL STRUT SUPPORTS AT ROOF
ELEVATION
PLAN AT BASE PLATE
PLAN
PLAN

10

11

12

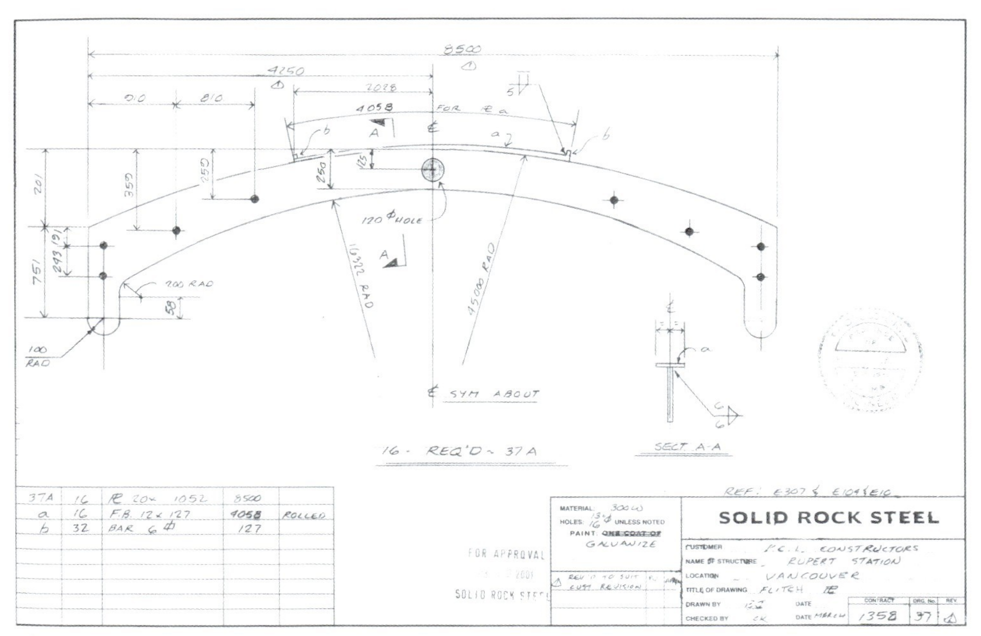

8500
4250
120 ϕ HOLE
SYM ABOUT
16 - REQ'D ~ 37A
SECT A-A
SOLID ROCK STEEL
FOR APPROVAL
CUSTOMER P.C.L. CONSTRUCTORS
NAME & STRUCTURE RUPERT STATION
LOCATION VANCOUVER

13

Pacific Mineral Museum
太平洋矿物博物馆

地　　点：温哥华
完成时间：2001年

The Pacific Mineral Museum was built in 1921 as the Ceperley Rounsefell Company Building for a local insurance company. In 1961 the building was renovated. The double height hall was filled in and a suspended masonite ceiling concealed the superb luminaire ceiling above. In 1998 Baker McGarva Hart/Henry Hawthorn Architect were commissioned by the Pacific Mineral Museum Society to create a showcase for the mining industry in B.C. The Society wished for a museum that complemented their superb mineral collection. The chairman of the Society, Ross Beaty, directed the architects to celebrate the heritage aspect of the building and he modified the architectural program to highlight those areas of the building where we could preserve and restore the original fabric.

The project involved lowering the infill second floor by 1'6" to the original mezzanine level in order to create a grand second floor gallery, exposing the original ceiling, the creation of 2 mezzanines within this gallery, and the restoration of the Hastings Street façade. The ground floor of the building was reconfigured to accommodate a gift shop, an introductory gallery, and support functions. Concrete shear walls were introduced to provide a stable base for the new second floor. The original vault door was uncovered behind a wall during construction and reused in the new museum vault exhibit on the second floor.The building was fully upgraded to meet the Vancouver building bylaw and a retail store, introductory gallery, workshop, lecture room, and service spaces were also incorporated into the building.

太平洋矿物博物馆建于1921年，是当地保险公司Ceperley Rounsefell的总部建筑。1961年建筑被重新装修改建，其两层高的大厅被填充，一个水泥吊顶掩盖了华丽的原始吊顶。1998年，太平洋矿物博物馆协会委托Baker McGarva Hart（VIA建筑设计事务所前身）及Henry Hawthorn建筑设计事务所设计一个展示不列颠哥伦比亚省矿业的空间。协会希望拥有一个可以展示不列颠哥伦比亚省灿烂的收藏品的博物馆。协会的主席Ross Beaty要求建筑师突出建筑的历史特征，并修改设计以强调建筑师应该保存和重建建筑原始的形态的部分。

改造工程包括，降低加层空间4cm，以

1

1~2 室内
3 入口构思草图

2

便于二层展示厅的使用和更好地展示原始的屋顶，加建第二个夹层空间，重修沿HASTINGS街的立面。太平洋矿物博物馆的首层设有纪念品商店，一个介绍厅和其他辅助设施。混凝土剪力墙为新的第二层提供结实的基础，原始的拱顶在施工时被遮盖，完工后成为新建二层的屋顶。

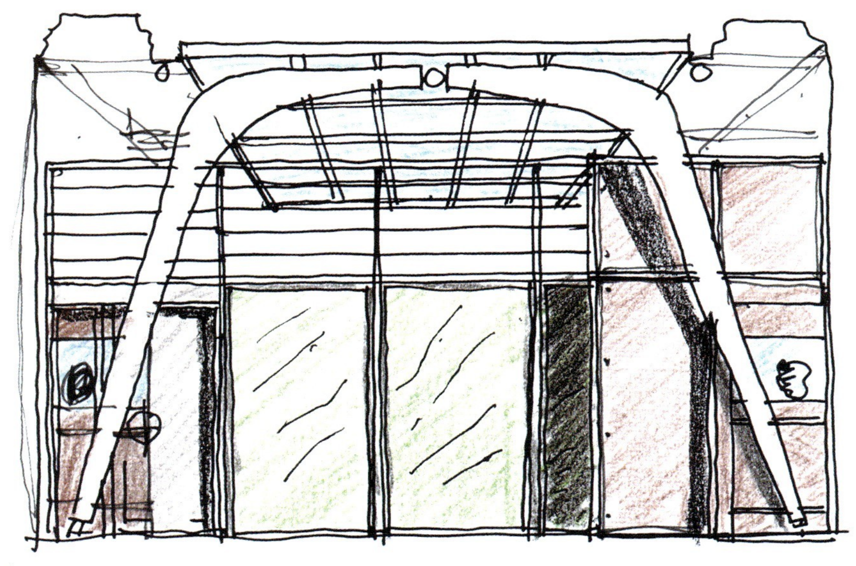

3

4

5

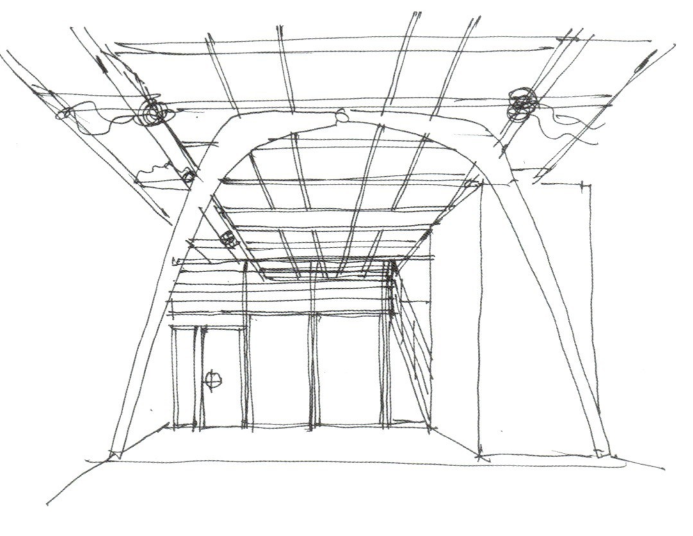

6

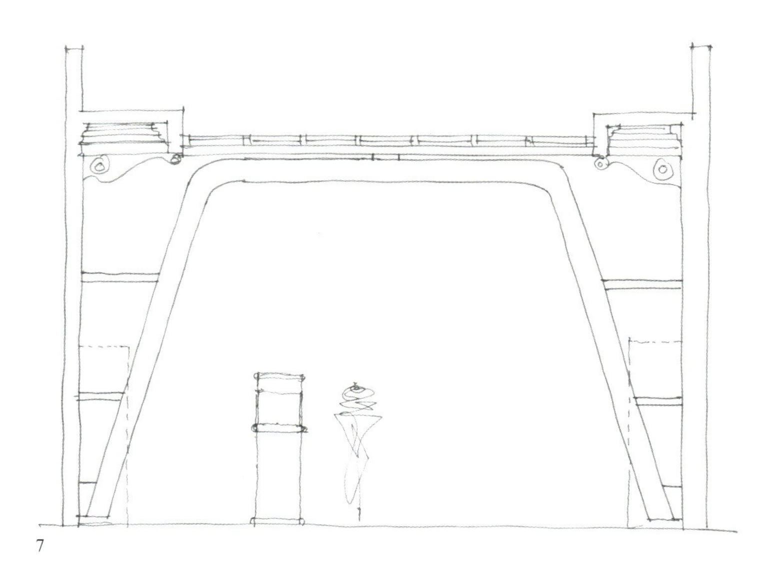

7

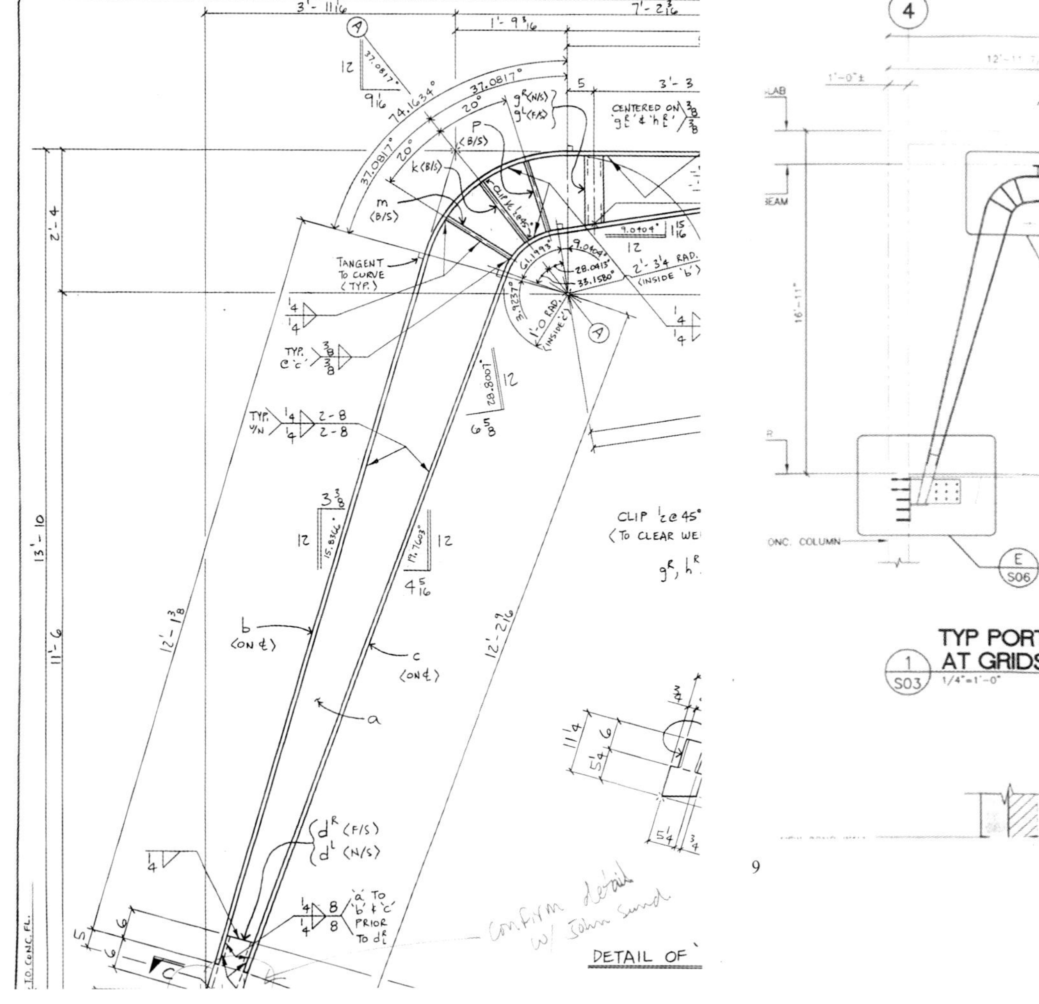

8

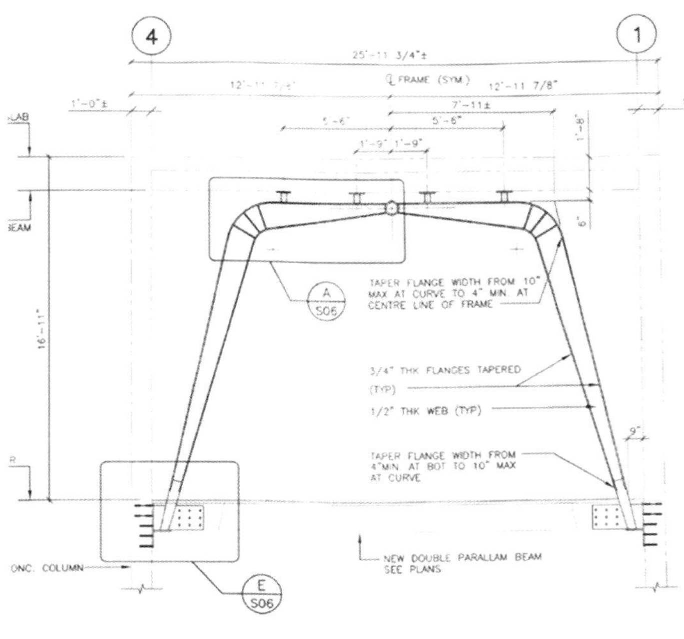

9

4 钢支撑
5 室内
6 室内支撑透视草图
7 室内支撑草图
8 屋顶细部
9 钢支撑施工详图

Club Intrawest at Palm Desert
西部俱乐部

地　　点：PALM，加利福尼亚州，美国
完成时间：1999年

Opened in early 1999, this 20 million ($US) multi-phased development was hailed as the first resort community to provide a truly authentic experience of this desert. The challenge was to capture the uniqueness of the local desert environment, while appropriately expressing the community and its lifestyle within a context of existing development that has relied heavily on irrigated lawns and extensive use of water features.

VIA ARCHITECT sought to create an exciting environment that was ecologically and culturally sensitive to the heritage and desert setting of the Coachella Valley. Our response was a classic, timeless architecture that enriches memories and expectations of the desert.

开业于1999年，造价两千万美金的西部俱乐部是分多期开发的，是第一个提供给沙漠爱好者的度假社区。设计的挑战性在于创造合宜的形态去适应当地的沙漠环境的同时，与当地现有的生活方式——极大程度上依赖灌溉的草坪和水的广泛使用——的融合。

VIA建筑设计事务所寻找并创造了一个令人兴奋的环境，在环境上、文化上都适应Coachella山谷沙漠的历史和特点。西部俱乐部是一个优雅的、超越了时间的建筑，丰富了参观者对沙漠的回忆和体验。

1

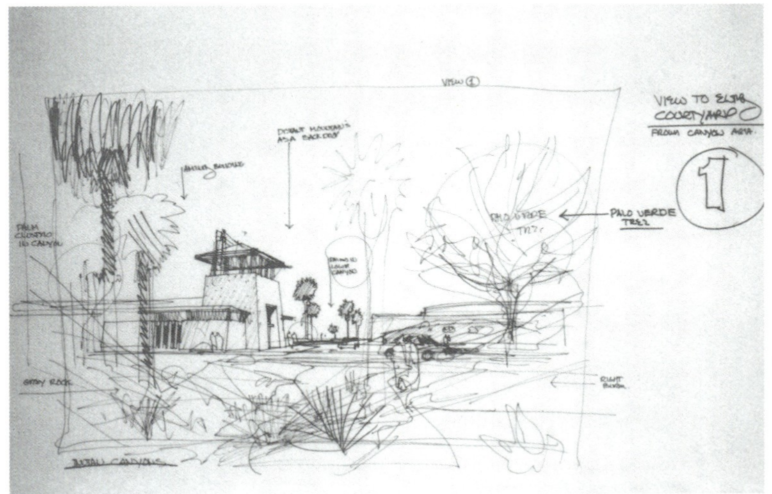

2

1 服务中心全景
2 构思草图
3 总平面图
4 立面渲染图

3

4

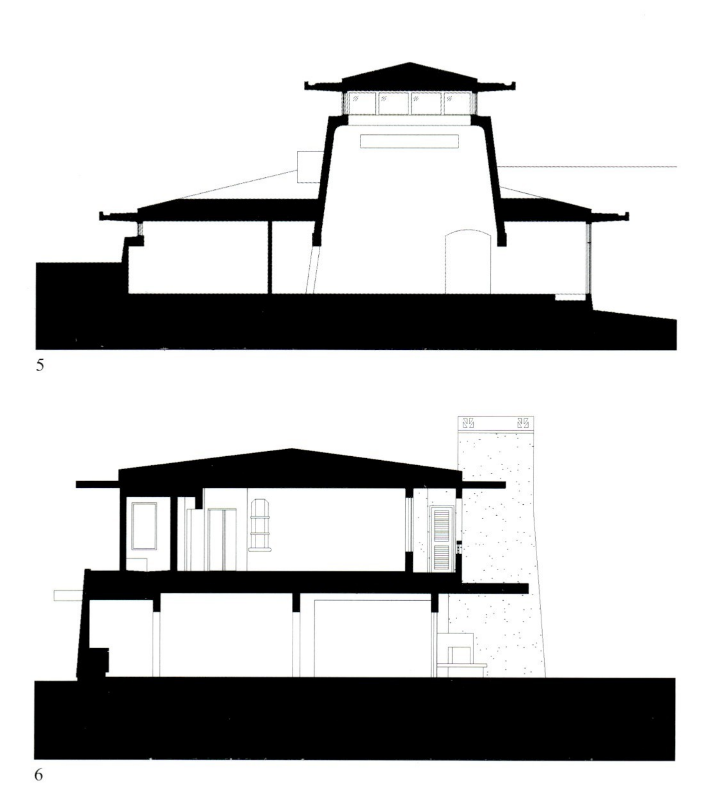
5

6

7

8

9

10

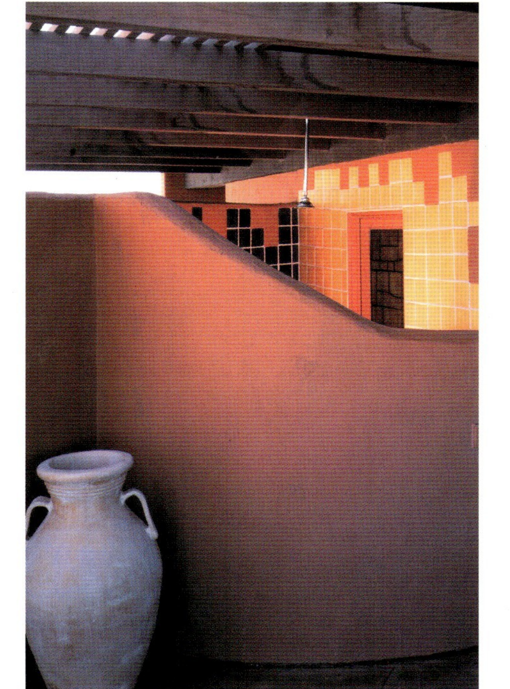

11

12

5 服务中心剖面图
6 标准单元剖面图
7 墙面细部
8 入口屋架细部和富有沙漠特色的院落
9 室内
10 远景
11 室内
12 庭院窗户

13　立面施工图
14　壁炉施工详图
15　室内楼梯和开口开窗详图
16　入口支架详图

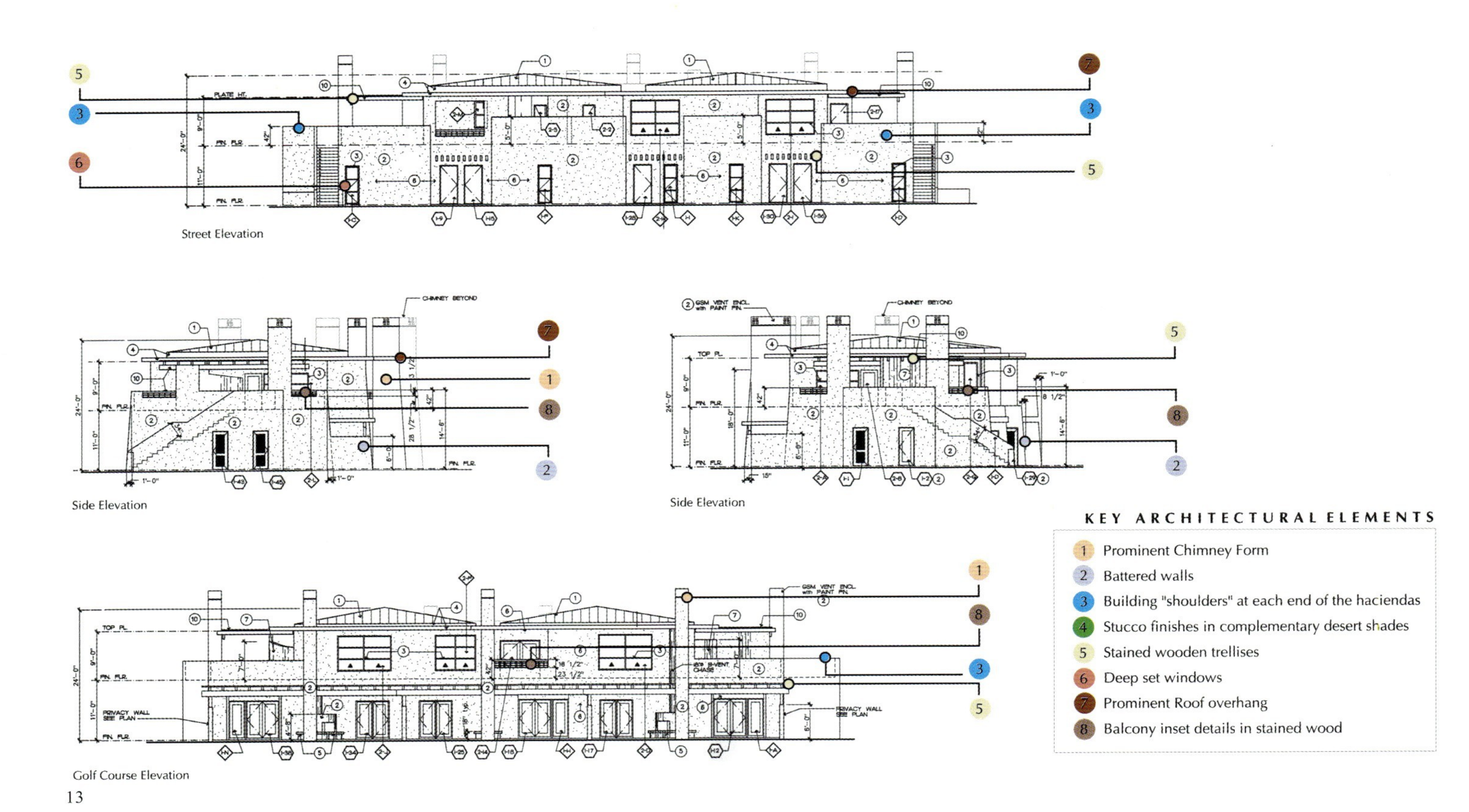

13

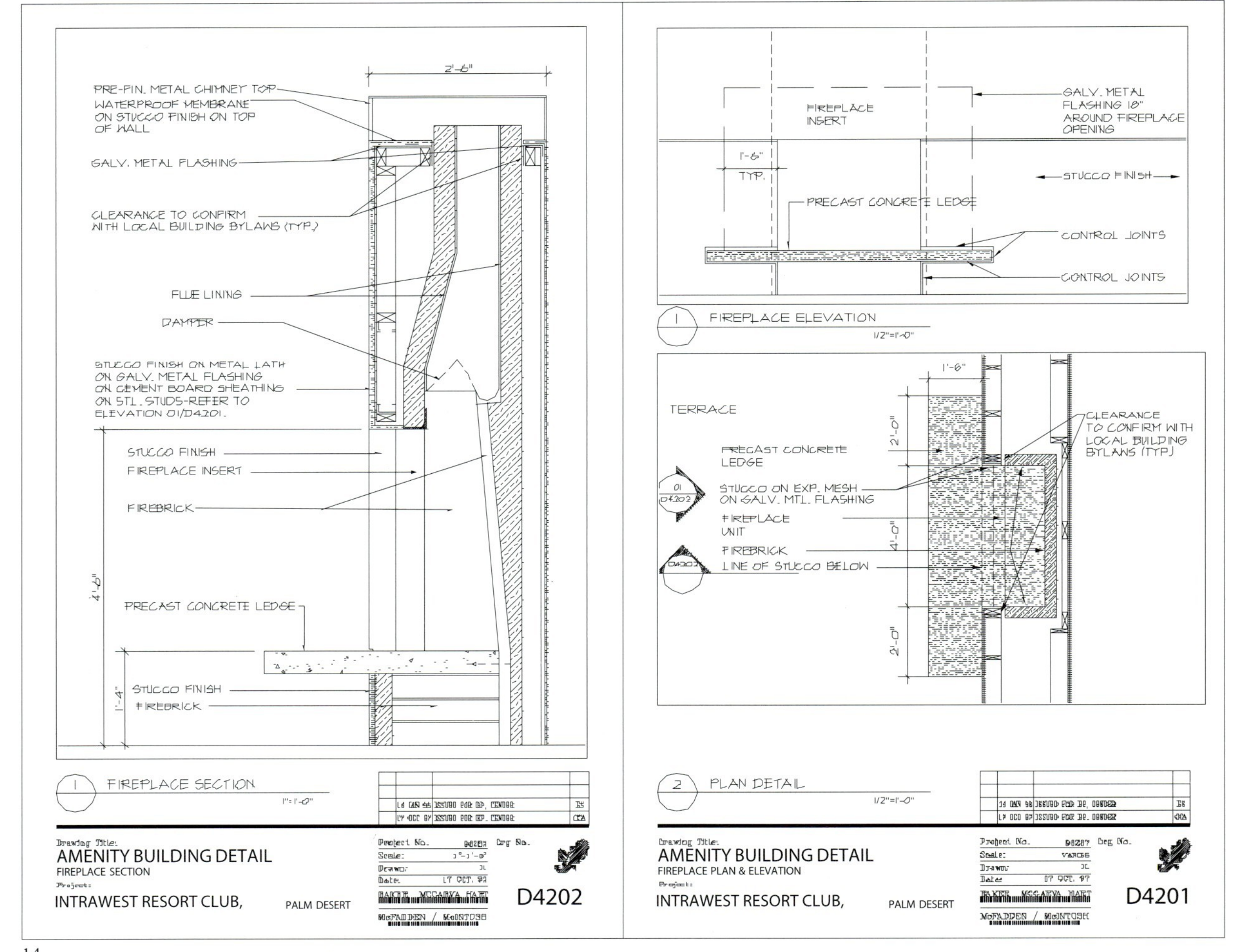

14

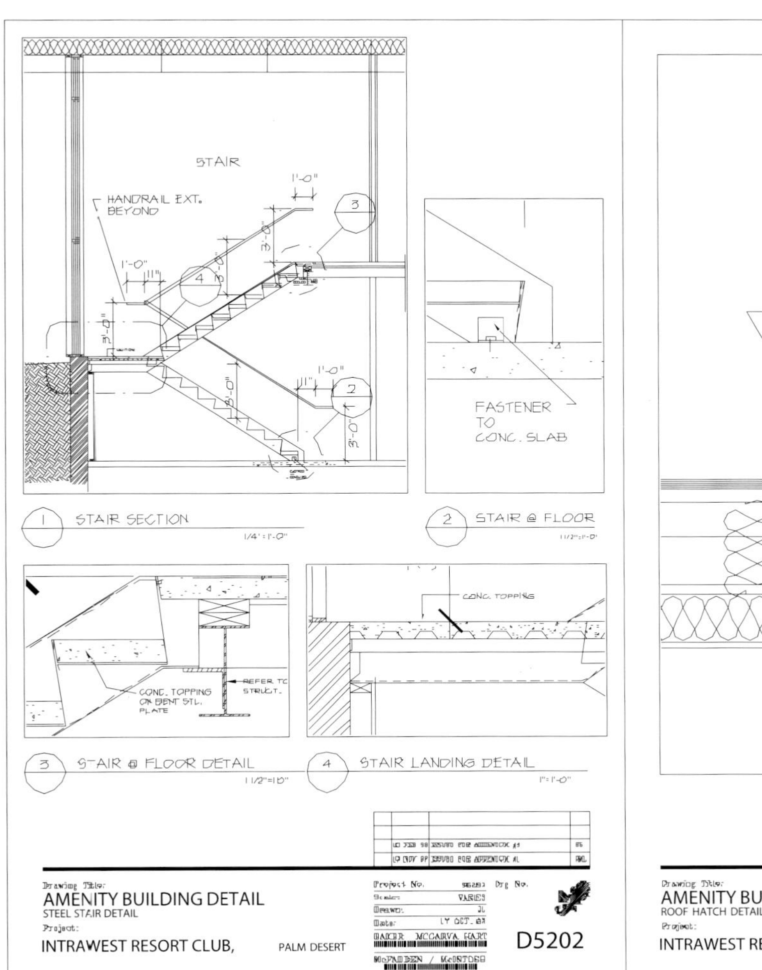
STAIR
HANDRAIL EXT. BEYOND
FASTENER TO CONC. SLAB
STAIR SECTION
1/4"=1'-0"
STAIR @ FLOOR
STAIR @ FLOOR DETAIL
STAIR LANDING DETAIL
CONC. TOPPING
Drawing Title:
AMENITY BUILDING DETAIL
STEEL STAIR DETAIL
Project:
INTRAWEST RESORT CLUB,
PALM DESERT
BAKER MCGARVA HART
McFADDEN / McINTOSH
D5202

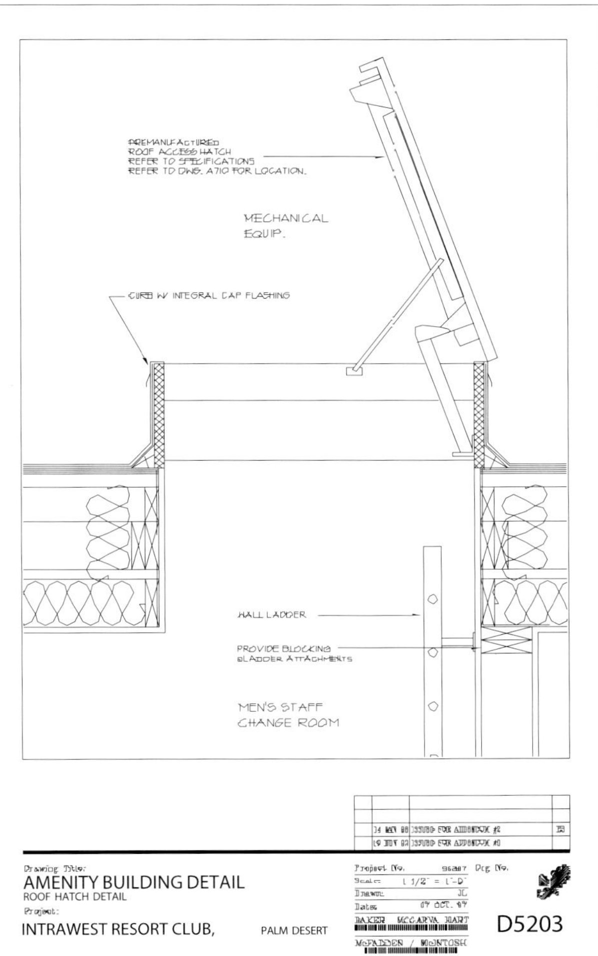
PREMANUFACTURED ROOF ACCESS HATCH
REFER TO SPECIFICATIONS
REFER TO DWG. A710 FOR LOCATION.
MECHANICAL EQUIP.
CURB W/ INTEGRAL CAP FLASHING
HALL LADDER
PROVIDE BLOCKING
BLADDER ATTACHMENTS
MEN'S STAFF CHANGE ROOM
Drawing Title:
AMENITY BUILDING DETAIL
ROOF HATCH DETAIL
Project:
INTRAWEST RESORT CLUB,
PALM DESERT
BAKER MCGARVA HART
McFADDEN / McINTOSH
D5203

15

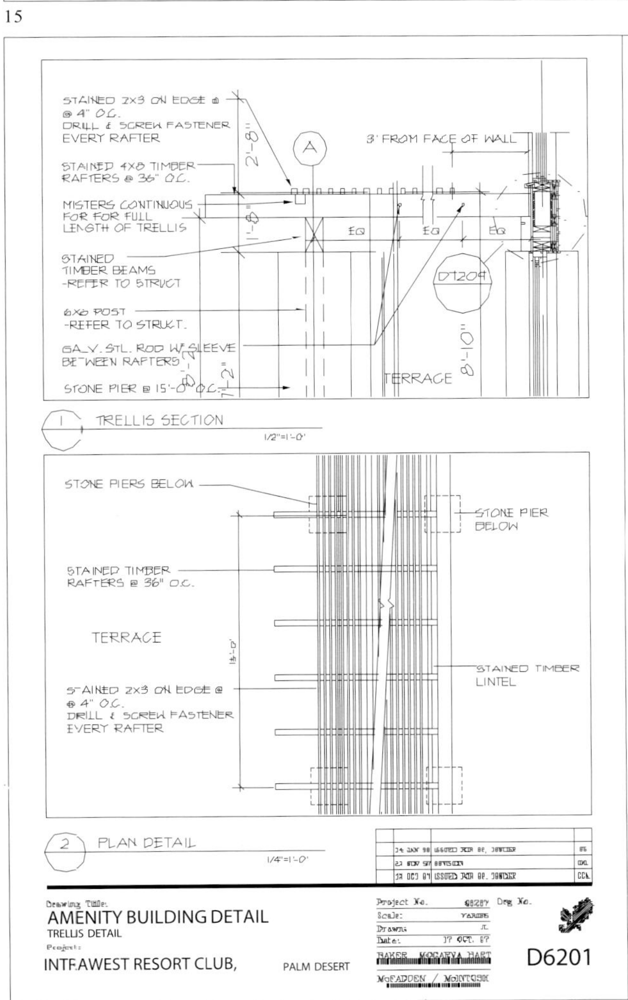
STAINED 2X3 ON EDGE @ 4" O.C.
DRILL & SCREW FASTENER EVERY RAFTER
STAINED 4X8 TIMBER RAFTERS @ 36" O.C.
MISTERS CONTINUOUS FOR FULL LENGTH OF TRELLIS
STAINED TIMBER BEAMS -REFER TO STRUCT
6X6 POST -REFER TO STRUCT.
GALV. STL. ROD W/ SLEEVE BETWEEN RAFTERS
STONE PIER @ 15'-0" O.C.
3' FROM FACE OF WALL
TERRACE
TRELLIS SECTION
1/2"=1'-0"
STONE PIERS BELOW
STONE PIER BELOW
STAINED TIMBER RAFTERS @ 36" O.C.
TERRACE
STAINED 2X3 ON EDGE @ 4" O.C.
DRILL & SCREW FASTENER EVERY RAFTER
STAINED TIMBER LINTEL
PLAN DETAIL
1/4"=1'-0"
Drawing Title:
AMENITY BUILDING DETAIL
TRELLIS DETAIL
Project:
INTRAWEST RESORT CLUB,
PALM DESERT
BAKER MCGARVA HART
McFADDEN / McINTOSH
D6201

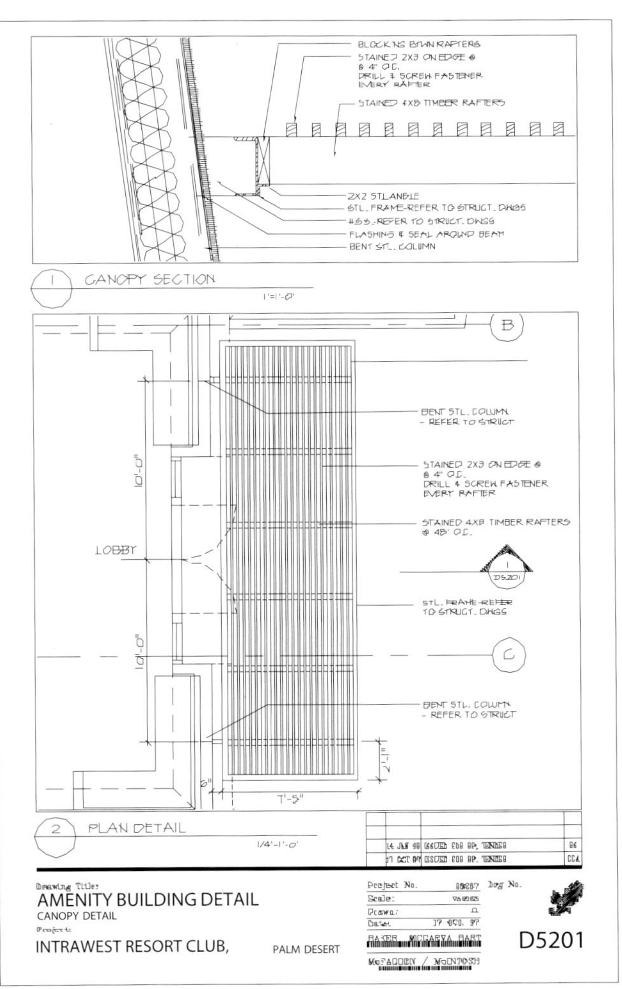
BLOCKING B/W RAFTERS
STAINED 2X3 ON EDGE @ 4" O.C.
DRILL & SCREW FASTENER EVERY RAFTER
STAINED 4X8 TIMBER RAFTERS
2X2 STL. ANGLE
STL. FRAME-REFER TO STRUCT. DWGS
FLASHING & SEAL AROUND BEAM
BENT STL. COLUMN
CANOPY SECTION
1"=1'-0"
BENT STL. COLUMN - REFER TO STRUCT
STAINED 2X3 ON EDGE @ 4" O.C.
DRILL & SCREW FASTENER EVERY RAFTER
STAINED 4X8 TIMBER RAFTERS @ 48" O.C.
STL. FRAME-REFER TO STRUCT. DWGS
LOBBY
PLAN DETAIL
1/4"=1'-0"
Drawing Title:
AMENITY BUILDING DETAIL
CANOPY DETAIL
Project:
INTRAWEST RESORT CLUB,
PALM DESERT
BAKER MCGARVA HART
McFADDEN / McINTOSH
D5201

16

Bing Thom Architects Inc.

谭秉荣建筑设计事务所

Since 1980, Bing Thom Architects has been collaborating with and advising cultural institutions, corporations, universities, governments, developers and communities around the world to help them successfully achieve their building aspirations. These projects have garnered a prodigious number of important design awards. More significantly they have all enjoyed broad popular appeal along with operational and financial success.

As principal of the firm, Bing Thom has been recognized with Canada's highest honour, the Order of Canada, for his contribution to architecture. Most recently, the year of 2002, he was awarded the Golden Jubilee Medal for services to his country. Bing Thom Architects undertakes only a selective few projects at any given time to ensure Bing Thom and at least one of his Directors are personally involved in all stages of every project. The core team of 45 experienced staff provides the foundation for larger teams that are assembled on a project-by-project basis. Their diversity is reflected in the fact that they come from 13 countries and can speak 14 languages. In addition, the majority of them possess at least one other area of expertise on top of architecture. The versatility that the team brings to each project ensures their clients receive an interdisciplinary perspective and an international focus.

谭秉荣建筑设计事务所成立于1980年，业主包括：文化机构、公司、大学、政府、开发商和社区。这些世界范围的工作带给谭秉荣建筑设计事务所众多的奖项，但是对于谭秉荣建筑设计事务所来说，更重要地是公众对其建筑的认可和使用上的成功。

谭秉荣是谭秉荣建筑设计事务所的总裁，因为谭秉荣对加拿大建筑界的特殊贡献，他于1995获得加拿大最高荣誉——加拿大勋章（THE ORDER OF CANADA）。谭秉荣还于2002年获得加拿大50周年特殊贡献勋章（GOLDEN JUBILEE）。

谭秉荣建筑设计事务所对设计任务的挑选极为严格，以保证谭秉荣本人或其合作者有充分的时间参与设计的每一个环节。谭秉荣建筑设计事务所的核心设计队伍拥有45个成员，来自13个国家，说14种语言。以多元文化背景设计师组成的设计团队使每一个设计项目得到多方位、多视角国际化的演绎。谭秉荣建筑设计事务所的另一个宗旨是保证设计至少有一方面在建筑界处于领先地位。

Central City - Surrey, British Columbia
不列颠哥伦比亚省萨里中心城

业　　主：ICBC PROPERTIES LTD
地　　点：萨里市
规　　模：1 700 000平方英尺
完成时间：2003年
摄 影 师：尼克·勒乌（NIC LEHOUX）

Conceived as a major urban intervention to develop a suburban town centre, this 1.7 million square foot mixed-use development includes an existing regional mall, a university campus and a major office building.

Architecturally, the Central City complex has four major components: a Tower, a Podium, a Galleria and an Atrium. Constructed above the mall, the Galleria is supported by nine concrete cruciform columns and is roofed over by an elegant timber roof and skylight. The university activities housed in this space overlook a fivestorey atrium space over the active shopping mall below. The Galleria is linked horizontally through a system of bridges to the Podium providing the university with 400,000 square feet of contiguous space on three floors. Central to the university space is the Atrium, the campus' "living room" which is roofed over by a space frame constructed of recycled timber. Designed to house high technology tenants, the curved tower has an offset core.

The building is conceived as part of a shifting and ever moving network of people and events from public transit to a complex of different programs and forms. The project celebrates this interconnectivity, which is critical to the health of a city, or any place where people live and work.

1

萨里市中心城是萨里市城市副中心开发的关键项目，总建筑面积1 700 000平方英尺。其功能包括一个区域性的商业中心，一个大学校园和一个办公楼。

建筑上，中心城的综合体包括四个主要部分：高层建筑、基座、中庭和前庭。中庭建造在商业中心的上部，由九个十字形的钢筋混凝土柱子支撑，上盖是一个优雅的木屋顶和天窗。位于中庭的大学活动区，俯视商业中心之上的五层前庭空间。一个水平方向的桥连接基座部分和中庭空间。三层的基座部分提供了400 000平方英尺的大学使用空间。前庭位于大学校区的中心，是学校的"起居室"。其屋顶是由可循环使用的木结构支撑。为满足高科技使用者的独特需求，弧形的高层办公楼具有一个错位的核心结构。

该建筑被设想为人们从公共交通系统转乘不同交通工具的永动的传送系统的一部分。萨里中心城的建成加速了这种内在联系，而这对于一座城市，或者说对于任何人们生活和工作在其中的场所的健康发展都是十分关键的。

2

3

1　社区内的视觉中心——高层塔楼
2　大厅——丰富的建筑材料运用
3　玻璃高窗和铝合金材质的通道

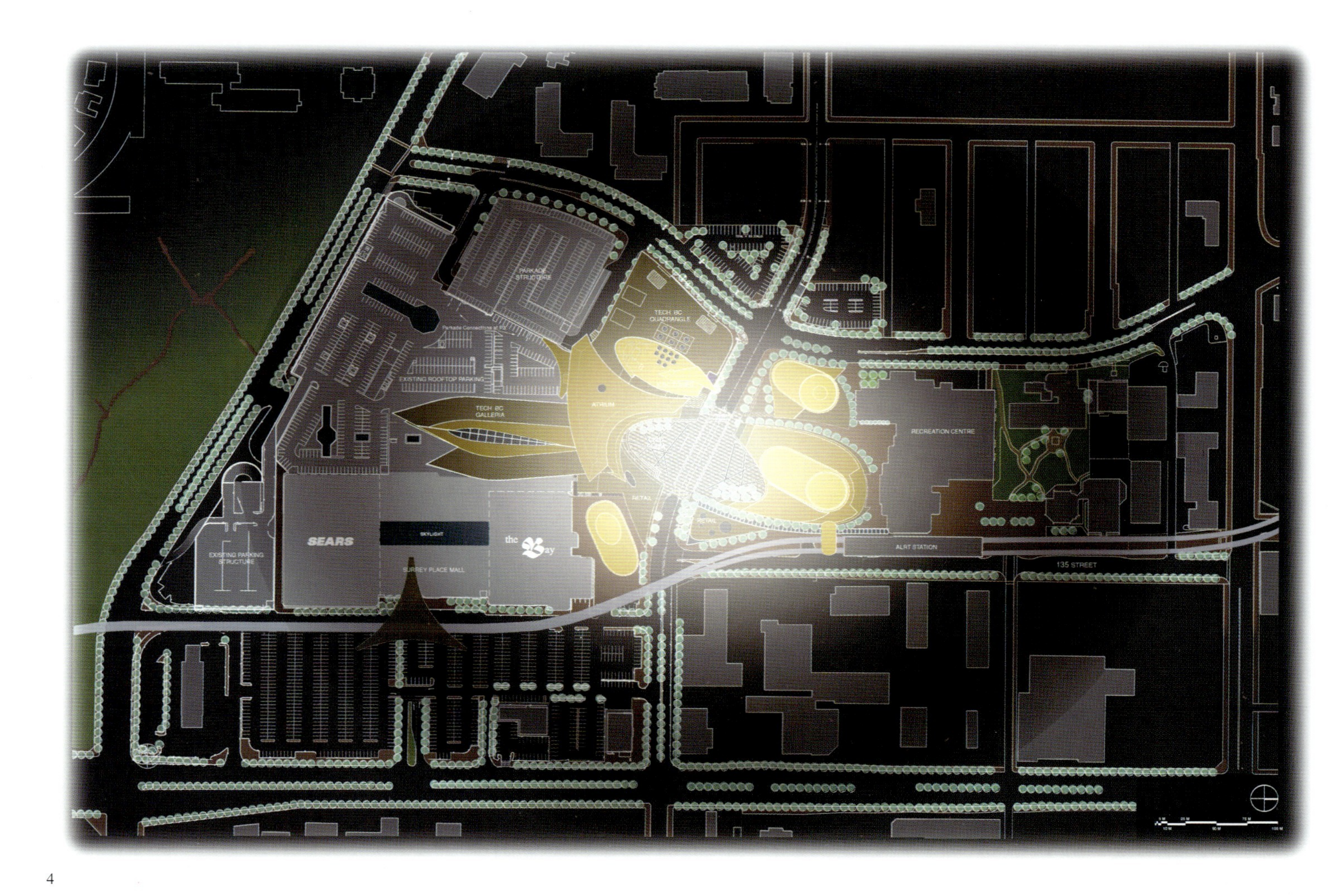
PARKADE STRUCTURE
Parkade Connections at F3
TECH BC QUADRANGLE
EXISTING ROOFTOP PARKING
TECH BC GALLERIA
ATRIUM
RECREATION CENTRE
RETAIL
RETAIL
SEARS
SKYLIGHT
the Bay
EXISTING PARKING STRUCTURE
SURREY PLACE MALL
ALRT STATION
135 STREET

4

5

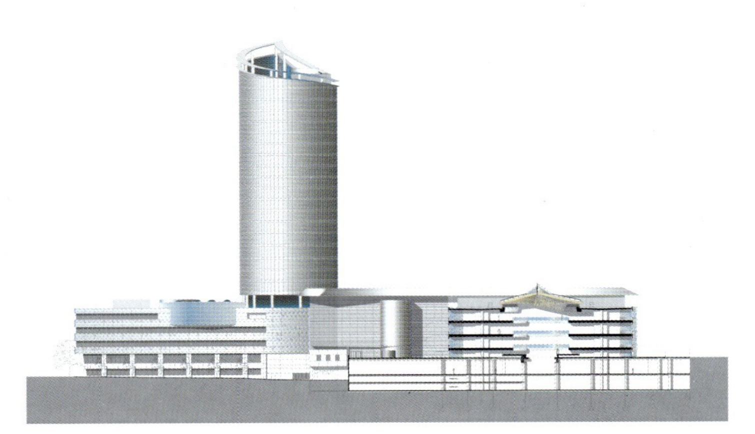

A-A cross section

6

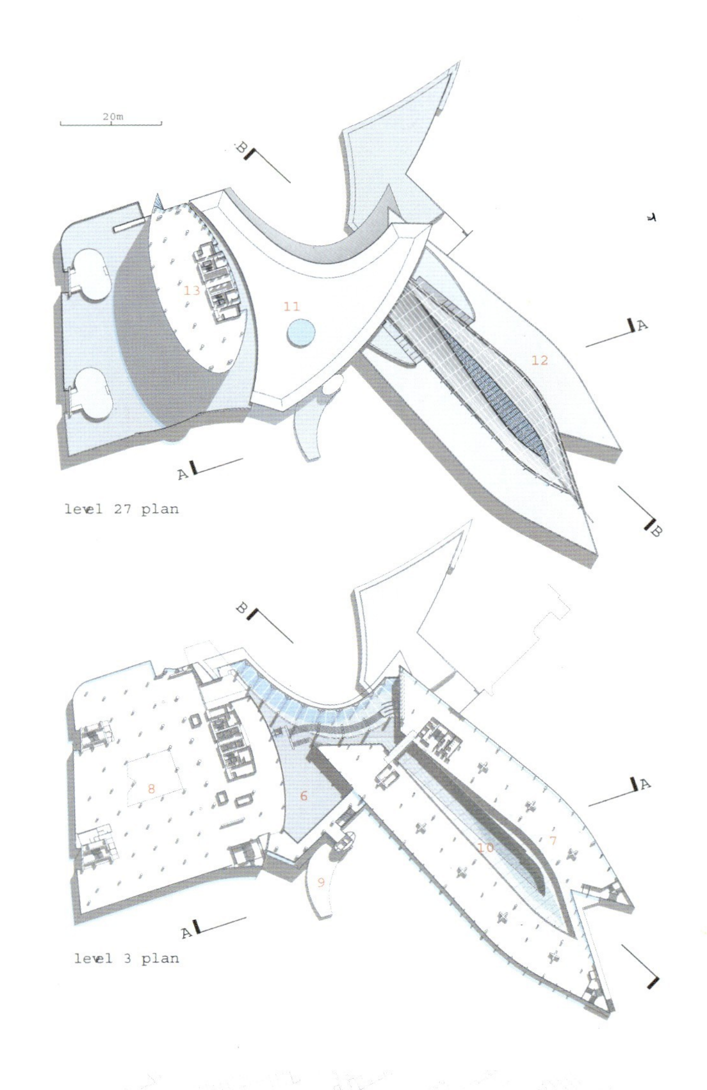

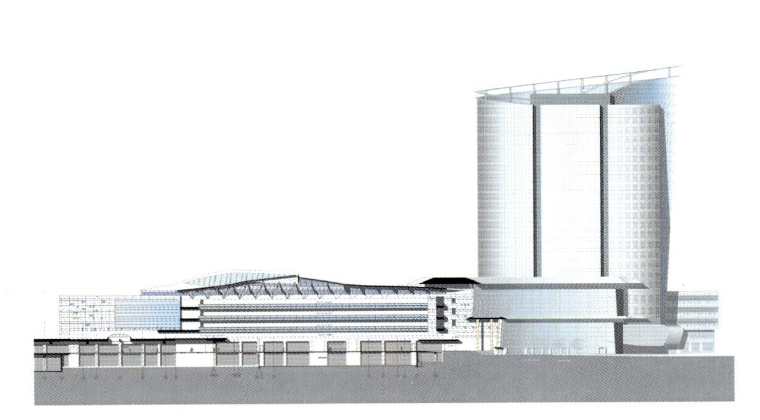

B-B long section

7

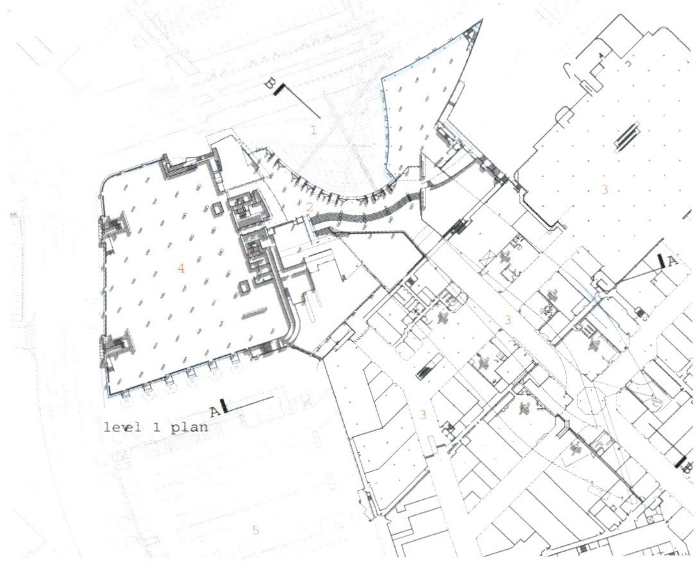

4　总平面图
5　建筑总体模型
6~7　建筑立面图
8　建筑平面图

8

9

9 玻璃墙面夜景
10 玻璃墙面细部
11 商业出租空间——木材的运用
12 建筑细部

10

11

12

13　室内细部
14~15　屋架细部
16　节点细部
17　屋架、节点详图
18　屋架详图

13

16

14

15

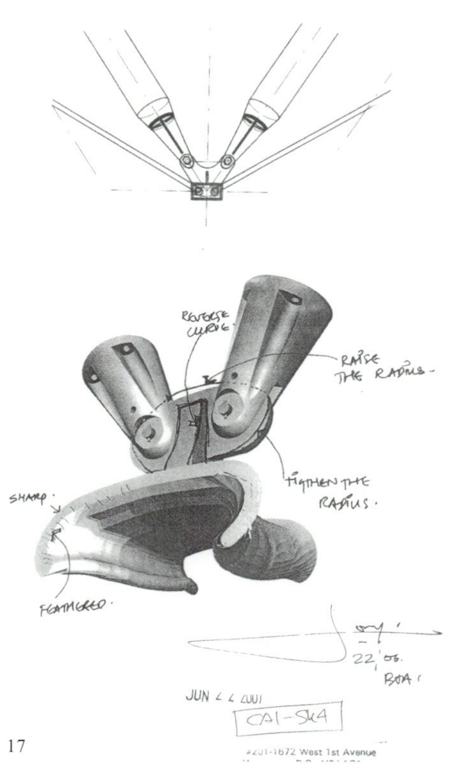

17

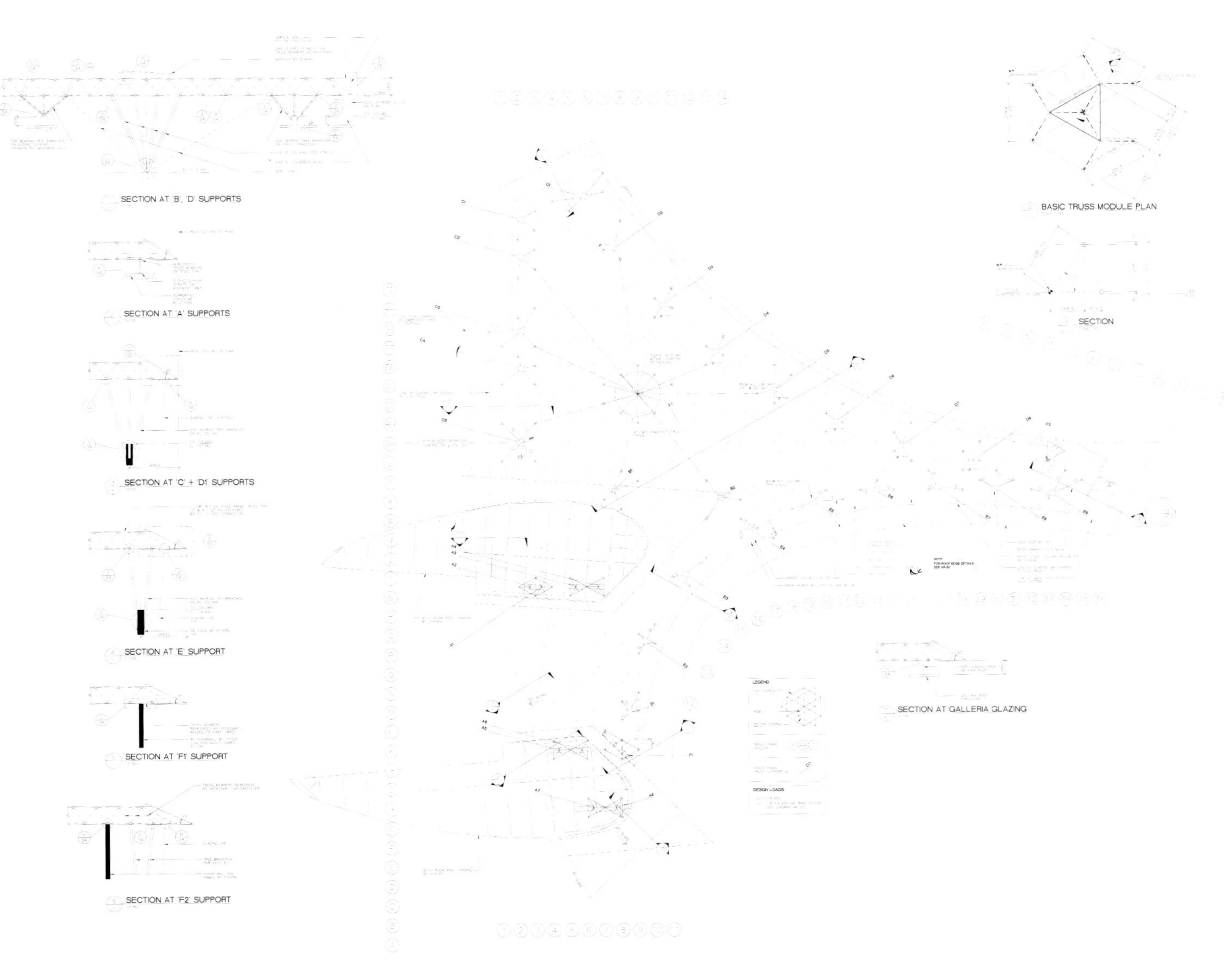
SECTION AT 'B', 'D' SUPPORTS
SECTION AT 'A' SUPPORTS
SECTION AT 'C' + 'D1' SUPPORTS
SECTION AT 'E' SUPPORT
SECTION AT 'F1' SUPPORT
SECTION AT 'F2' SUPPORT
BASIC TRUSS MODULE PLAN
SECTION
SECTION AT GALLERIA GLAZING
LEGEND
DESIGN LOADS
TETRAHEDRAL SPACE-FRAME TRUSS ROOF SUPPORT PLAN

18

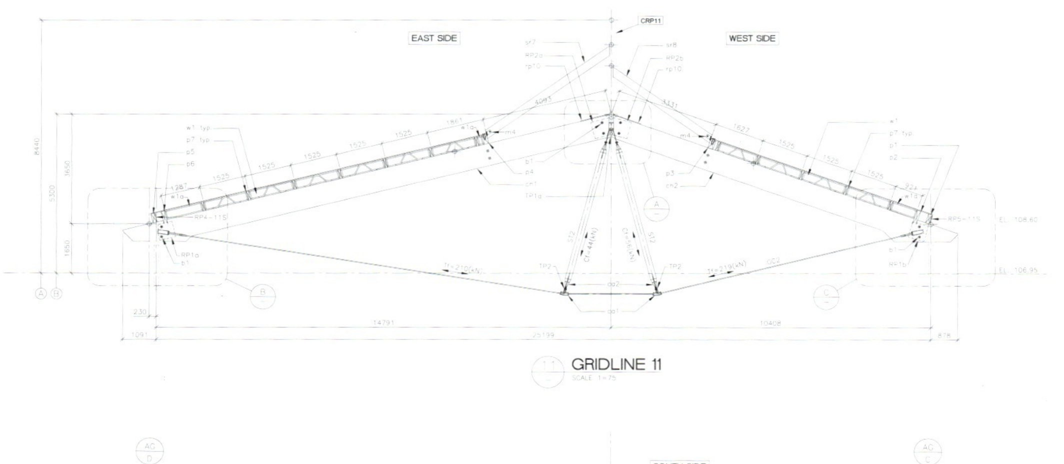
EAST SIDE
WEST SIDE
GRIDLINE 11

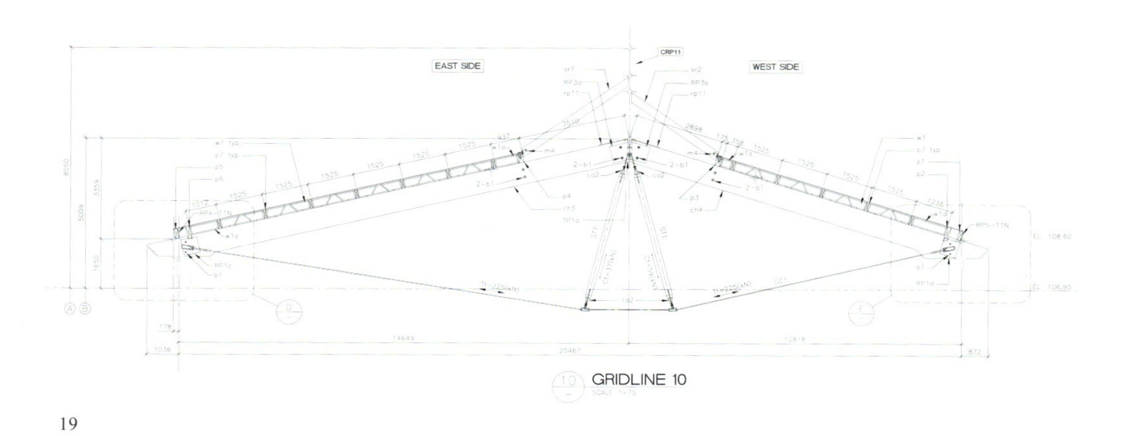
SOUTH SIDE
EAST SIDE
WEST SIDE
NORTH SIDE
TRUSS 11

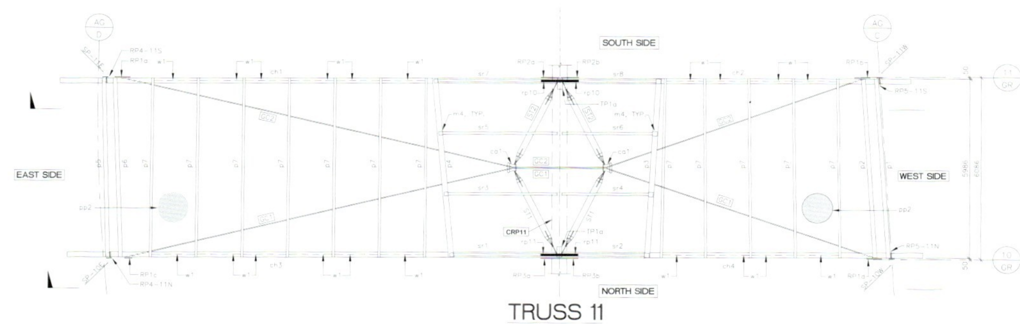
EAST SIDE
WEST SIDE
GRIDLINE 10

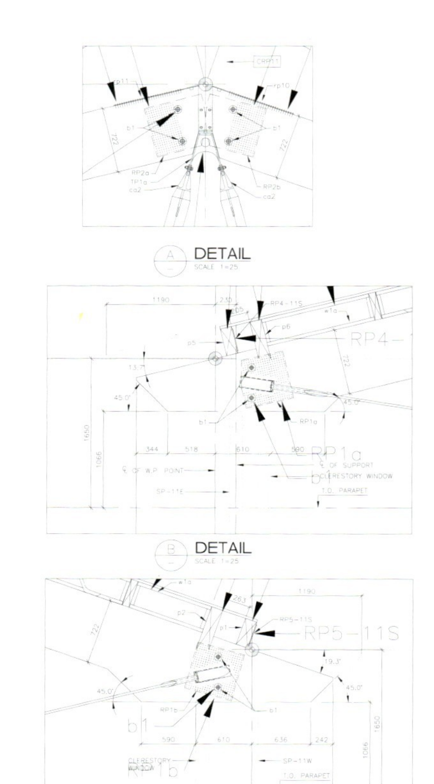
DETAIL
DETAIL

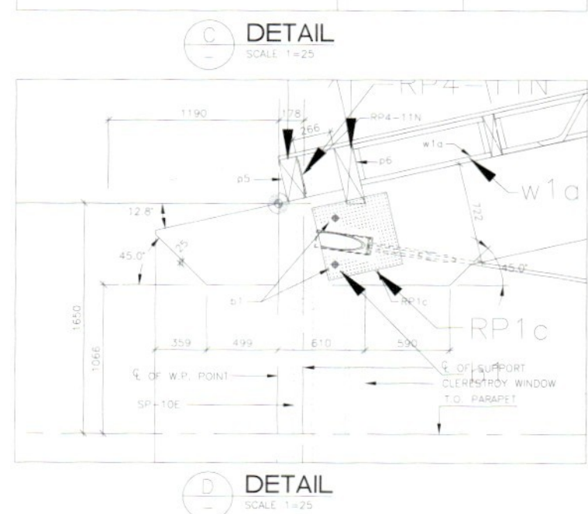
DETAIL
DETAIL

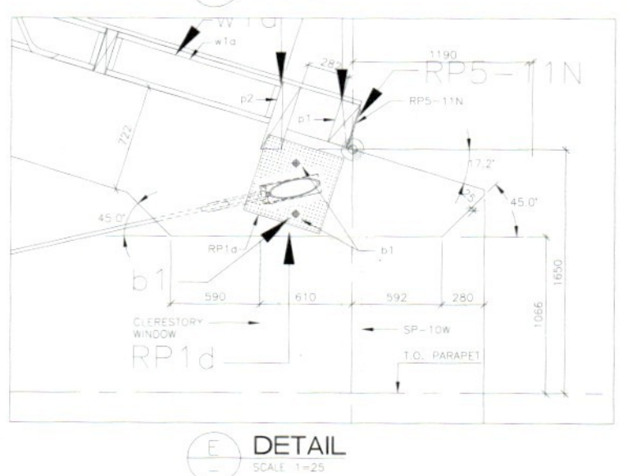
DETAIL
ROOF TRUSS LAYOUT

19

20

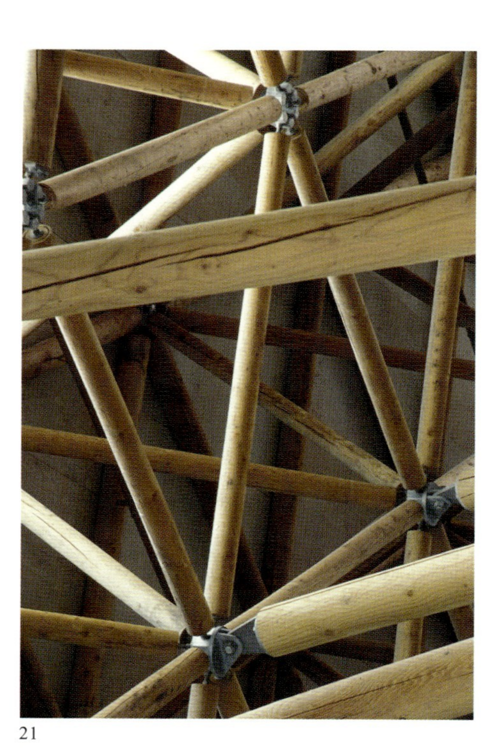

21

19 屋顶详图
20~21 屋顶细部
22~23 室内细部
24 木构件细部
25 木构件及玻璃幕墙细部（室内）
26 木构件细部

22

24

25

23

26

27

28

29

27~31　玻璃幕墙细部（室外）
32　玻璃幕墙详图

30

31

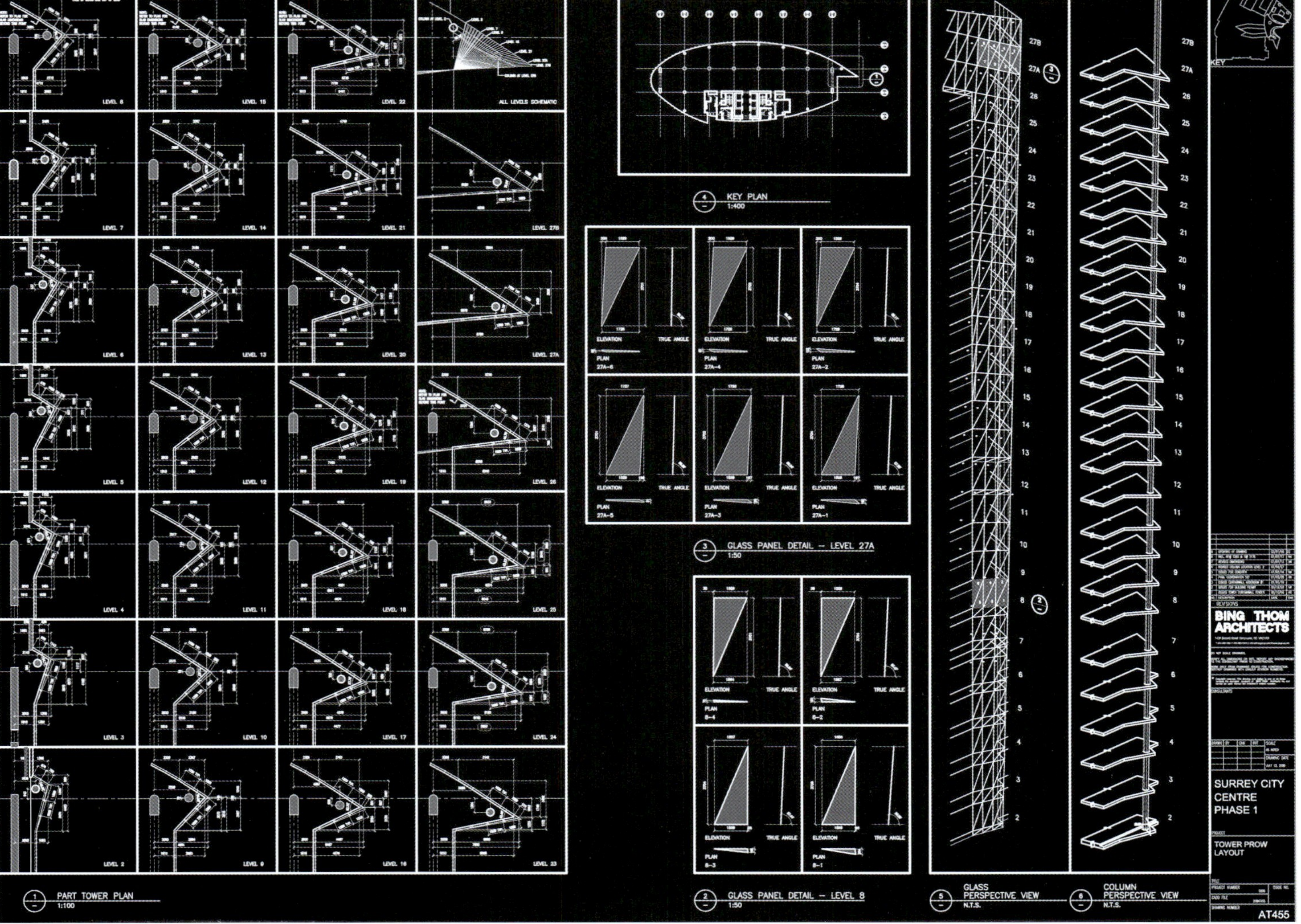

PART TOWER PLAN
1:100
KEY PLAN
1:400
GLASS PANEL DETAIL - LEVEL 27A
1:50
GLASS PANEL DETAIL - LEVEL 8
1:50
GLASS PERSPECTIVE VIEW
N.T.S.
COLUMN PERSPECTIVE VIEW
N.T.S.
BING THOM ARCHITECTS
SURREY CITY CENTRE PHASE 1
TOWER PROW LAYOUT
AT455

32

Aberdeen Mall - Richmond, British Columbia

Aberdeen商业中心

业　　主：FAIRCHILD DEVELOPMENT
地　　点：里士满市
规　　模：38 000m²
完成时间：2003年
摄 影 师：NIC LEHOUX

The essential idea of this project is to create a retail/entertainment centre that engages its context and brings the life of the Centre to the streets. Conceived as a glass lantern, the mall is wrapped by an undulating and luminous mural of glass, flowing with the curves of the streets and displaying the active life of the shops, market place, entertainment and restaurants. Unlike conventional North American malls with department stores anchoring the plan, this shopping centre creates a new more urbane typology that is a synthesis of western and Asian sensibilities.

This exterior wall of glass, sweeps, curves and carves its presence in the three main entranceways located on three major streets in Richmond. Composed of a multitude of glazed glass panels, this wall envelops the entire building. The pixilated, horizontal patterns of the movement formed by the glass are achieved by the play of light, colour and transparency. Like a movie screen, this glass mural will project outside the activities within the Centre while at the same time reflecting inside the ever-changing movement of the streets and weather.

Aberdeen商业中心的最重要的构思是创造一个零售和娱乐中心可以融合其各项功能，让街道成为生活的中心。整个Aberdeen商业中心的外墙覆盖着波浪形明亮的玻璃，这个设计灵感来自玻璃灯笼。随着街道弧线的改变，商业中心展示着其丰富的内容：商店、市场、娱乐和餐饮活动。与传统的北美商业中心以大型商场为中心不同，Aberdeen商业中心创造了一种新的城市型商业中心的模式——一种融合东西方商业中心的综合体。

外墙的玻璃所展示和装饰的三个主要出入口位于里士满市的三条主要街道上。由不同彩色玻璃板组成的外墙覆盖了整个商业中心。玻璃组成的引人入胜的水平动感图案由光线、色彩和透明玻璃共同演绎。像电影屏幕一样，玻璃墙把外部的活动内容带入室内的同时，也反射街道的活动和天气的变化。

1

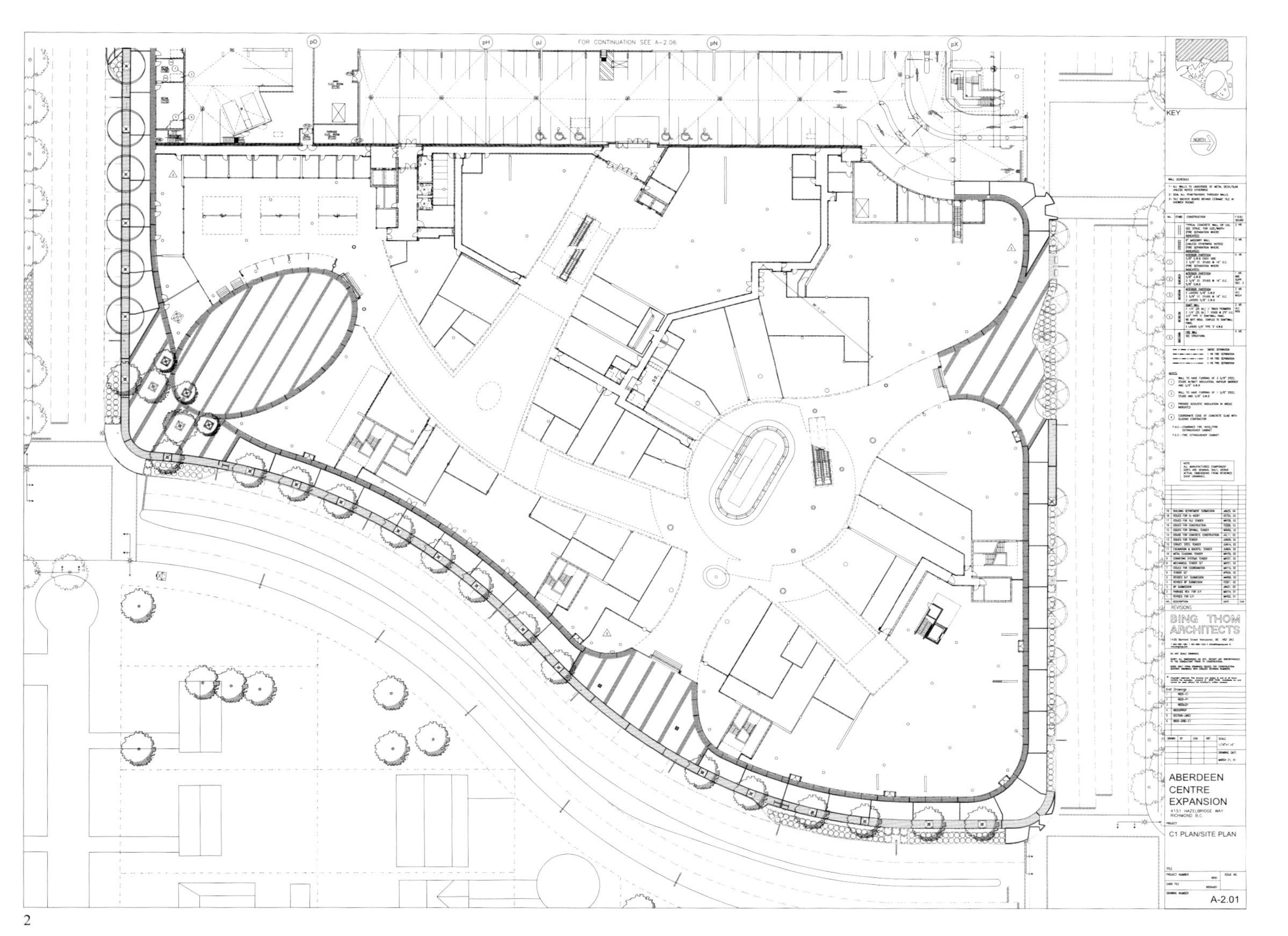

2

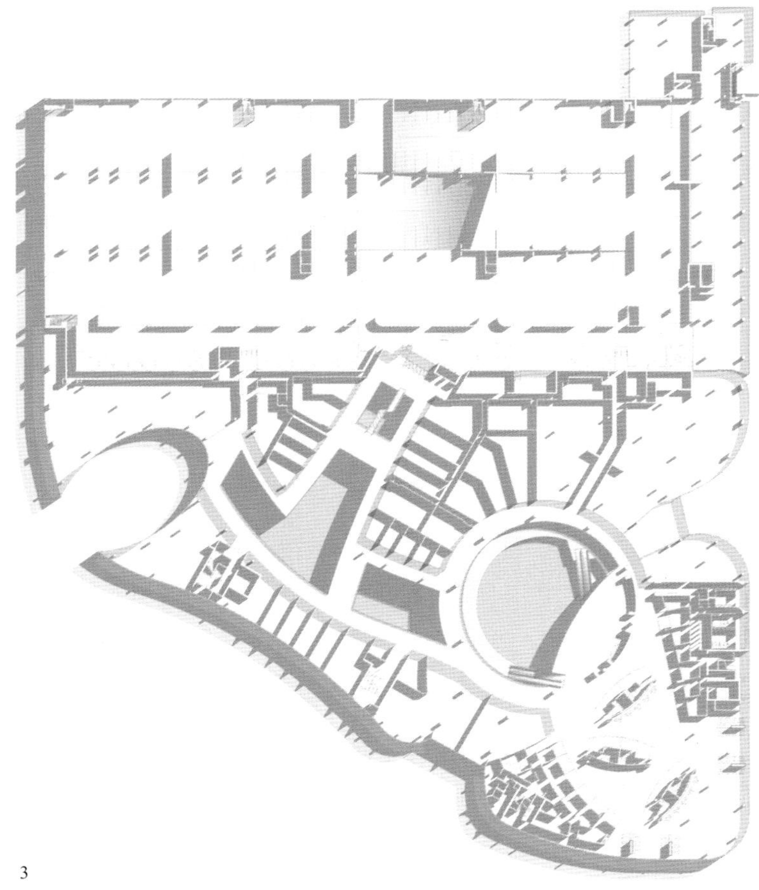

3

1 Aberdeen 商业中心彩色玻璃幕墙
2 总平面图
3 建筑平面图

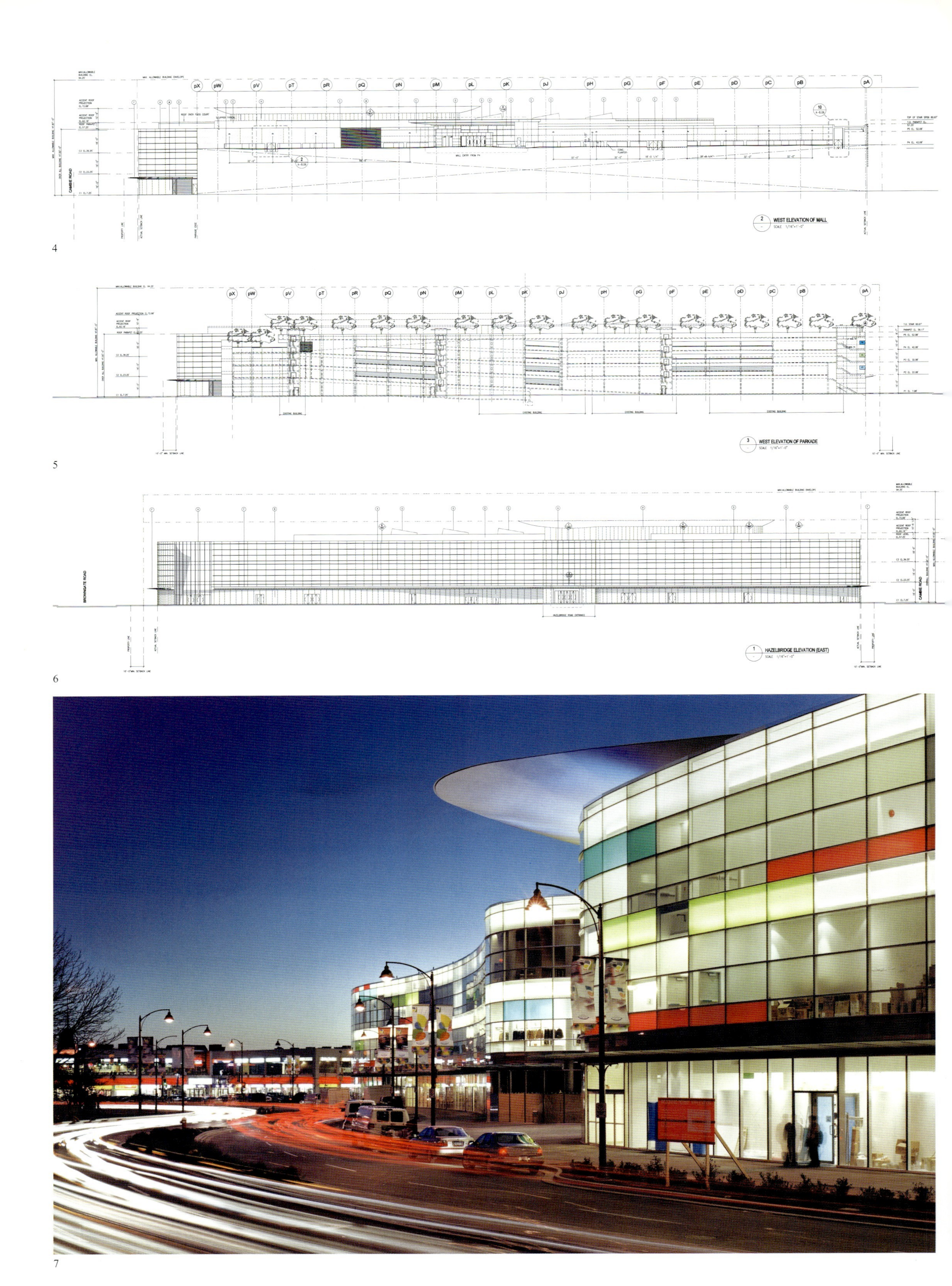
WEST ELEVATION OF MALL
CAMBIE ROAD
WEST ELEVATION OF PARKADE
HAZELBRIDGE ELEVATION (EAST)
BROWNGATE ROAD
CAMBIE ROAD

4

5

6

7

8

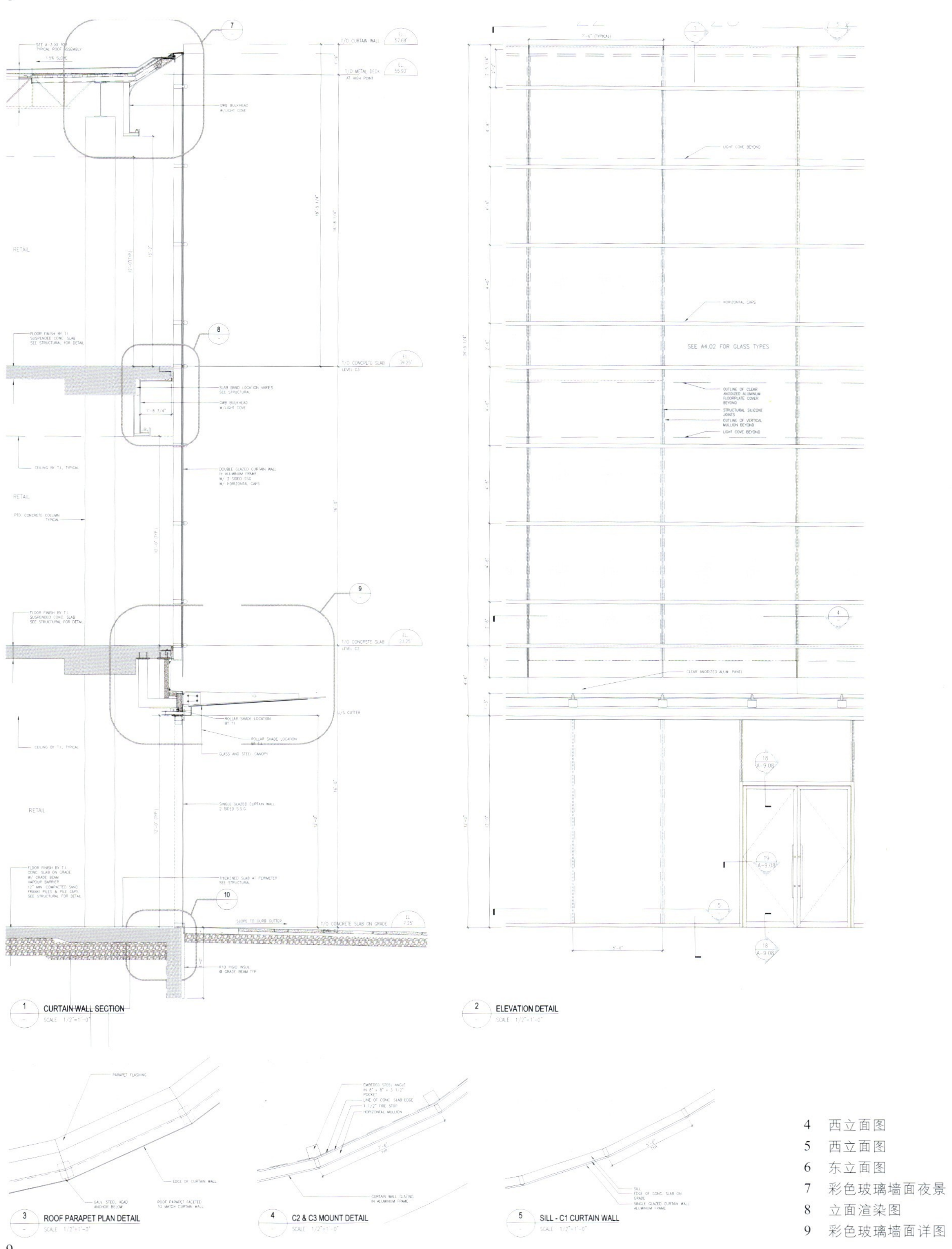

9

4 西立面图
5 西立面图
6 东立面图
7 彩色玻璃墙面夜景
8 立面渲染图
9 彩色玻璃墙面详图

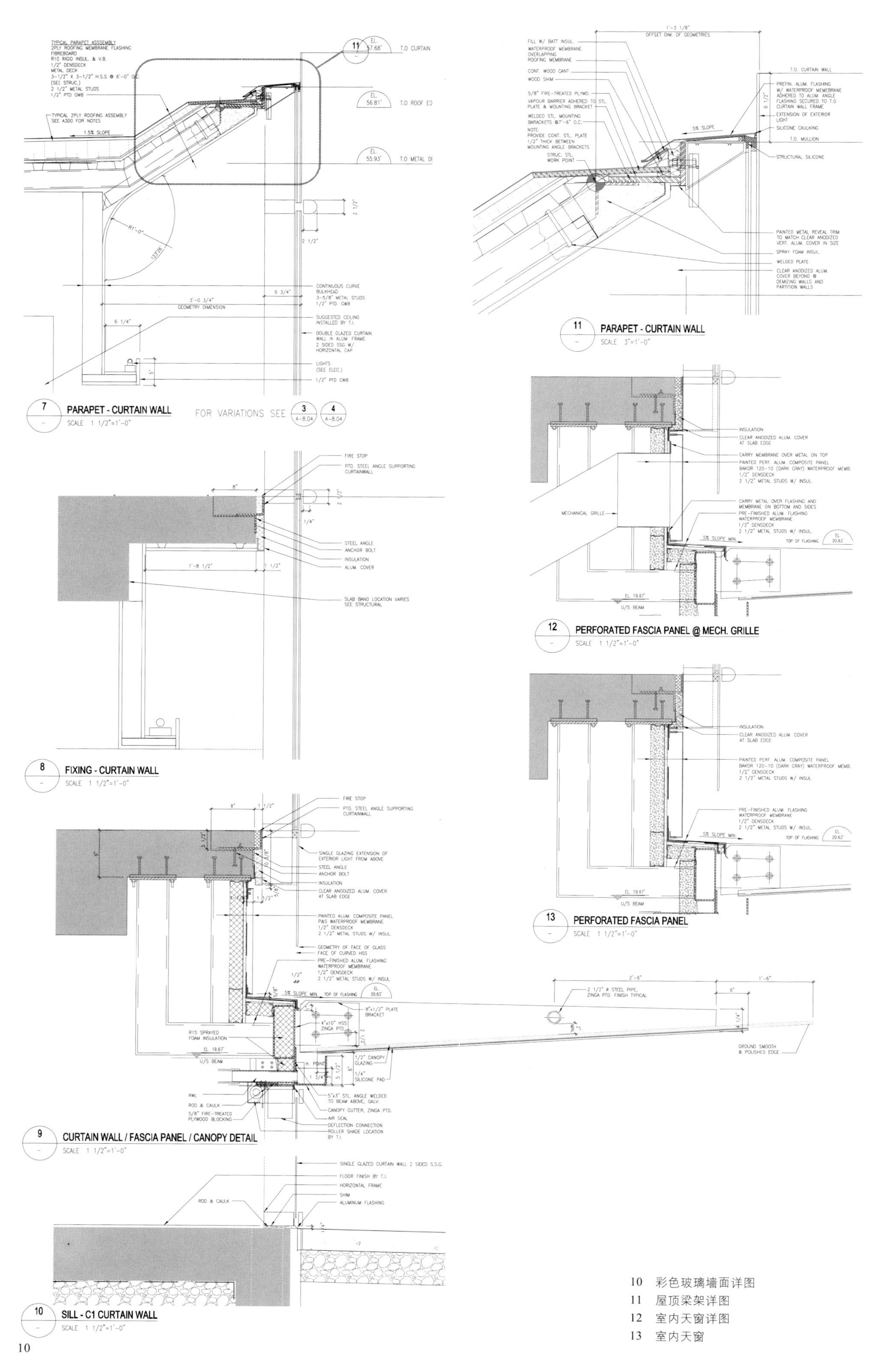

10 彩色玻璃墙面详图
11 屋顶梁架详图
12 室内天窗详图
13 室内天窗

10

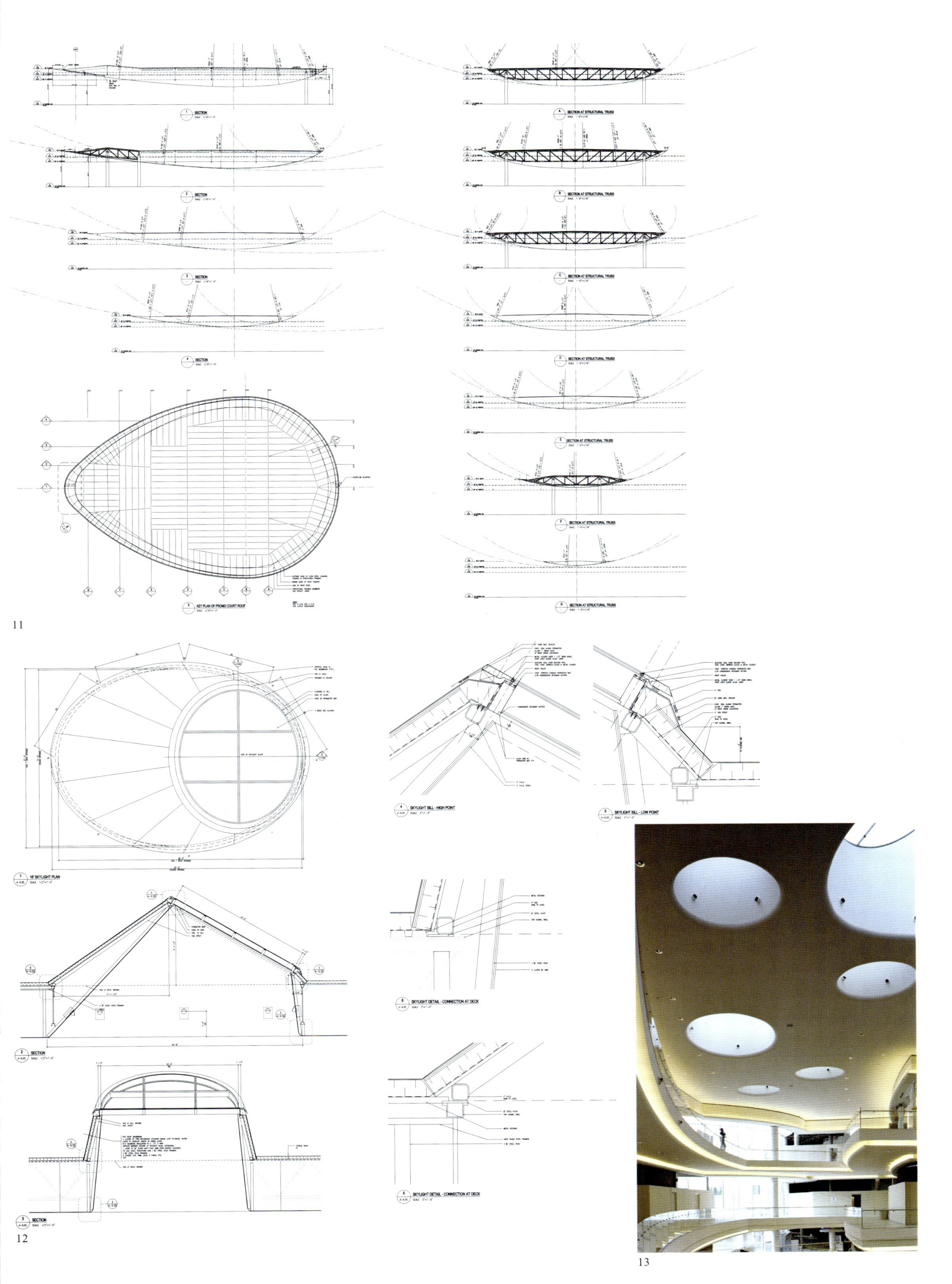

11

12

13

The Chan Centre for the Performing Arts - Vancouver, British Columbia
陈氏演艺中心

业　　主：不列颠哥伦比亚大学
地　　点：温哥华市
规　　模：76 000平方英尺
完成时间：1997年
摄 影 师：马丁·泰斯莱斯（Martin Tessles）

The Chan Centre at the University of British Columbia is designed as a group of three linked facilities including the 1,400-seat concert hall, a 250-seat experimental theatre and a 160-seat cinema. Envisioned by BTA as a performing venue in a forest setting, the building is integrated into an existing evergreen grove and acts as a teaching facility, a ceremonial complex and a venue for the broader communty.

A glazed two-storey curvilinear lobby connects all of the facilities, visually drawing in the surrounding forest and heightening the sense of dramatic discovery for the visitor.

Recalling a musical instrument in shape and detailing, the concert hall's curved concrete walls are supplemented with the warmth of wood and painted plaster. The featured element of the room is the 27 ton adjustable acoustic canopy, which together with the sound absorbing cloth banners can tune the room to fit the acoustic requirements for every performance. The acoustic canopy is designed to look like a chandelier to bring sparkle and awe to the room.

1

陈氏演艺中心包括了一个1 400座的音乐厅，一个250座的实验剧场和一个160座的电影院。

Bing Thom很重视陈氏演艺中心的独特选址——海岸边的树林。陈氏演艺中心的建筑成为当地常绿树丛的一部分，并作为教学设施、综合庆典场所和多团体社区成员的活动场地。

全玻璃的二层弧形的大厅连接所有的设施。视觉上，将建筑和周围环境融为一体，并为使用者提供逐步探索环境的惊喜和愉悦。

为满足形态和细部的乐器效果，音乐厅采用圆弧形的混凝土墙，并装饰以温暖的木材和粉刷的水泥灰。音乐厅内的独特装饰是27t重的可调节的隔声板。通过对隔声板和吸音材料的运用，可以满足不同的音乐效果。隔声板设计为一个巨大的吊灯形状，可以给大厅带来闪光和崇高的感觉。

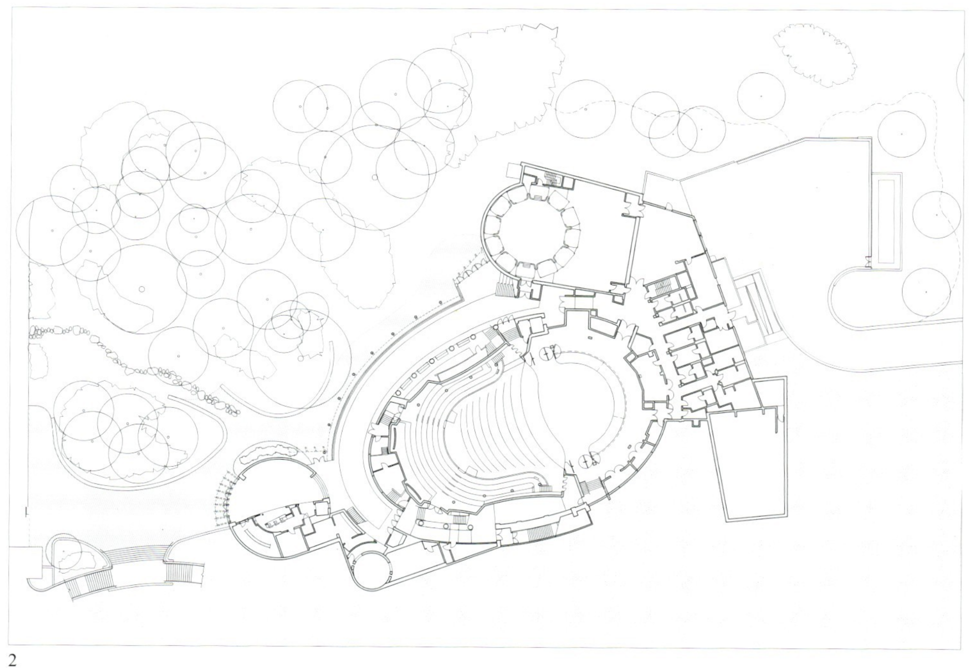

2

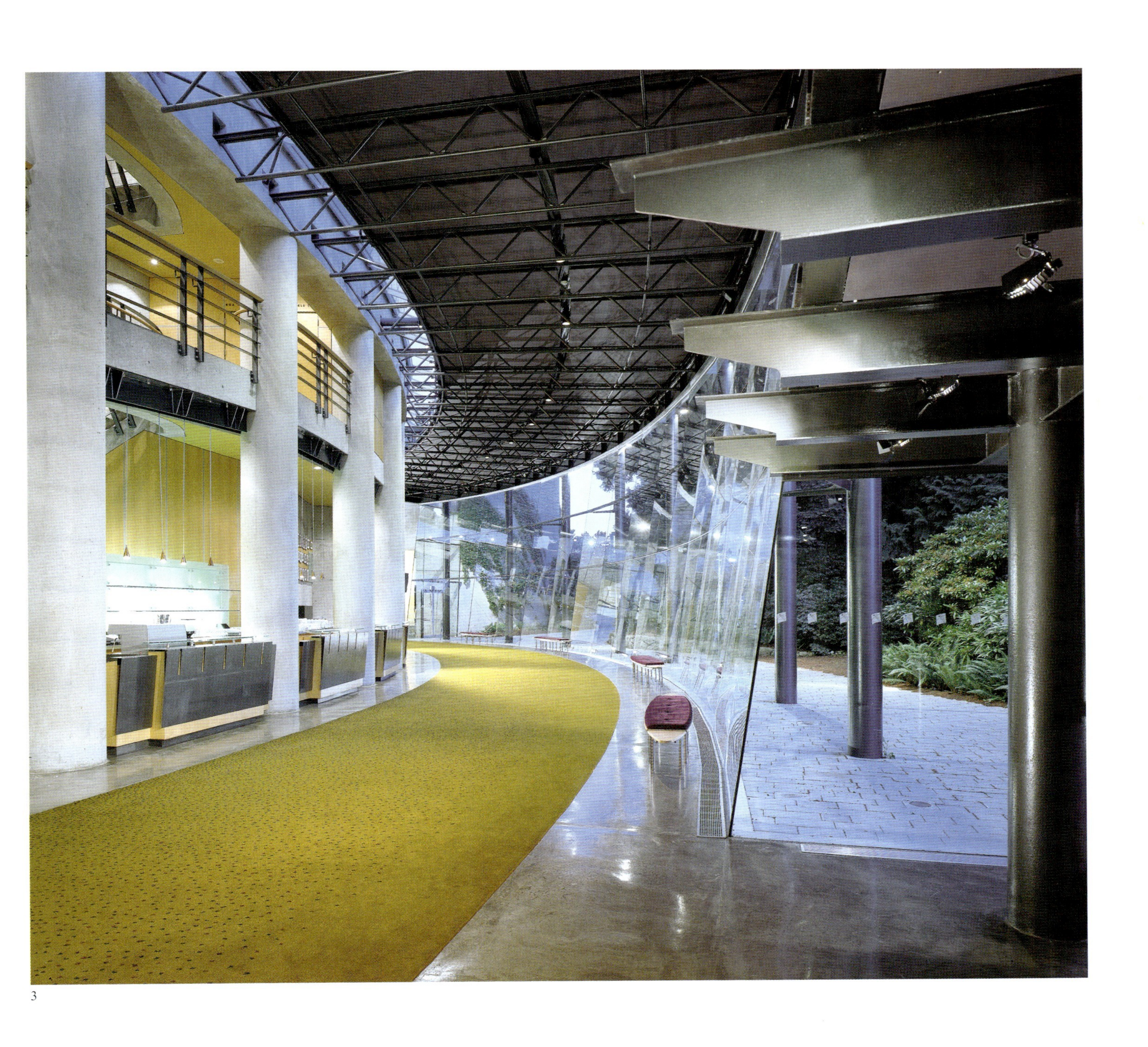

3

1　演艺中心全景
2　演艺中心平面图
3　演艺中心室内夜景

4

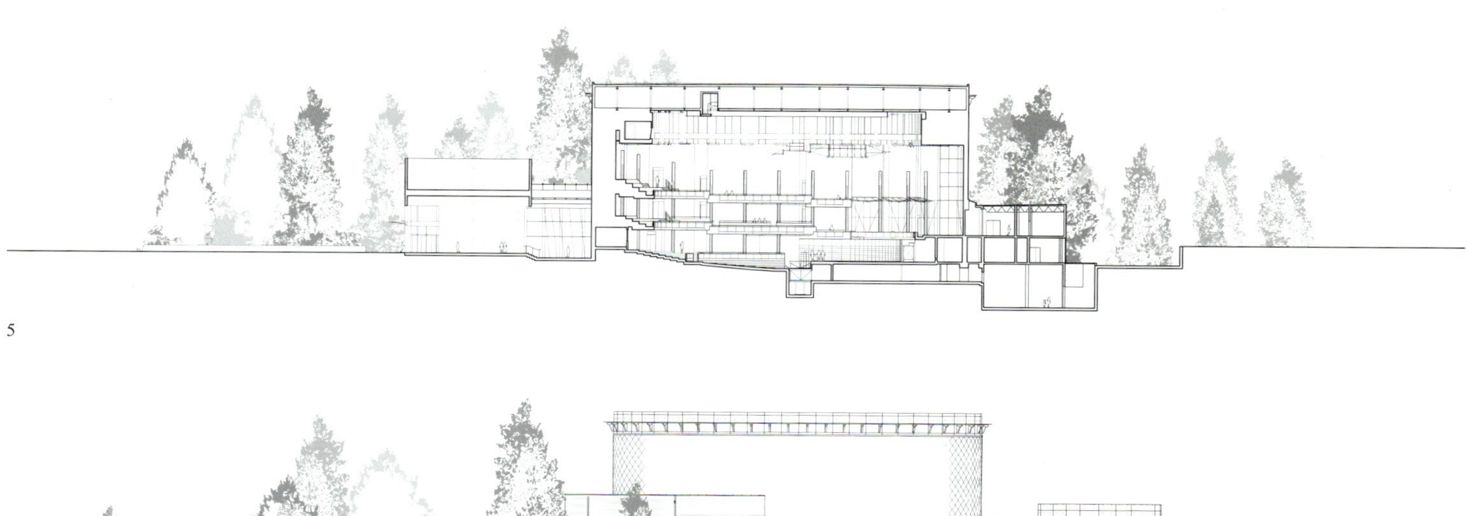

5

6

7

8

9

10

4　室外的景观
5　演艺中心剖面图
6　演艺中心立面图
7　演艺中心入口
8　演艺中心雨篷和玻璃细部
9~10　演艺中心室内细部

11

12

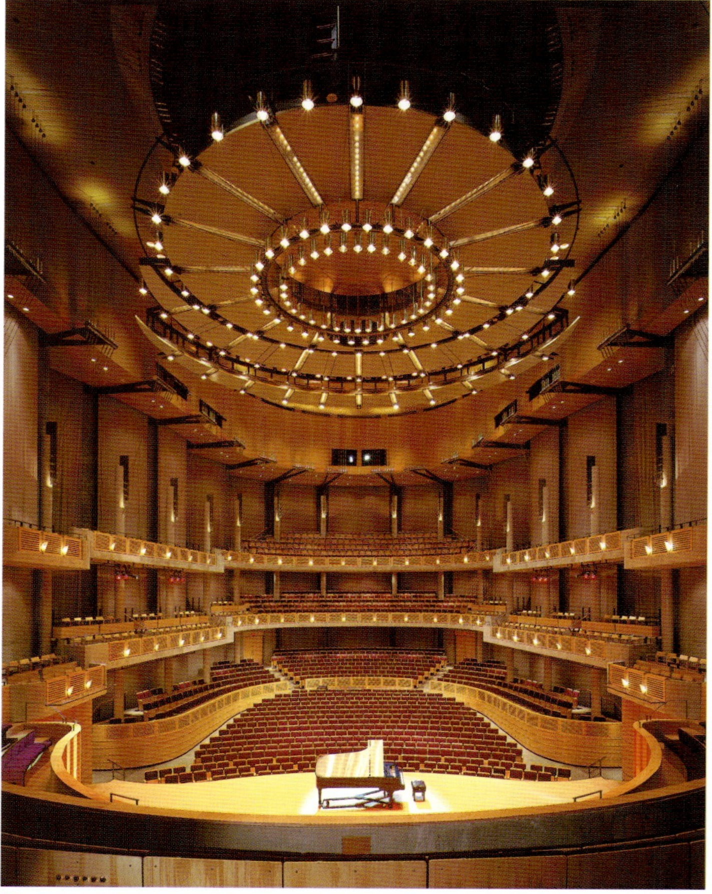
13

11　表演大厅顶棚细部
12　表演大厅演奏现场
13　表演大厅
14　顶篷详图
15　表演大厅细部
16　屋顶细部
17　表演大厅
18　洗手间细部

15

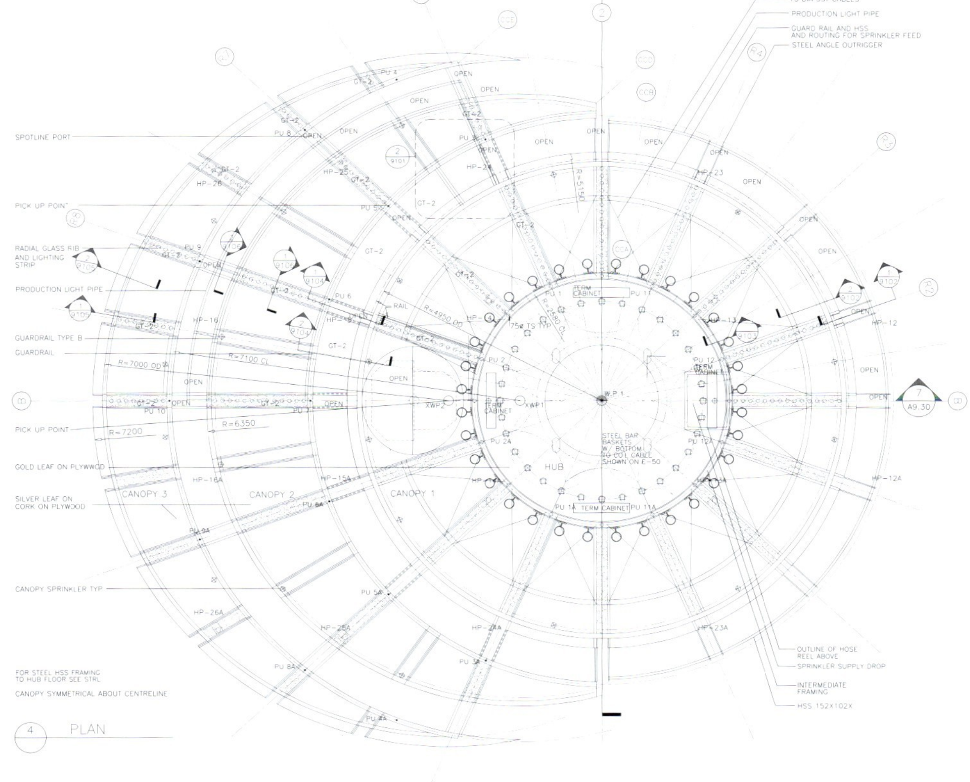

14

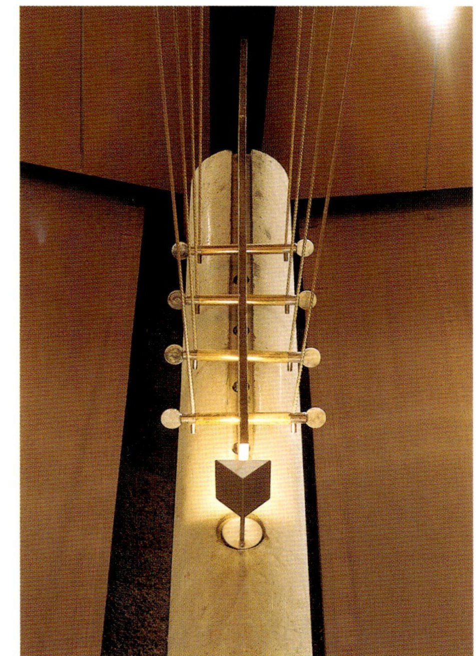

16

17

18

Acton Ostry Architects Inc.

Acton Ostry 建筑设计事务所

Established in 1992, Acton Ostry Architects is an practice with a track record of producing high-quality and technically outstanding public, institutional and cultural projects. Their experience encompasses a wide range of facility programming, feasibility studies, master planning, building design, interior design and public consultation services.

Their projects have been recognized for their sensitivity in responding to the uniqueness of their particular location and client requirements through respected awards such as; Canadian Architect Awards of Excellence, Lieutenant Governor of British Columbia Awards in Architecture, American Wood Council Awards, and Canadian Wood Council Awards–including the prestigious Ron Thom Award which was awarded to the Har-El Synagogue project.

Acton Ostry 建筑设计事务所创立于1992年，致力于创造高质量和技术卓越的公共和文化建筑。他们的工作范围涉及广泛，包括：可行性研究、城市规划、室内设计、建筑设计、公共咨询服务。

Acton Ostry建筑设计事务所强调与环境的和谐和充分反映客户的利益，这使得他们的建筑得到众多的奖项，其中包括：加拿大杰出建筑师奖，不列颠哥伦比亚省杰出建筑政府奖，美国杰出木建筑设计奖和加拿大杰出木建筑设计奖，其中包括Har-EI礼拜堂所赢得的建筑最高奖项—— Ron Thom 奖。

Acton Ostry 建筑设计事务所相信建筑的质量影响到居住空间、工作空间，以及两者之间大量的公共空间的质量。真正意义上的建筑包括完美的视觉、环境因素、建筑技术、结构、设备设计、建筑预算和工期的各个方面。建筑应反映文化的进步和现实生活。折射社会、文化和技术的现代生活和传统因素是建筑师的灵感。建筑设计手法、环境因素、建筑技术以及对建筑材料及建筑本身的理解使建筑师对复杂的建筑设计有准确地认识。

Acton Ostry建筑设计事务所一贯重视可持续性设计。可持续性设计包括平衡环境因素，充分利用能源、文化和社会因素等方面。其目标是创造和再利用建筑物，以满足使用者最理想的使用要求。可持续性设计对业主及使用者都长期有利。包括业主在内的设计队伍共同完成设计，以减低建筑能源的消耗以及对环境的冲击，同时保护室内环境的质量。这种设计理念贯穿了从最初的项目分析到最终的设计的全过程。

Chief Matthews Primary School

Chief Matthews 小学

地点：Old Massett

The site is flanked on the west and north by forest and on the south by a gravel road and dwellings that follow the line of the beach in traditional Haida fashion. The community hall, the only other large institutional building in the village, is to the east.

Principles of Haida culture and architecture are reflected and reinterpreted in the design of the school, which consists of three independent wings surrounding a focal language and library center. Two of the wings house learning activities while the third contains administrative and service functions.

The traditional axes of supernatural power are utilized to lie out the plan of the school as well as to serve to channel natural elements along their paths. Each of the wings has been placed at a different grade level and is served by a circulation ramp, which descends to the language and library center. This focal area is depressed relative to the exterior grade and recalls the spatial quality of the interior of a longhouse building and pit. Heavy timber trusses support a large central skylight where the essence of a traditional longhouse merges with the concept of the Haida cosmos - which the longhouse represents. The language and library center are used by all members of the community to further the rebirth of Haida culture. Elders assist with documenting and preserving the Haida language. Ceremonial songs and dances are performed and taught to the children in the central space.

1

2

1 学校全景
2 学校活动场地
3 总平面图

Chief Matthews 小学建筑的西侧和北侧是树林，而南面为一条石子路和以传统Haida方式沿海滨展开的居住区。东面则是社区会堂，是除学校外村庄内惟一的公共建筑。

按照印第安人Haida文化和建筑的传统，Chief Matthews 小学由一个集中的语言和图书中心，以及围绕着这一中心的三个独立的侧翼组成。其中两个为教学空间，第三个为管理和服务空间。

传统超自然的斧被运用以组织空间，并以此连接沿途的自然因素。所有的侧翼位于不同高度的地面上，并有交通斜坡向下通向语言和图书中心。中心区域低调地与室外场地呼应，并与印第安人长屋建筑和坑建筑的室内空间组合。像传统的印第安长屋建筑表现出的Haida宇宙观，沉重的杉木屋架支撑一个大中心天窗。语言和图书中心被社区的成员用来拯救Haida文化，长者在此记录和保存Haida语言，仪式歌曲和舞蹈在此表演并传授给儿童。

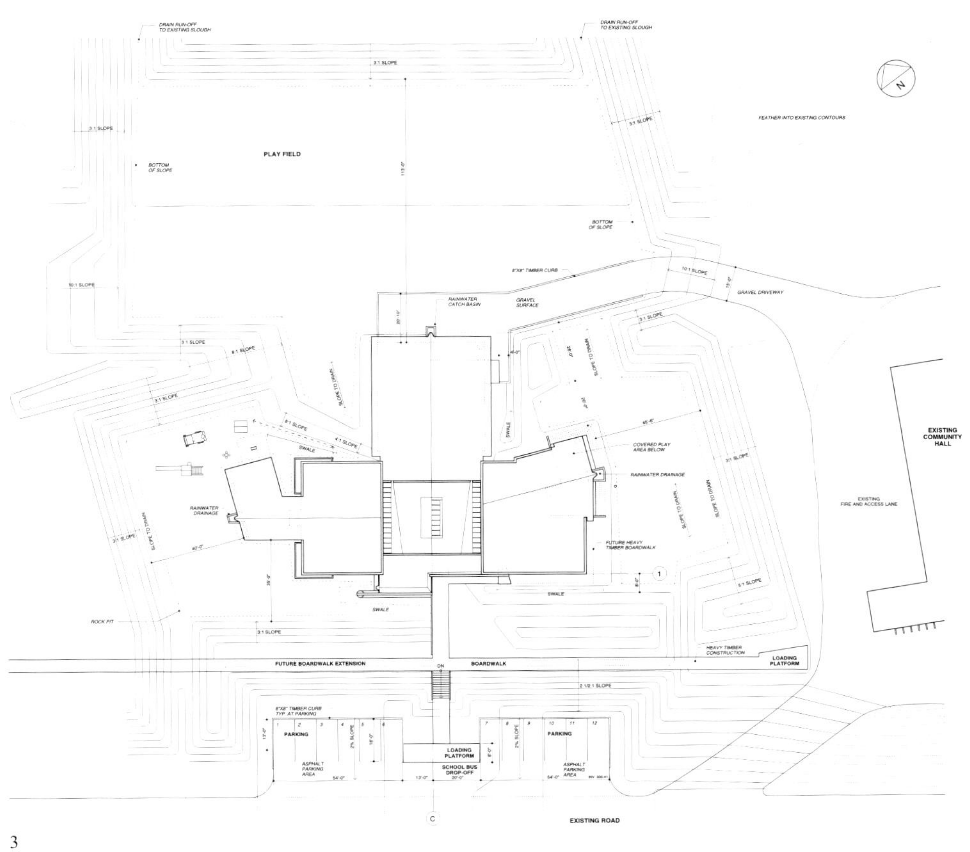

3

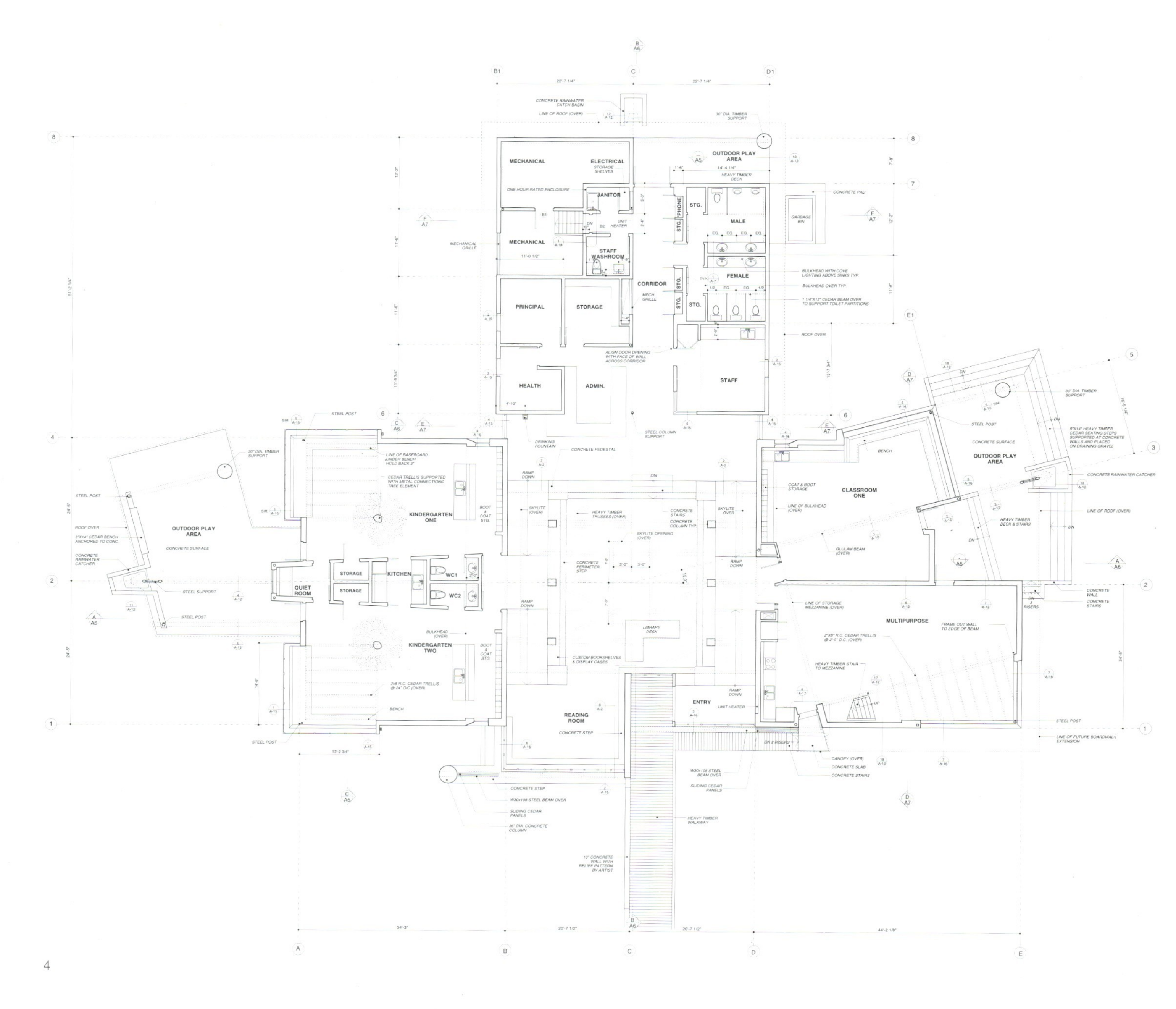

MECHANICAL
ELECTRICAL
JANITOR
STAFF WASHROOM
OUTDOOR PLAY AREA
STG.
PHONE
MALE
FEMALE
CORRIDOR
PRINCIPAL
STORAGE
HEALTH
ADMIN.
STAFF
OUTDOOR PLAY AREA
KINDERGARTEN ONE
QUIET ROOM
STORAGE
KITCHEN
WC1
WC2
KINDERGARTEN TWO
CLASSROOM ONE
OUTDOOR PLAY AREA
MULTIPURPOSE
LIBRARY DESK
READING ROOM
ENTRY

4

5

6

4 建筑平面图
5 仪式墙
6 儿童活动场地
7 建筑室外细部
8 墙体细部
9 建筑墙体细部
10 建筑雨篷细部
11 入口雨篷细部

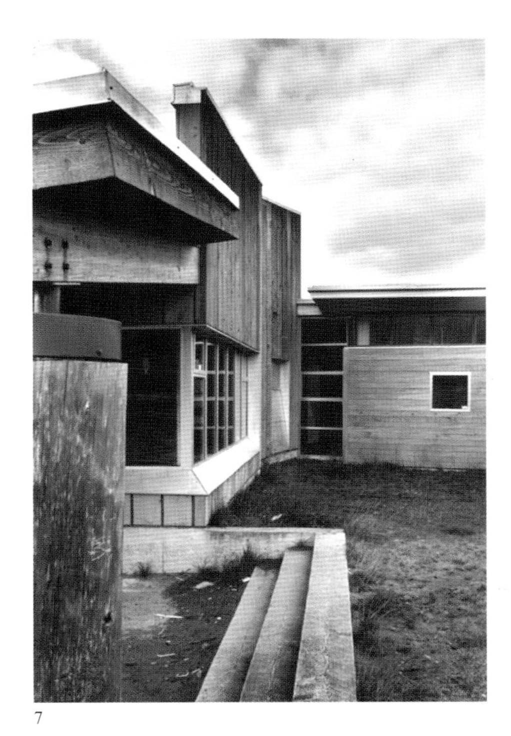
7

8

9

10

11

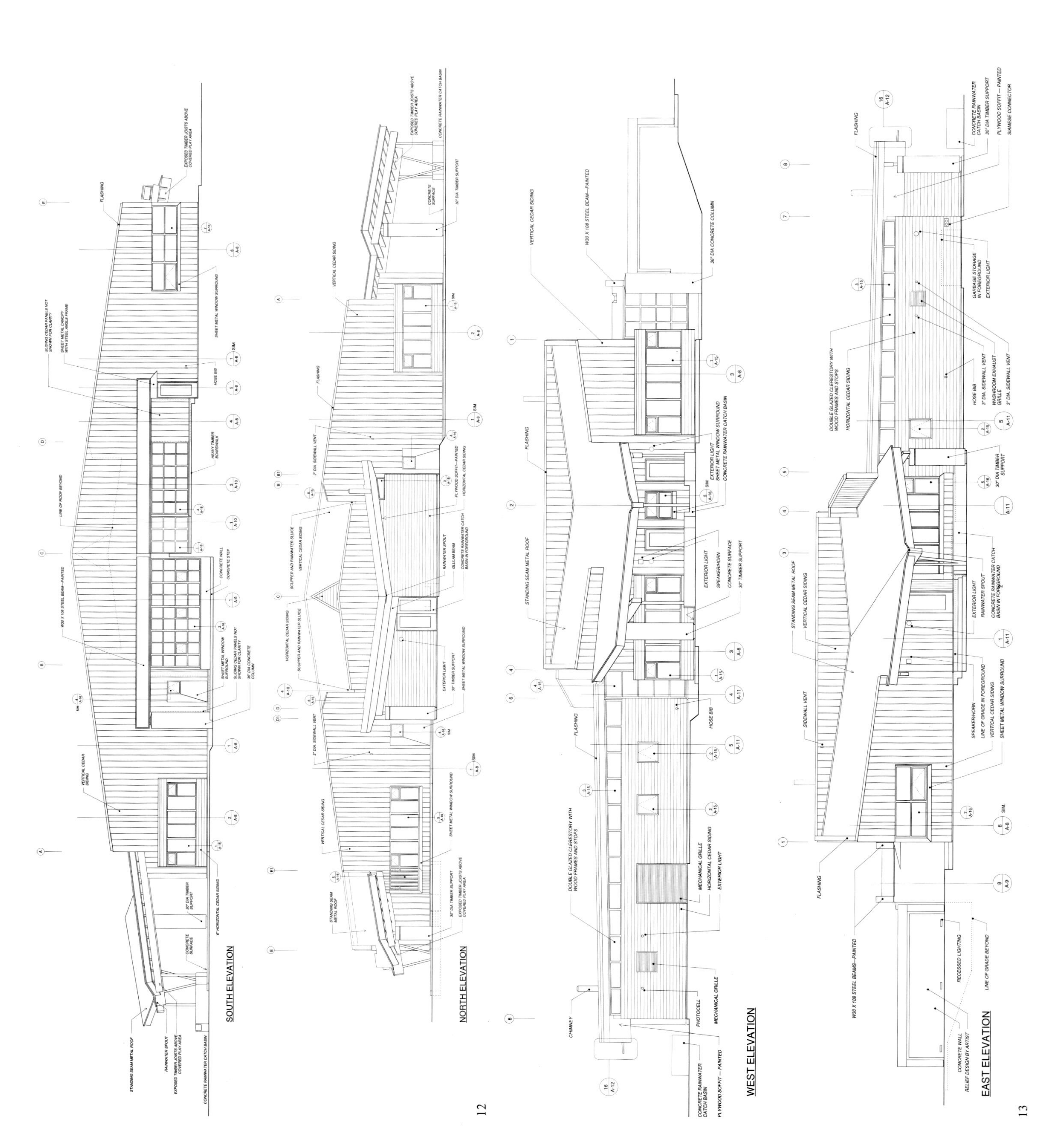
SOUTH ELEVATION
NORTH ELEVATION
WEST ELEVATION
EAST ELEVATION
12
13

12　建筑南、北立面
13　建筑西、东立面
14　屋顶细部
15　屋顶细部
16　建筑剖面详图

14

15

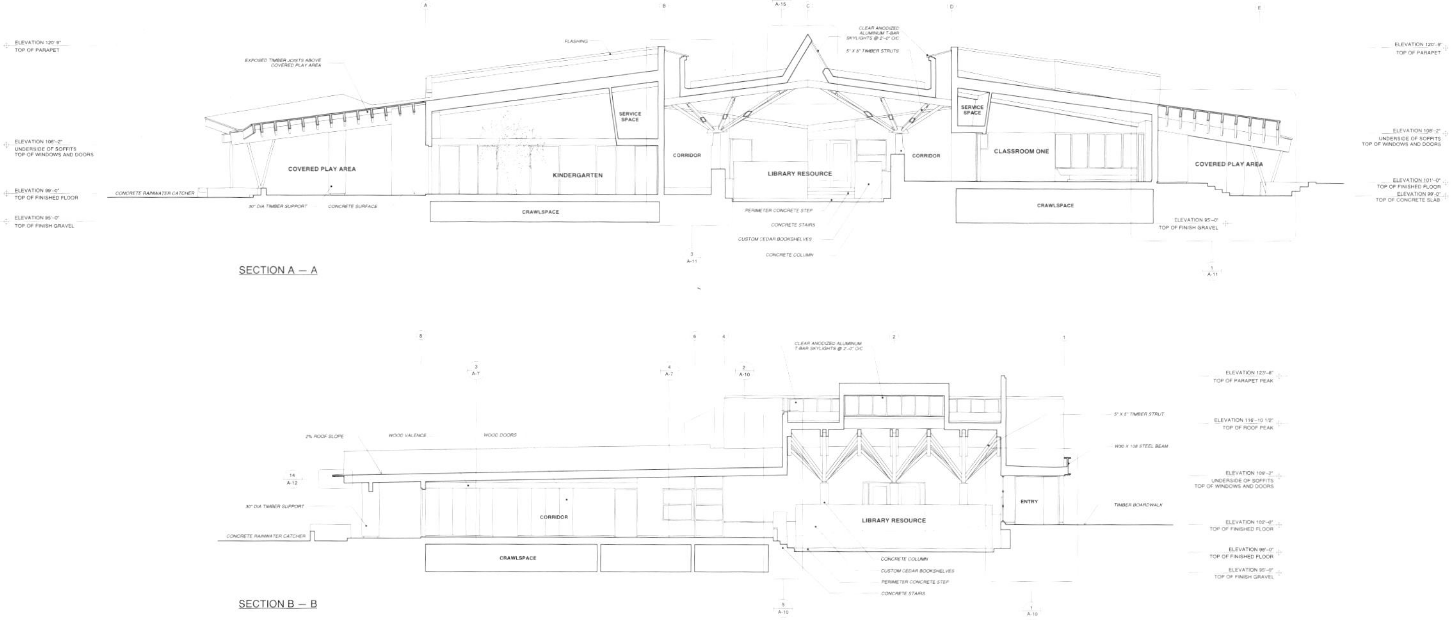

16

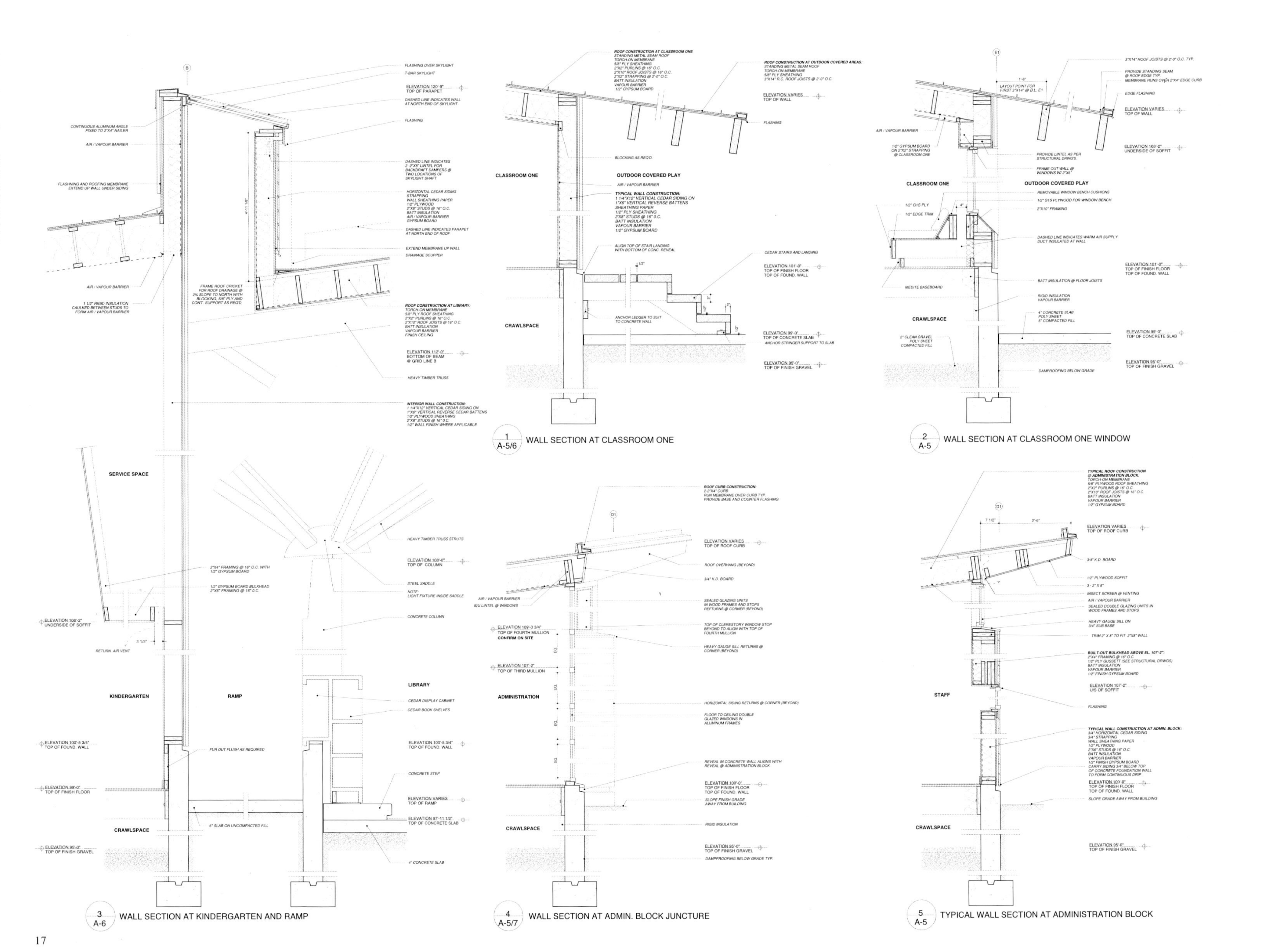

1 A-5/6 WALL SECTION AT CLASSROOM ONE
2 A-5 WALL SECTION AT CLASSROOM ONE WINDOW
3 A-6 WALL SECTION AT KINDERGARTEN AND RAMP
4 A-5/7 WALL SECTION AT ADMIN. BLOCK JUNCTURE
5 A-5 TYPICAL WALL SECTION AT ADMINISTRATION BLOCK
CLASSROOM ONE
OUTDOOR COVERED PLAY
CRAWLSPACE
SERVICE SPACE
KINDERGARTEN
RAMP
LIBRARY
ADMINISTRATION
STAFF

17

17　屋顶详图
18　墙体剖面详图

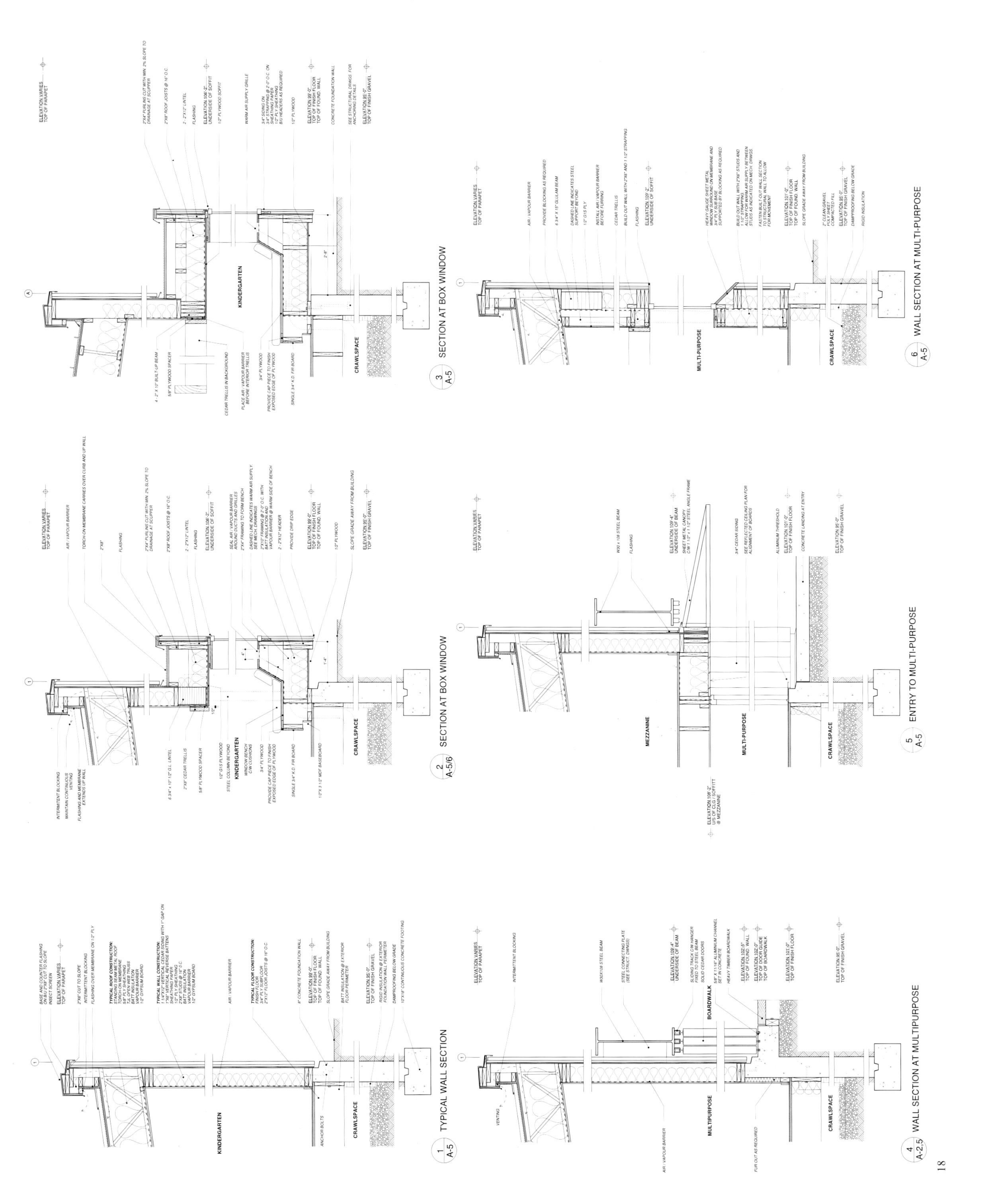

Har-El Synagogue

Har-El礼拜堂

地点：西温哥华市（West Vancouver）

The congregation Har-El, Mountain of God, envisioned building the first synagogue to be constructed in West Vancouver in which to accommodate their growing membership of one hundred-fifty families, and located at the intersection of the Trans-Canada highway and a busy arterial thoroughfare on the North Shore Mountains in West Vancouver.

The site is covered with second growth forest and is traversed by a small creek which swells to river proportions during the spring thaw runoff from the mountains. Flood plain requirements impose restrictions for occupied floor levels. Vehicular access is restricted by provincial highway authority to a single location at the southeast side of the site. Ministry of Environment covenants restrict the proximity of construction to the salmon-spawning creek. Municipal by-laws impose further setback restrictions in relation to the surrounding thoroughfares, senior care facility, and residential neighborhood.

The synagogue, positioned at the northeast corner of the site, is anchored by heavy masonry walls and concrete elements, which insulate the congregation from the highway traffic. The building opens toward and embraces the creek and surrounding wooded landscape, forming a terrace to the south. The sanctuary is oriented east toward Jerusalem. The school bridges the creek linking the synagogue to the community.

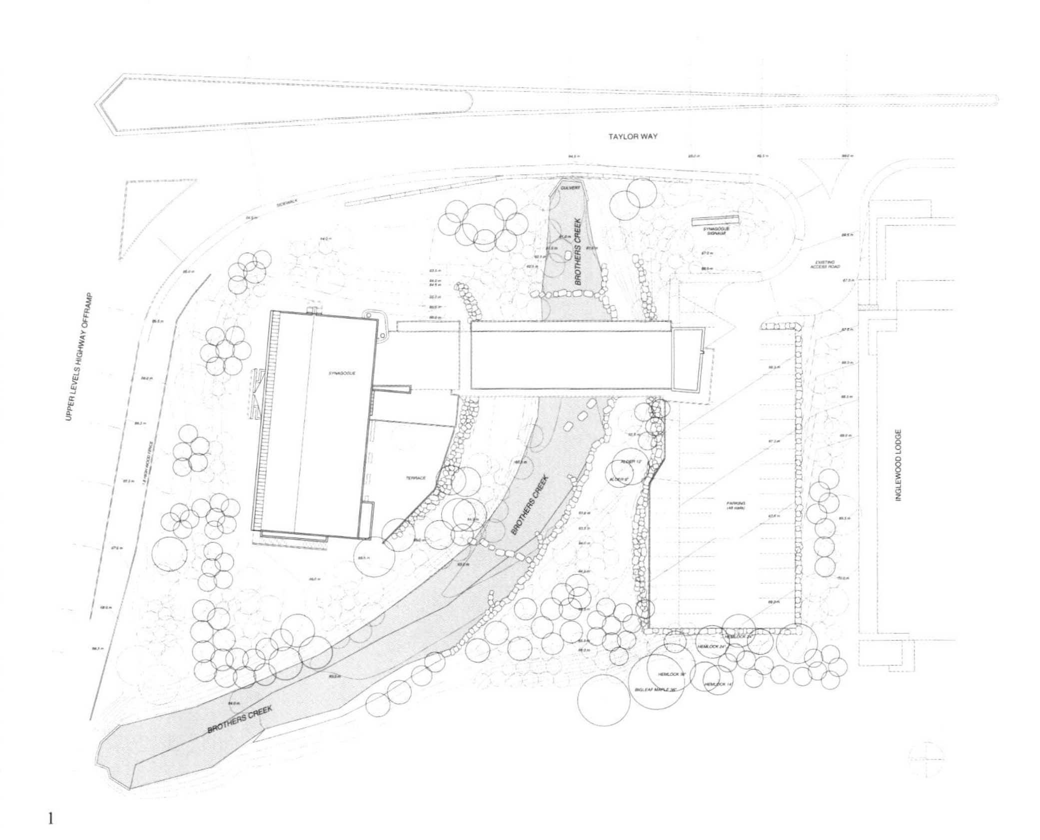

1

Har-El礼拜堂是西温哥华市域第一个礼拜堂，服务于150个家庭，并不断增长。其地理位置十分显要——位于加拿大穿境高速公路和一个通向北岸山脉的繁忙支路的交叉口上。

整个场地覆盖着再生树林，有一条穿越而过的小溪。春天小溪会因山脉上的积雪融水而汇成一条小河，因此首层有防洪设计的要求。由于不列颠哥伦比亚省高速公路法规的限制，机动车出入口位于场地的西南方。环境部门也因其位于三文鱼保育场地内，而对设计提出了限制条件。市政府也对建筑的退红线有所限制，此外，限制条件还包括周围环境、长者医疗设施和居民区等因素。

建筑位于场地的西北角，由沉重的石墙和混凝土支撑，以隔绝高速公路的噪声。通过一个南向的花园，建筑面向小溪和环绕的树林，并与其连为一体。圣坛向东面对基督。学校是连接礼拜堂和社区的桥梁。

1　总平面图
2　礼拜堂全景

2

3

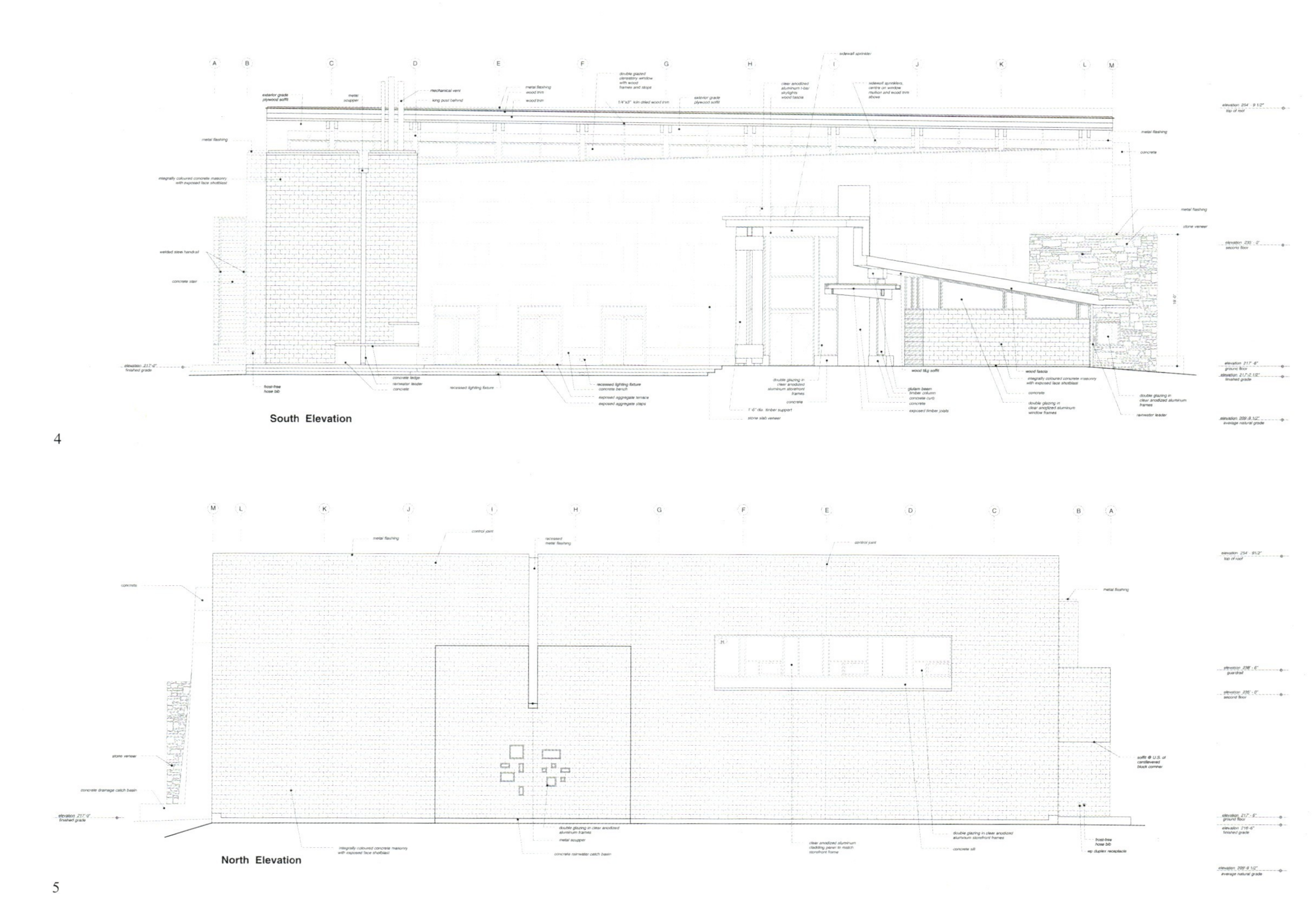

4

5

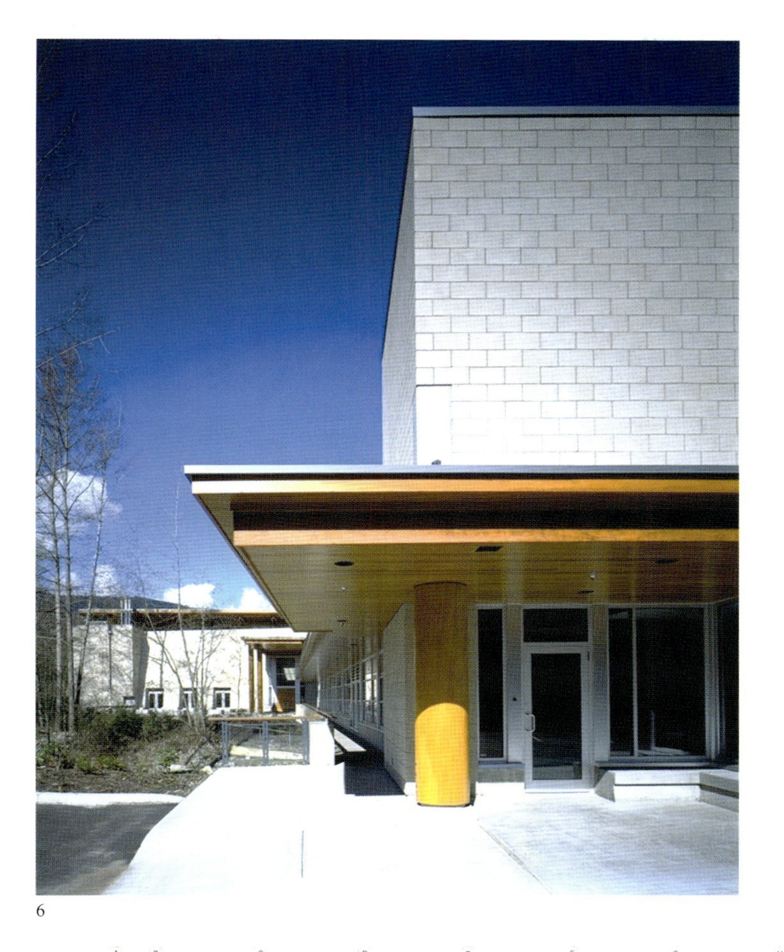

6

3 南向立面景观
4 建筑南立面图
5 建筑北立面图
6 挑檐细部
7 E–E 剖面图
8 F–F 剖面图

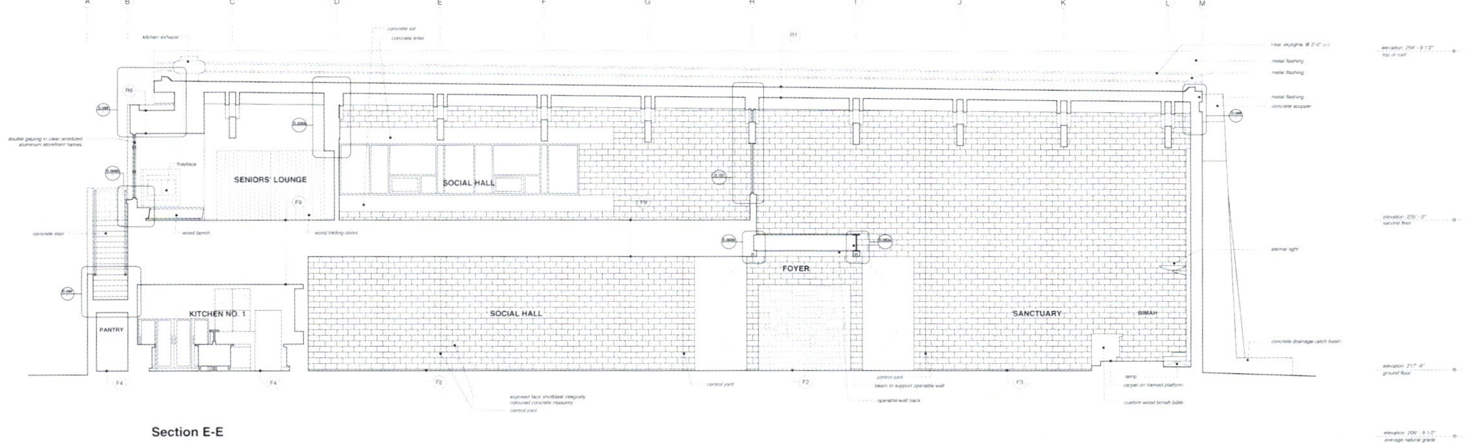

Section E-E

7

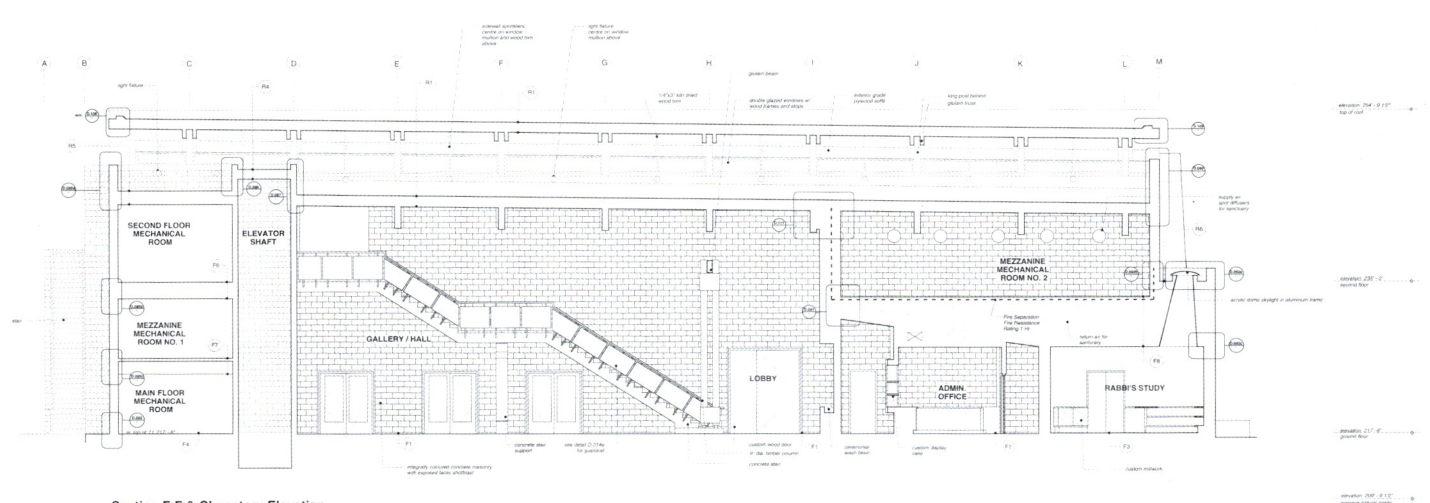

Section F-F & Clerestory Elevation

8

9

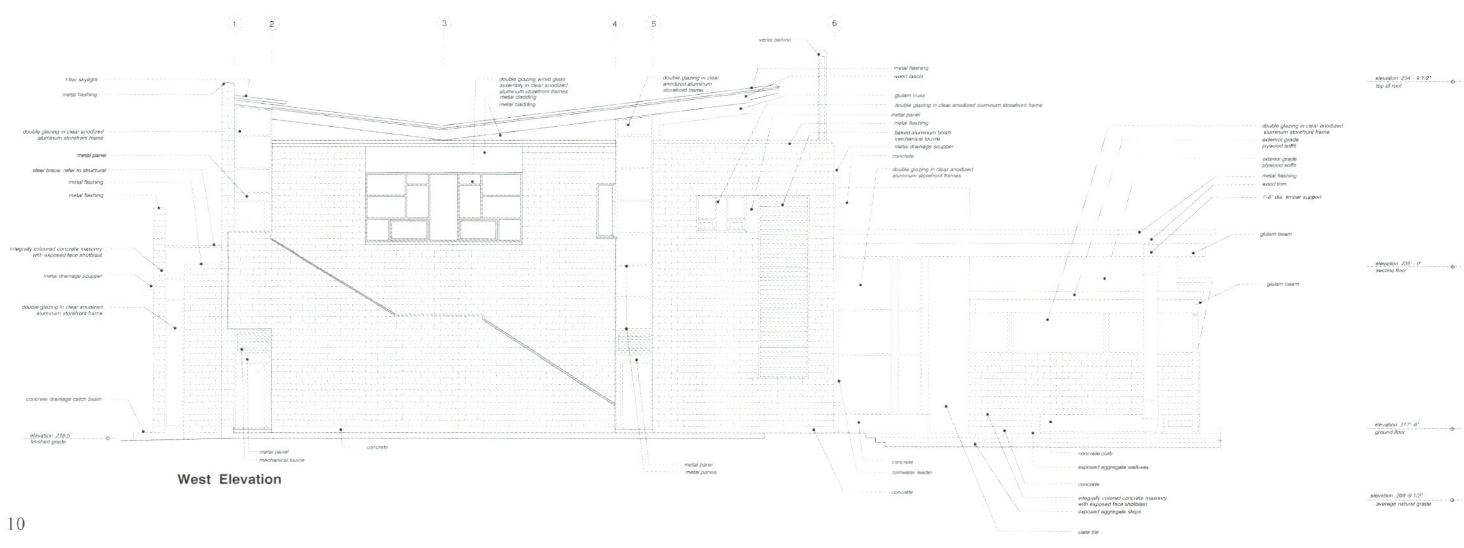

10

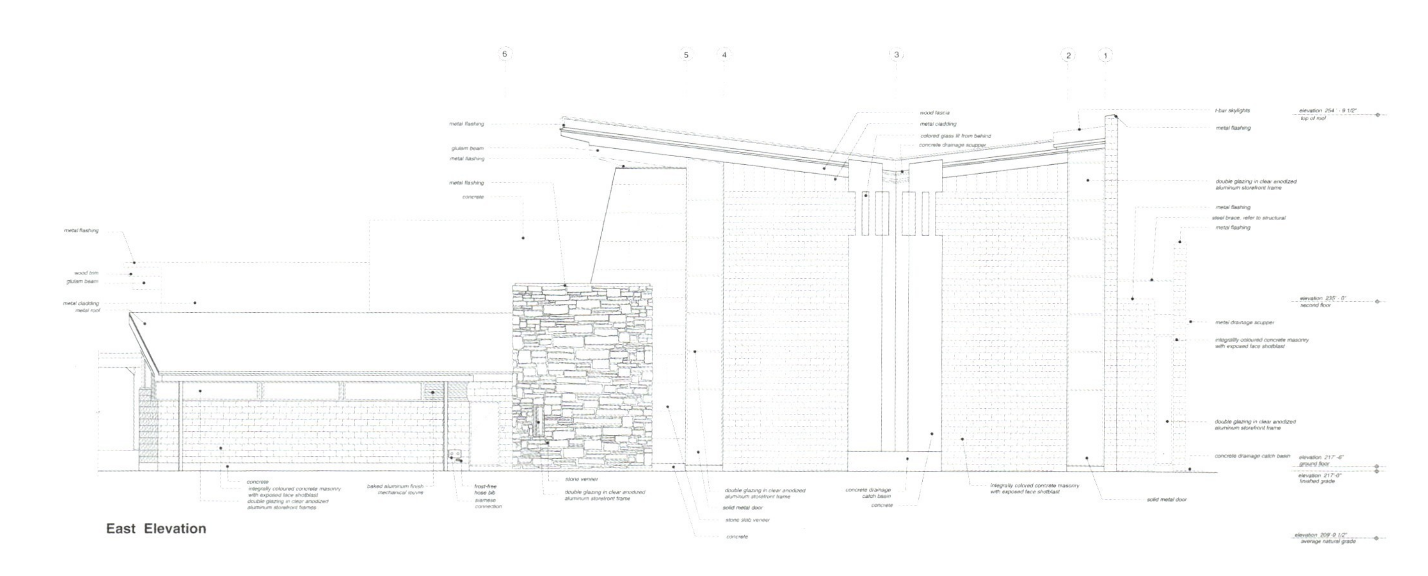

11

9 沿快速通道一侧礼拜堂的景观
10 西立面图
11 东立面图
12 室内细部
13 室内
14 剖面及室内立面详图

12　　13

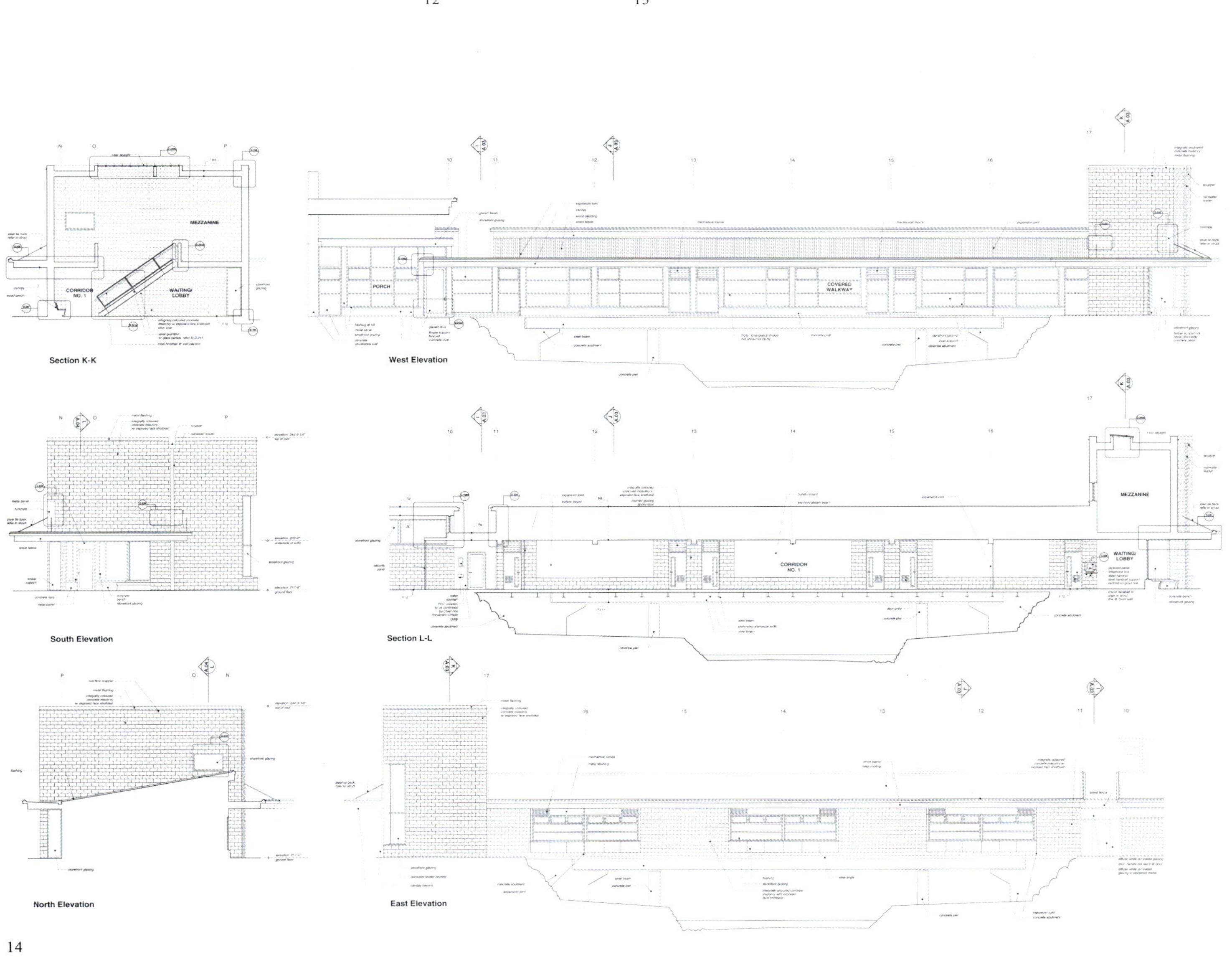

14

15

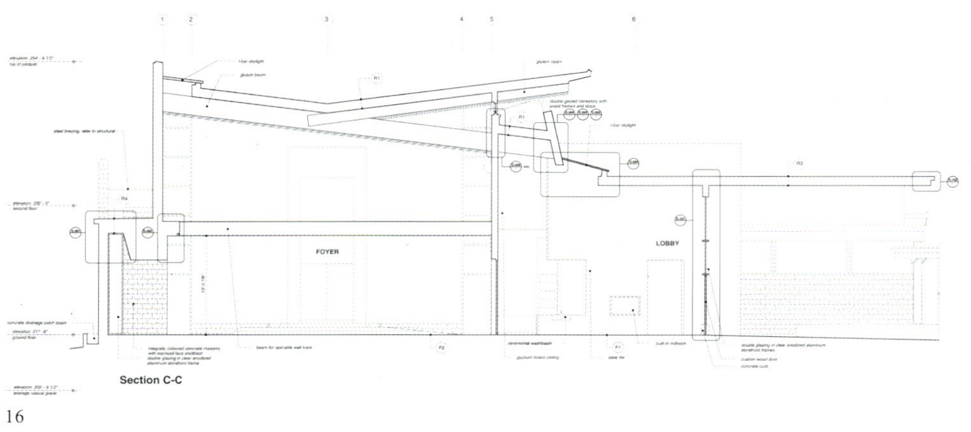

16

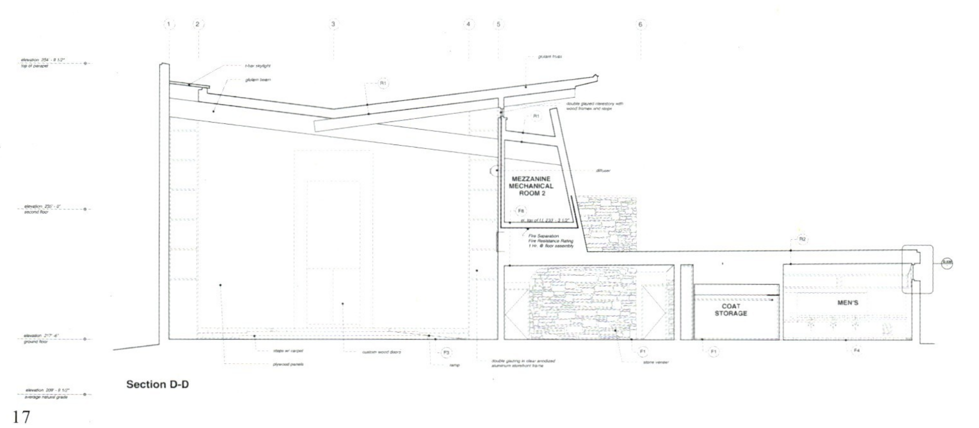

17

15　入口细部
16　C–C 剖面图
17　D–D 剖面图
18　屋顶剖面详图
19　屋顶剖面详图
20　屋顶剖面详图

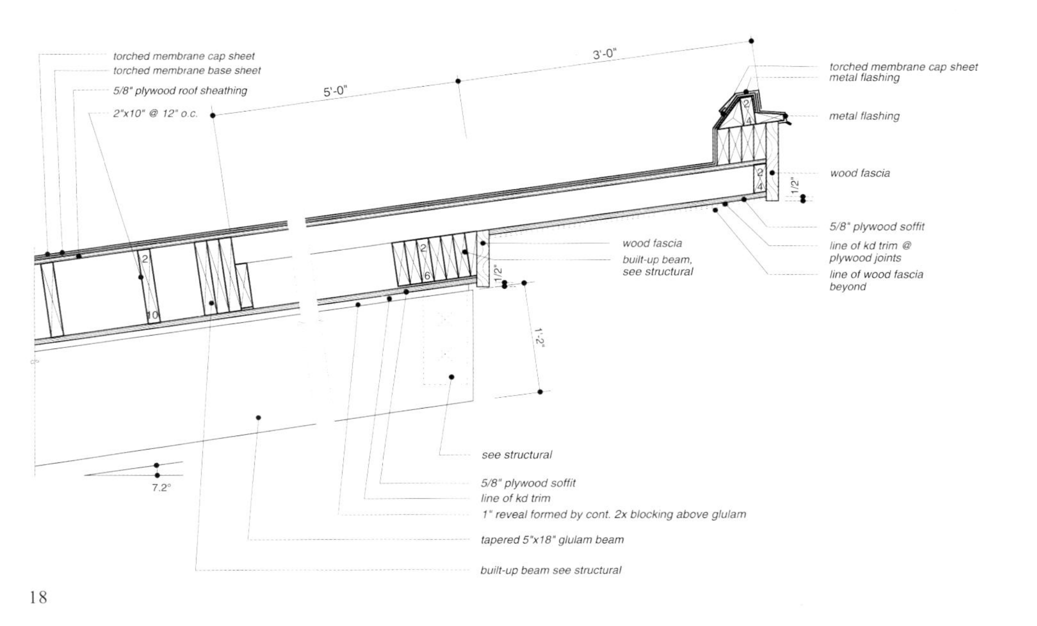

torched membrane cap sheet
torched membrane base sheet
5/8" plywood roof sheathing
2"x10" @ 12" o.c.
5'-0"
3'-0"
torched membrane cap sheet
metal flashing
metal flashing
wood fascia
1/2"
5/8" plywood soffit
line of kd trim @ plywood joints
line of wood fascia beyond
wood fascia
built-up beam, see structural
1/2"
1'-2"
see structural
5/8" plywood soffit
line of kd trim
1" reveal formed by cont. 2x blocking above glulam
tapered 5"x18" glulam beam
built-up beam see structural
7.2°

18

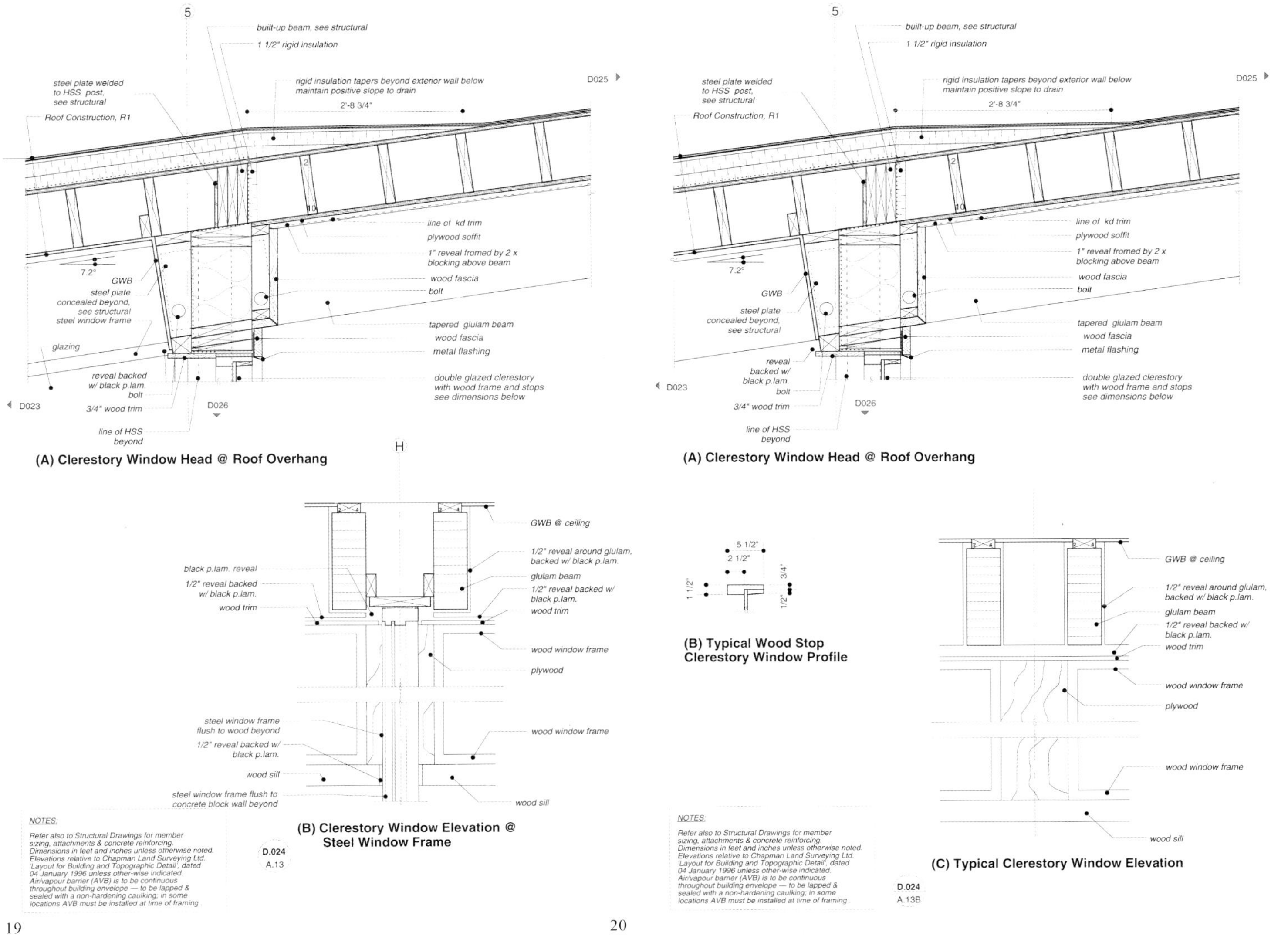

5
built-up beam. see structural
1 1/2" rigid insulation
steel plate welded to HSS post, see structural
rigid insulation tapers beyond exterior wall below maintain positive slope to drain
D025
2'-8 3/4"
Roof Construction, R1
7.2°
line of kd trim
plywood soffit
1" reveal fromed by 2 x blocking above beam
GWB
wood fascia
steel plate concealed beyond, see structural
bolt
steel window frame
tapered glulam beam
glazing
wood fascia
metal flashing
reveal backed w/ black p.lam.
double glazed clerestory with wood frame and stops see dimensions below
bolt
D023
3/4" wood trim
D026
line of HSS beyond
(A) Clerestory Window Head @ Roof Overhang
H
GWB @ ceiling
black p.lam. reveal
1/2" reveal around glulam, backed w/ black p.lam.
1/2" reveal backed w/ black p.lam.
glulam beam
wood trim
1/2" reveal backed w/ black p.lam.
wood trim
wood window frame
plywood
steel window frame flush to wood beyond
wood window frame
1/2" reveal backed w/ black p.lam.
wood sill
steel window frame flush to concrete block wall beyond
wood sill
NOTES:
Refer also to Structural Drawings for member sizing, attachments & concrete reinforcing.
Dimensions in feet and inches unless otherwise noted.
Elevations relative to Chapman Land Surveying Ltd. 'Layout for Building and Topographic Detail', dated 04 January 1996 unless other-wise indicated.
Air/vapour barrier (AVB) is to be continuous throughout building envelope — to be lapped & sealed with a non-hardening caulking; in some locations AVB must be installed at time of framing
D.024
A.13
(B) Clerestory Window Elevation @ Steel Window Frame
5
built-up beam, see structural
1 1/2" rigid insulation
steel plate welded to HSS post, see structural
rigid insulation tapers beyond exterior wall below maintain positive slope to drain
D025
2'-8 3/4"
Roof Construction, R1
7.2°
line of kd trim
plywood soffit
1" reveal fromed by 2 x blocking above beam
wood fascia
GWB
bolt
steel plate concealed beyond, see structural
tapered glulam beam
wood fascia
metal flashing
reveal backed w/ black p.lam.
double glazed clerestory with wood frame and stops see dimensions below
D023
bolt
3/4" wood trim
D026
line of HSS beyond
(A) Clerestory Window Head @ Roof Overhang
5 1/2"
2 1/2"
1 1/2"
3/4"
1/2"
(B) Typical Wood Stop Clerestory Window Profile
GWB @ ceiling
1/2" reveal around glulam, backed w/ black p.lam.
glulam beam
1/2" reveal backed w/ black p.lam.
wood trim
wood window frame
plywood
wood window frame
wood sill
NOTES:
Refer also to Structural Drawings for member sizing, attachments & concrete reinforcing.
Dimensions in feet and inches unless otherwise noted.
Elevations relative to Chapman Land Surveying Ltd. 'Layout for Building and Topographic Detail', dated 04 January 1996 unless other-wise indicated.
Air/vapour barrier (AVB) is to be continuous throughout building envelope — to be lapped & sealed with a non-hardening caulking; in some locations AVB must be installed at time of framing
D.024
A.13B
(C) Typical Clerestory Window Elevation

19

20

21

22

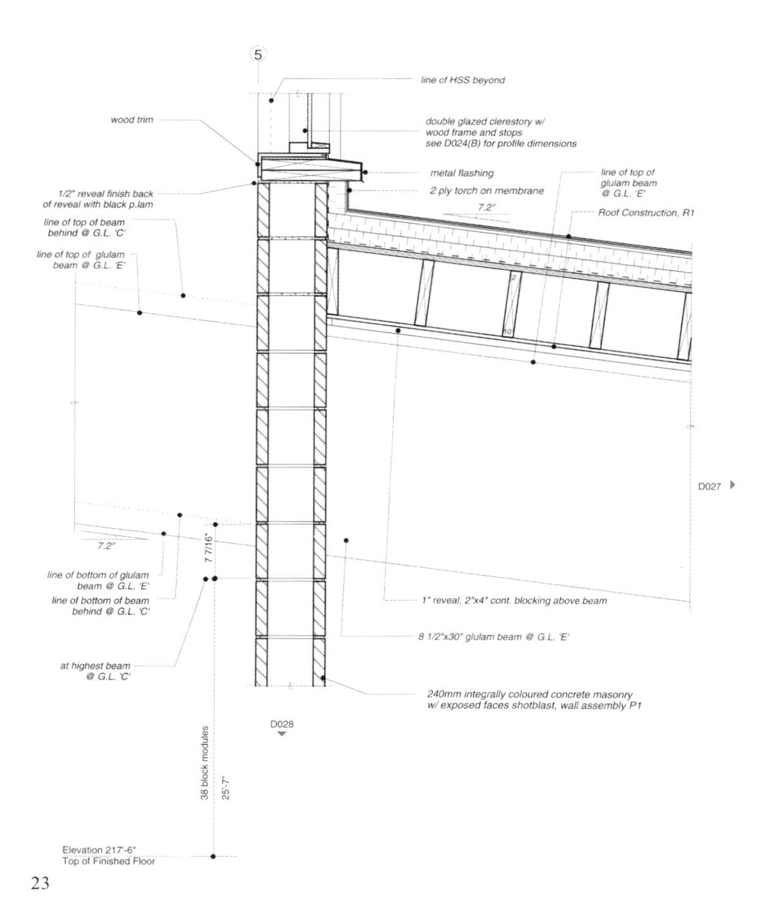

5
line of HSS beyond
wood trim
double glazed clerestory w/
wood frame and stops
see D024(B) for profile dimensions
metal flashing
2 ply torch on membrane
line of top of
glulam beam
@ G.L. 'E'
1/2" reveal finish back
of reveal with black p.lam
line of top of beam
behind @ G.L. 'C'
7.2°
Roof Construction, R1
line of top of glulam
beam @ G.L. 'E'
D027
7.2°
line of bottom of glulam
beam @ G.L. 'E'
line of bottom of beam
behind @ G.L. 'C'
7 7/16"
1" reveal, 2"x4" cont. blocking above beam
8 1/2"x30" glulam beam @ G.L. 'E'
at highest beam
@ G.L. 'C'
240mm integrally coloured concrete masonry
w/ exposed faces shotblast, wall assembly P1
D028
38 block modules
25'-7"
Elevation 217'-6"
Top of Finished Floor

23

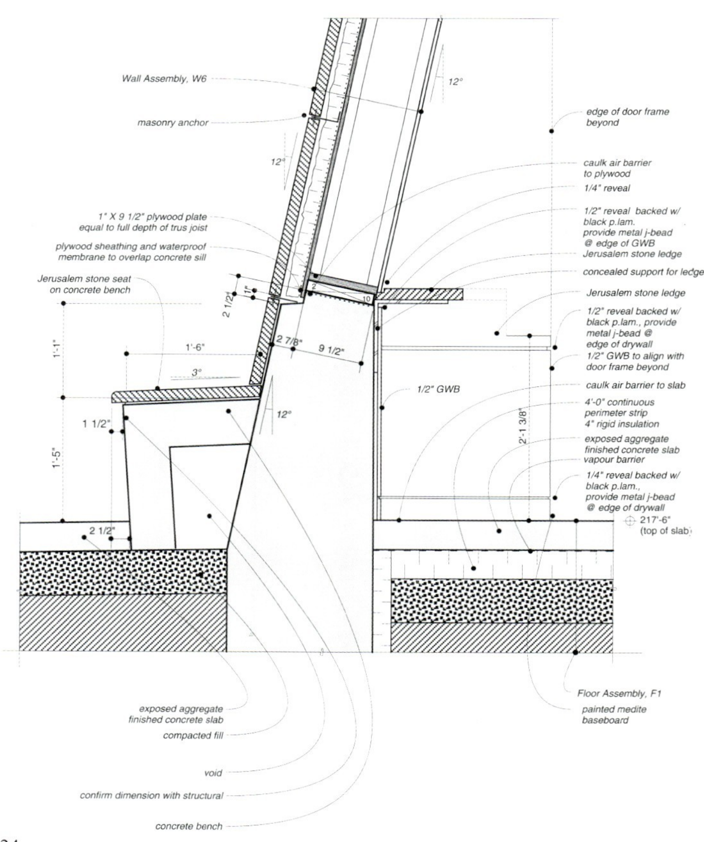

Wall Assembly, W6
12°
masonry anchor
edge of door frame
beyond
12°
caulk air barrier
to plywood
1/4" reveal
1" X 9 1/2" plywood plate
equal to full depth of trus joist
1/2" reveal backed w/
black p.lam.
provide metal j-bead
@ edge of GWB
plywood sheathing and waterproof
membrane to overlap concrete sill
Jerusalem stone ledge
concealed support for ledge
Jerusalem stone seat
on concrete bench
Jerusalem stone ledge
2 1/2"
1/2" reveal backed w/
black p.lam., provide
metal j-bead @
edge of drywall
1'-1"
1'-6"
2 7/8"
9 1/2"
1/2" GWB to align with
door frame beyond
3"
1/2" GWB
caulk air barrier to slab
12°
4'-0" continuous
perimeter strip
4" rigid insulation
1 1/2"
2'-1 3/8"
exposed aggregate
finished concrete slab
vapour barrier
1'-5"
1/4" reveal backed w/
black p.lam.,
provide metal j-bead
@ edge of drywall
2 1/2"
217'-6"
(top of slab)
Floor Assembly, F1
exposed aggregate
finished concrete slab
painted medite
baseboard
compacted fill
void
confirm dimension with structural
concrete bench

24

21 过道细部
22 过道细部
23 墙体详图
24 墙体细部
25 净身池细部
26 浴室墙面详图
27 浴室墙面详图

25

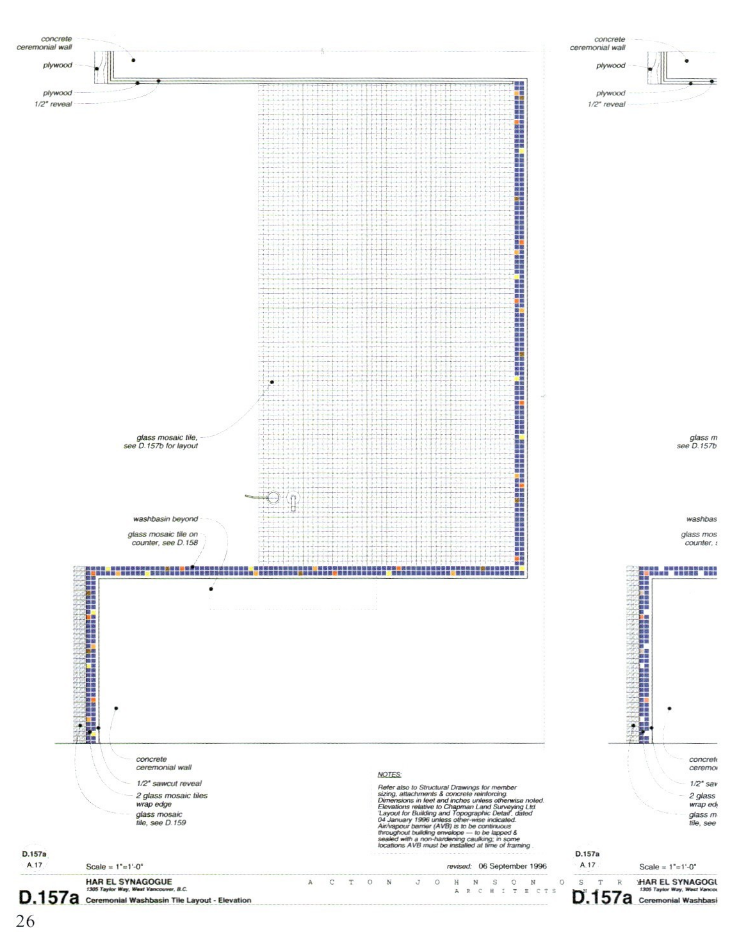

26

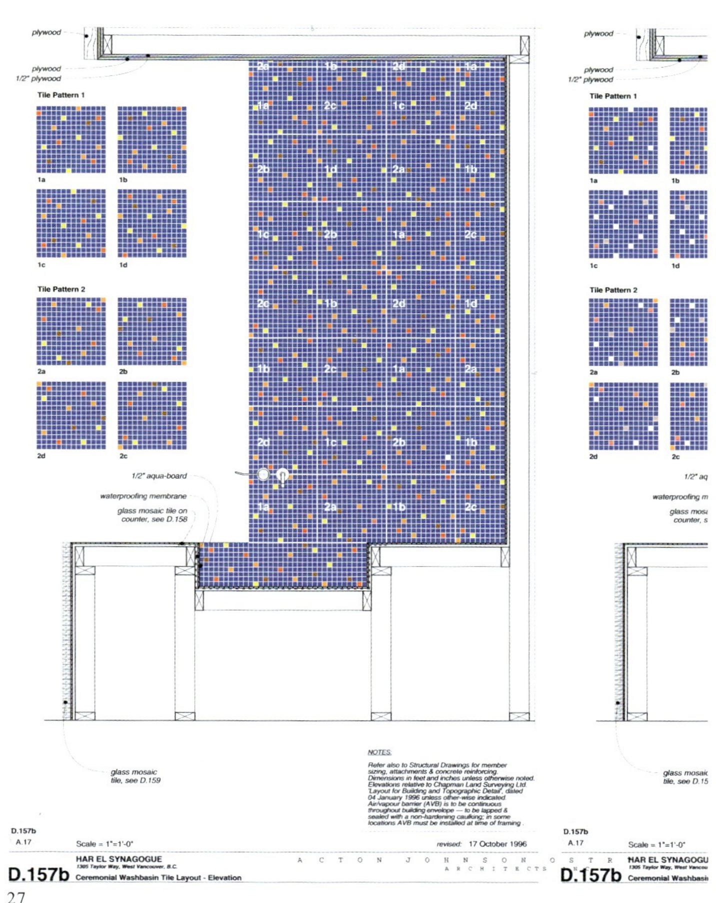

27

28

28　楼梯细部
29~31　楼梯节点详图

5

steel guardrail w/ glass panels see D.095

D014h(A) sim.

Wall Assembly P1

1 1/2"

concrete

sprinkler line

6"

12"

16"

1/2"

recessed lighting

16"

1/2" plywood panel

u/s stair beyond

(A) Section Through Stair Landing @ Gallery/Hall

D.095a
A.07

29

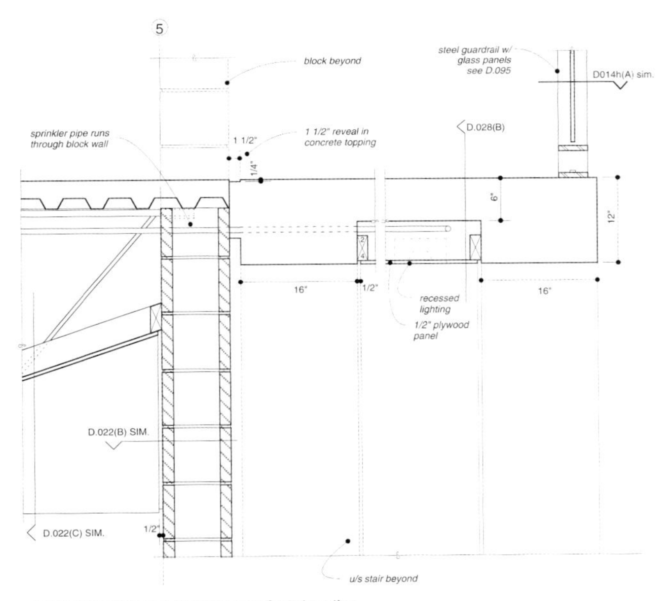

(A) Section Through Cantilevered Stair Landing

D.028a
A.13

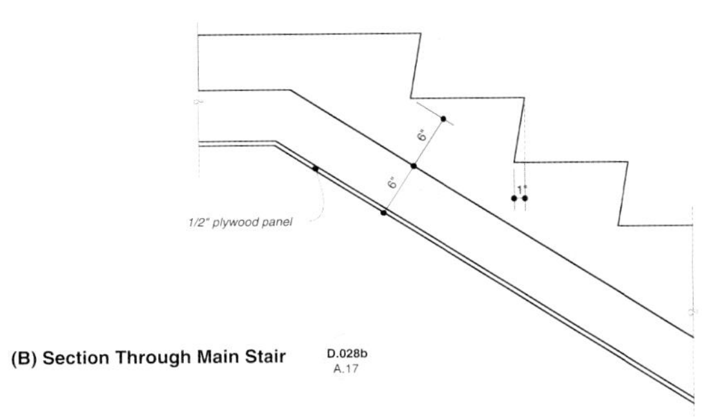

(B) Section Through Main Stair

D.028b
A.17

30

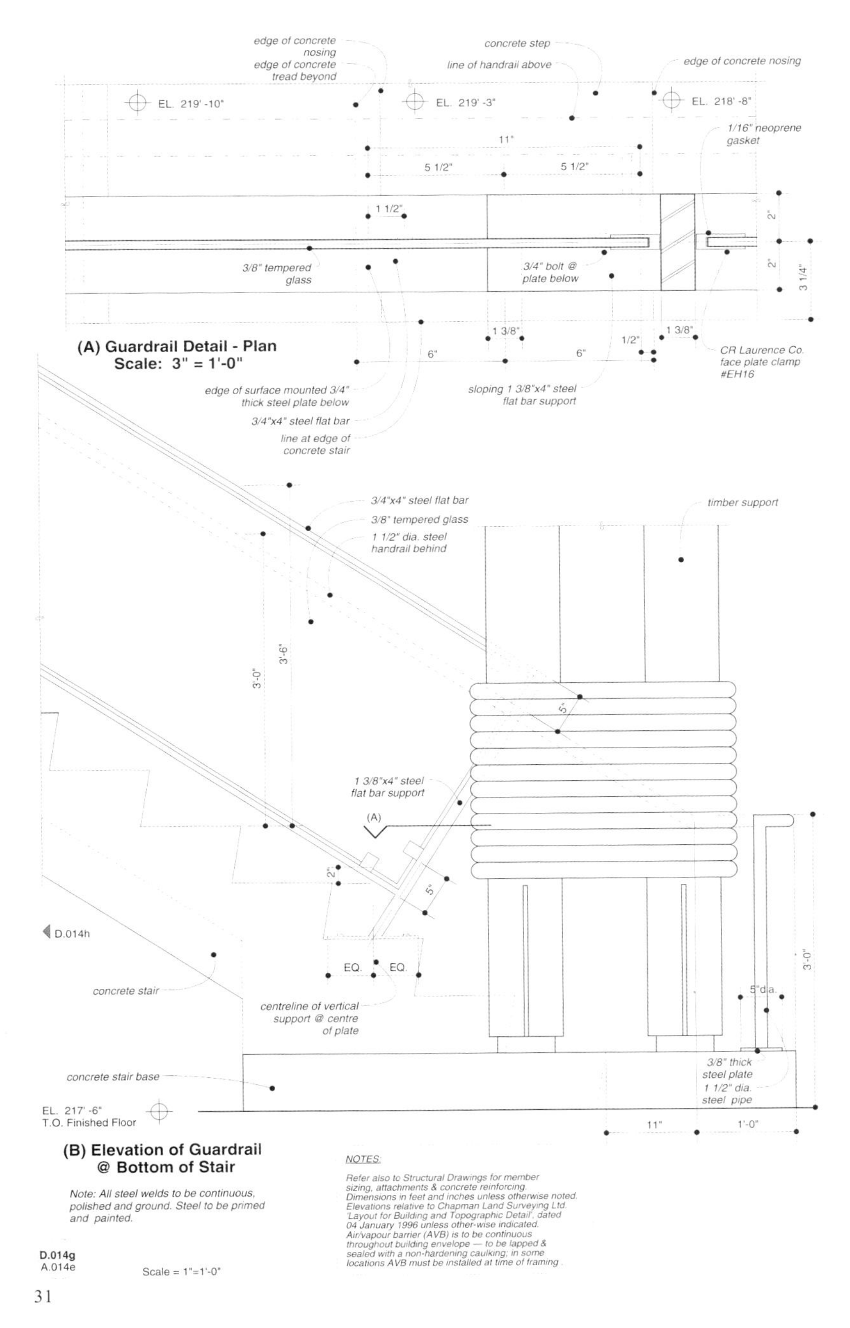

31

The Siple Residence
Siple 住宅

地点：北温哥华

The client is an independent filmmaker partially paralyzed as the result of an automobile accident. The project is the total renovation and addition to a 1950's modernist spec-built rancher located in North Vancouver.

The objective was to design an independent living and working environment free of the institutional overtones typically associated with universal design. The existing plan and basic organization of the original house have been retained, although virtually every interior partition has been relocated to increase and enhance the relationship of the interior with the outdoors - as well as to meet specific accessibility requirements of the client.

The robust and unselfconscious character of the original house with its planar surfaces, punched openings and horizontal wood frame canopies has been reinterpreted and redefined through the introduction of a natural palette of bold, rich red mahogany panels and box windows complemented by warm yellow clear-finished Douglas Fir panels and exposed wood-framed glazed canopies. Many windows have been added or relocated to delineate views in a way that is both client and site specific. Vertical channel cedar siding is painted black to provide a neutral background to accentuate the simple, but effective, vocabulary of architectural details.

The interior utilizes similar architectural devices to order space and create a consistent architectural character. In the living room, a portion of the ceiling is peeled back to reveal the roof joists to define the central seating area. At a smaller scale, timber baffles modulate light from new and existing skylights. In the northwest corner of the living room, a glass-enclosed nook is pushed out to the adjacent trees to overlook the ravine that borders the property. Large sliding storefront windows have been added at the south to provide access to a large terrace.

A film-editing studio has been added west of the living area. A tall floor-to-ceiling window frames the trunk of a large Douglas fir that anchors the studio to the site. A double garage, utility room and entry foyer has been added to the east of the site to replace an existing carport and garage that were not functional for use by the client.

1

Siple住宅的业主是一个因车祸伤残的独立电影制作人，设计工程为全面改建和扩建20世纪50年代的乡村住宅。

设计目标是提供一个独立的生活和工作的环境，与常规建筑不同，要求有残障设计。现有的和基本的住宅平面得以保留。重新安排室内的隔墙以加强室内外的联系，并可以满足业主行动的需要。

粗壮的和不自信的原住宅的外立面，包括平凡的表面，突兀的开口和水平木架雨篷，被有自然纹理的红色石板，干净的杉木装饰的窗和裸露的玻璃木制雨篷所代替。许多的窗户因业主和场地的要求而被改动。竖向开启的雪松窗被漆成黑色的原因是提供一个中性的背景，以突出简单但却有效的建筑细部。

室内设有类似的建筑设施以安排空间和显示建筑的个性。起居室内，部分顶棚裸露出来，显示屋顶的交接处，并以此标识中心生活区。小一些的尺度上，木制遮阳板调节新的和已有的天窗光线。起居室的西北角，一个玻璃小室伸出与树木接近，并可以俯瞰场地的界限——山涧。巨大的推拉窗位于南侧，增加面向花园的通道。加建的一个电影编辑室位于生活区的西侧。一个高大的从地板到顶棚的杉木窗框支撑着编辑室的结构。一个双车位的车库，设备间和入口厅位于场地的东侧，以代替现有的已无法使用的停车棚和车库。

1 住宅入口
2 平面图
3 入口夜景
4 适应使用者需求的宽敞空间

3

4

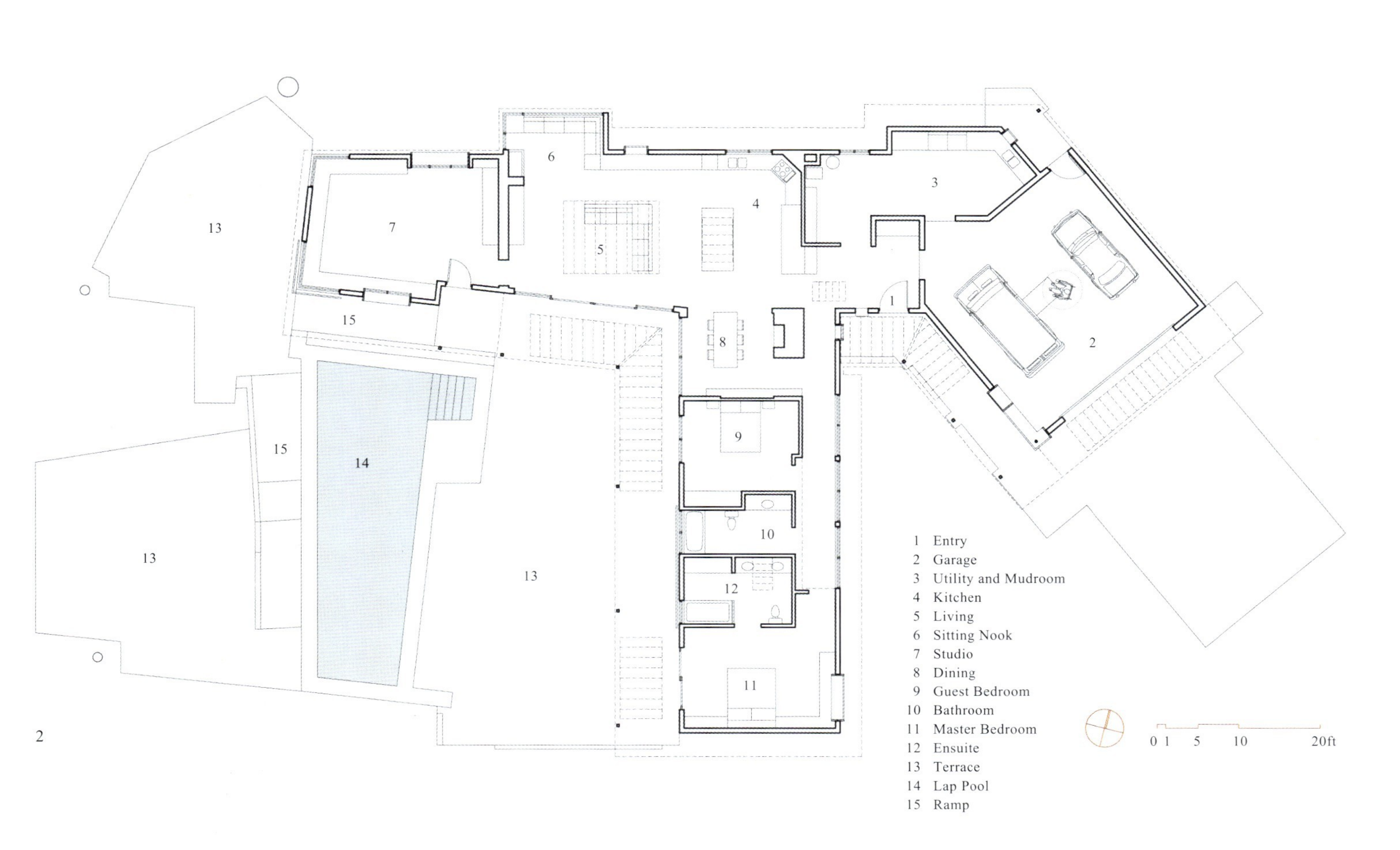
1 Entry
2 Garage
3 Utility and Mudroom
4 Kitchen
5 Living
6 Sitting Nook
7 Studio
8 Dining
9 Guest Bedroom
10 Bathroom
11 Master Bedroom
12 Ensuite
13 Terrace
14 Lap Pool
15 Ramp
0 1 5 10 20ft

2

5~6 挑檐细部
7,9~10 室内细部
8,11~12 剖面图

5

6

7

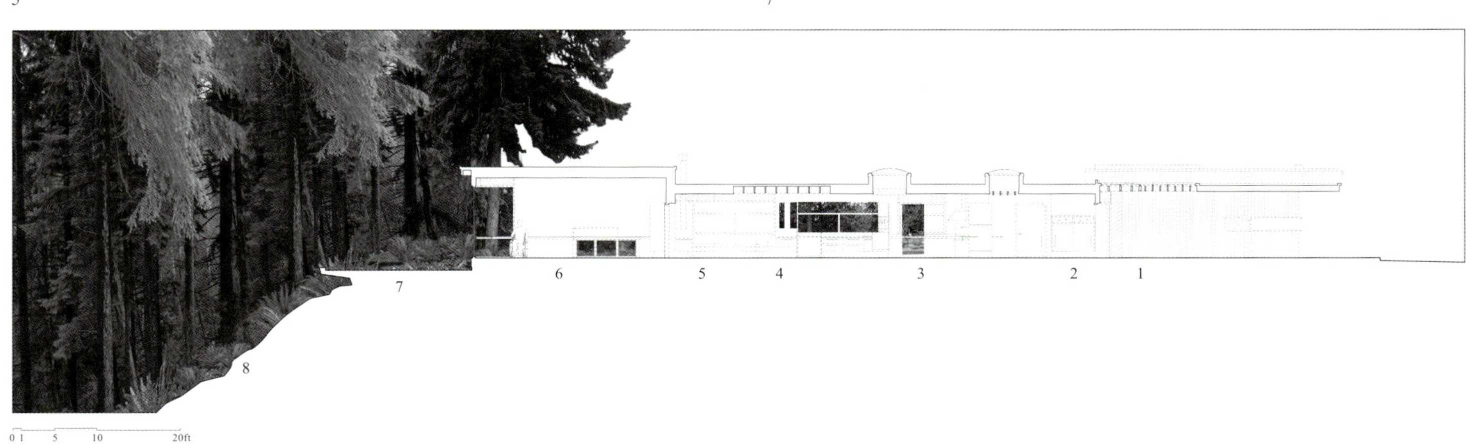

1 approach
2 entry
3 kitchen
4 living room
5 sitting nook
6 studio
7 terrace
8 Mosquito Creek Park

8

9

10

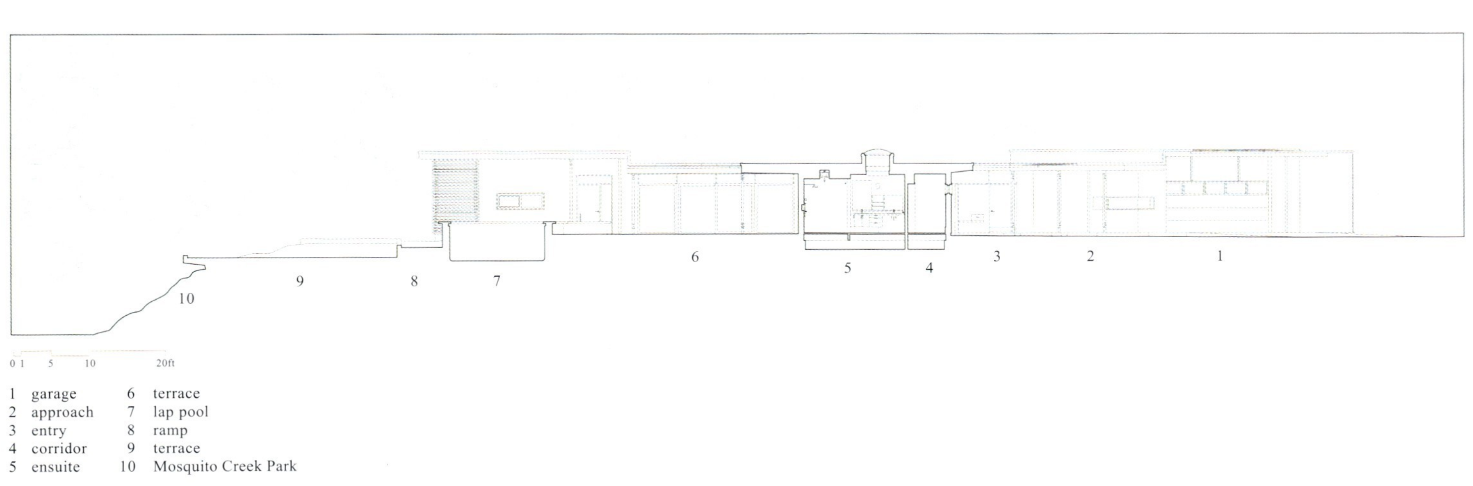

1 garage
2 approach
3 entry
4 corridor
5 ensuite
6 terrace
7 lap pool
8 ramp
9 terrace
10 Mosquito Creek Park

11

1 lap pool
2 covered exterior
3 sitting nook
4 studio
5 terrace
6 Mosquito Creek Park

12

Skidegate Elementary School
斯基德盖特小学

地点：斯基德盖特

The Sk'aadgaa Naay Elementary School, located in Skidegate, BC, was designed and built to serve the two adjacent communities of Queen Charlotte City and Skidegate. Drawing from the technologies of the logging and fishing industries, as well as the First Nation culture, the building incorporates extensive use of heavy timber framing, exposed to the interior in the public spaces. In addition, some of the rough cedar columns provided an opportunity for the local Haida artists to incorporate their work.

The site on which the school is located is topographically quite varied, with a blanket of second growth forests and marshes, traversed by a significant watercourse and with spectacular views over Skidegate Inlet in the distance. The form and orientation of the building takes advantage of these natural site characteristics, minimizing major disruptions to the landscape, while reinforcing the varied natural wildlife habitats as educational opportunities for the students. The angled wings of the building also provided for age segregated outdoor play areas protected from the prevailing winds. Disrupted areas have been replanted with indigenous plant materials.

Situated in a region of high precipitation, the control of water on the site became a significant element in the design. The path of the existing watercourse with its rich ecosystem was reinforced by the site development, with all surface runoff returned to the landscape by natural site grading via a detention pond. Roof water falls to the ground into a series of exposed basins, troughs and gravel trenches and is also returned naturally to the landscape. The occupants are often in direct visual contact with these water flows.

Since an elementary school usually provides the child's first exposure to a public institution, the building was planned in two intersecting blocks to provide a clear distinction between the communal more public spaces and the smaller, more residential scale classroom spaces. The blocks are skewed slightly to accommodate the constraints of the site: the watercourse, topographic conditions, forested areas and views. Ancillary functions are housed in smaller scale volumes attached to these larger blocks.

1

Circulation spaces were viewed as potential areas for small group instruction, so naturally lit alcoves with benches are provided at various locations, in particular in the classroom corridor with its exposed "vertebral" roof structure, and in the larger assembly space below the pyramidal skylight. Students, as well as Elders from the community, are encouraged to gather here and make use of the resources of the adjacent Haida Studies Room, in a manner reflecting the traditional teaching methods.

Because of the high latitude and overcast climate, the introduction of natural light into the spaces was an important design factor. An extended building perimeter permitted extensive glazing, and in conjunction with top lighting from skylights, introduced natural light deep into the interior spaces.

1 学校全景
2 总平面图
3 建筑构成结构

斯基德盖特小学位于斯基德盖特市，服务两个相邻的社区Queen Charlotte和Skidegate。因当地的特殊条件，包括木材的使用技术、渔业和印第安人的文化，建筑使用了沉重的木结构，并裸露于室内的公共空间。实际上，一些表面粗糙的松木柱是当地印第安人——Haida艺术家的作品。

建筑场地起伏多变：茂密的再生树林、沼泽地、横穿而过的小河和远处海湾的美丽景色。建筑的形态和朝向是综合场地因素的结果，并尽可能不破坏环境，同时也是学生良好的自然课堂。有角度的翼状物是按年龄组别的室外活动场地的标志。破坏的区域,原场地的植物被植回。

因为当地的强降雨量，场地内的雨水控制十分重要。通过设立一个停留池以汇聚地表水来加强现有水道和其丰富的生态系统。屋面雨水由地面上露天的盆汇聚，再通过暗沟回归自然生态系统。使用者可以观察到这些水的流动过程。

一般来说，小学是儿童接触的第一个公共教育建筑，因此建筑设计为两个明显的区域以提供清晰的指引——开敞的公共空间和相对较小有居住建筑尺度的教室空间。建筑物的两片的组合受场地因素的影响，包括：水道，地面条件，森林和景观因素。辅助的房间是附着于大体量建筑上的小尺度空间。

交通空间被当作可能的小型课堂设计，因此有自然光线的凹室——特别是有暴露屋脊的教室间的通道内和大一些位于金字塔形天窗下的集会空间——设有长凳。学生和社区内的长者可以聚集在此，使用相邻的Haida印第安人研究设施，并反映传统的教育方式。

由于高纬度和多变的天气，自然光线的引入十分重要：宽大的玻璃窗和天窗可以给室内充足的自然光。

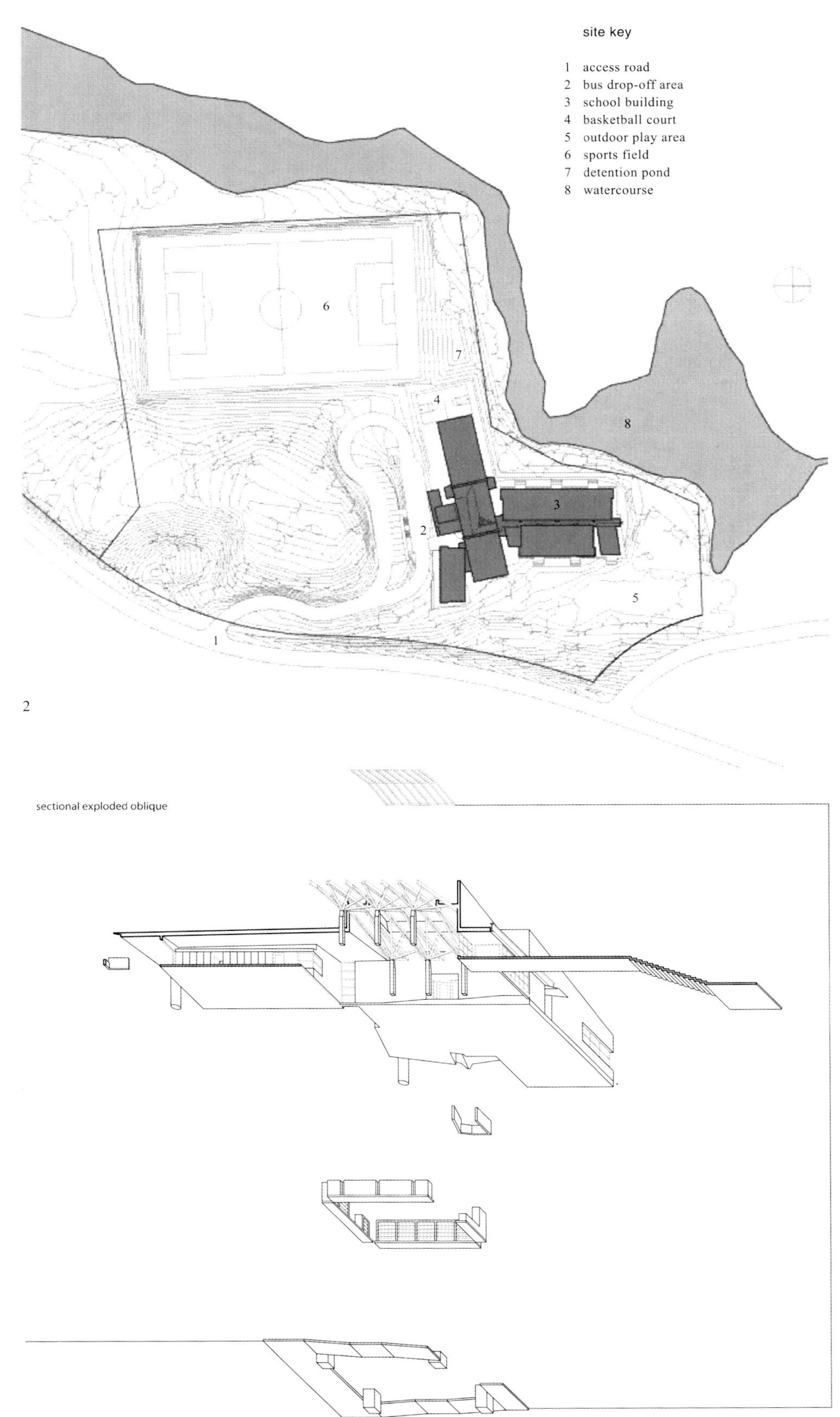

2

3

4

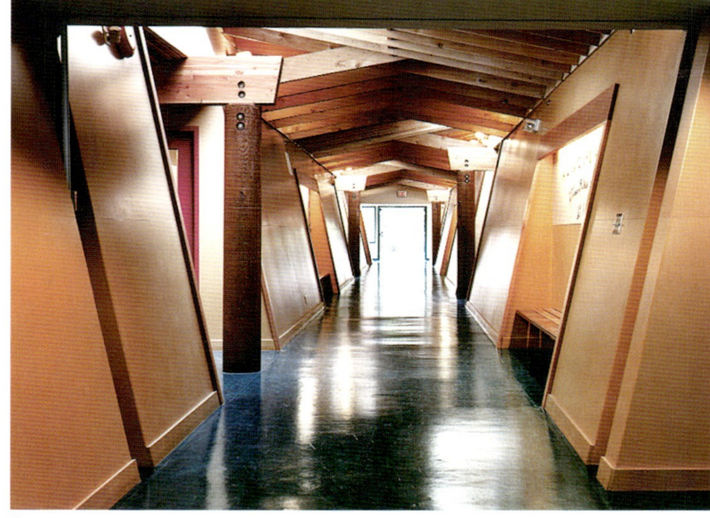

5

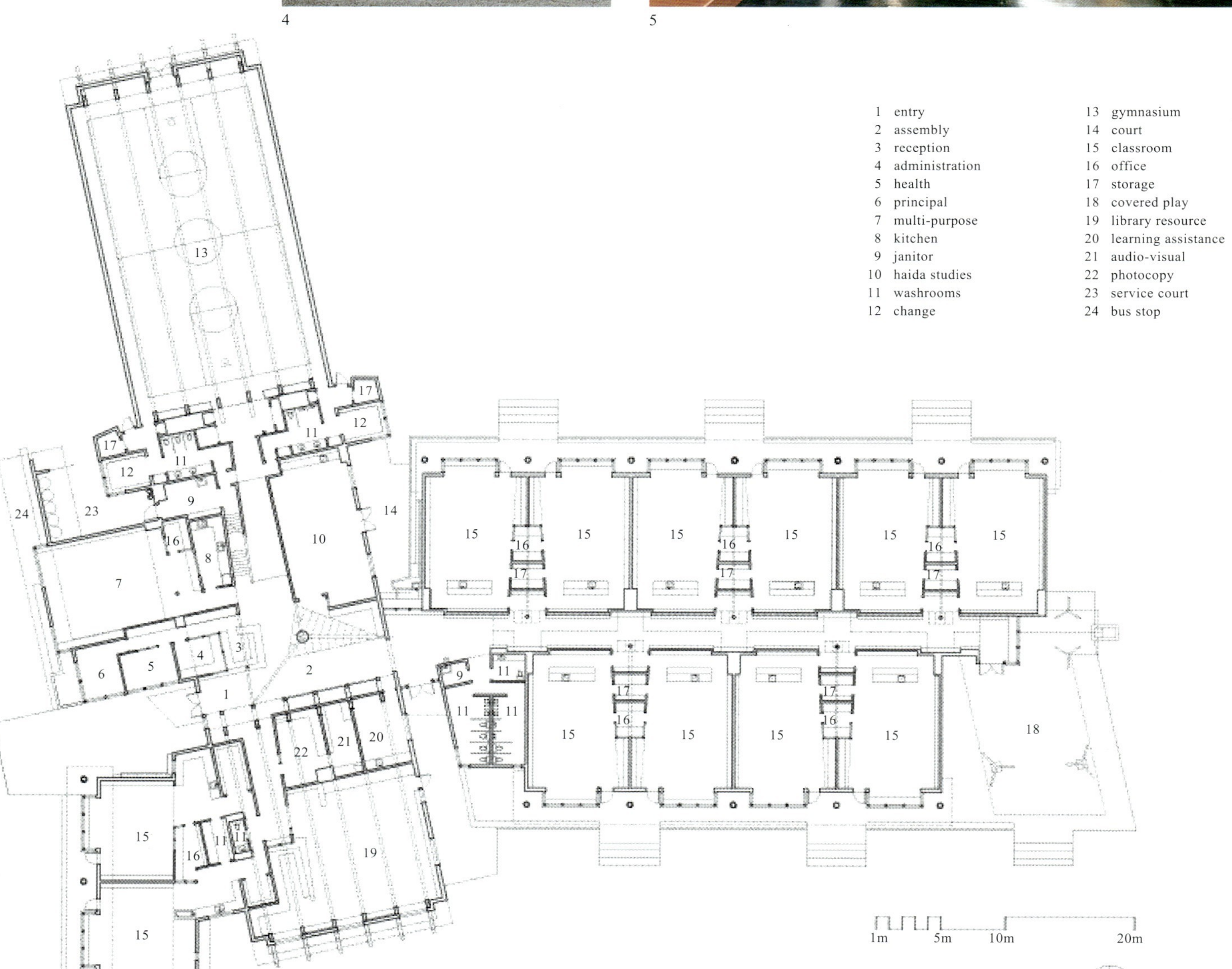
1 entry
2 assembly
3 reception
4 administration
5 health
6 principal
7 multi-purpose
8 kitchen
9 janitor
10 haida studies
11 washrooms
12 change
13 gymnasium
14 court
15 classroom
16 office
17 storage
18 covered play
19 library resource
20 learning assistance
21 audio-visual
22 photocopy
23 service court
24 bus stop
1m
5m
10m
20m

6

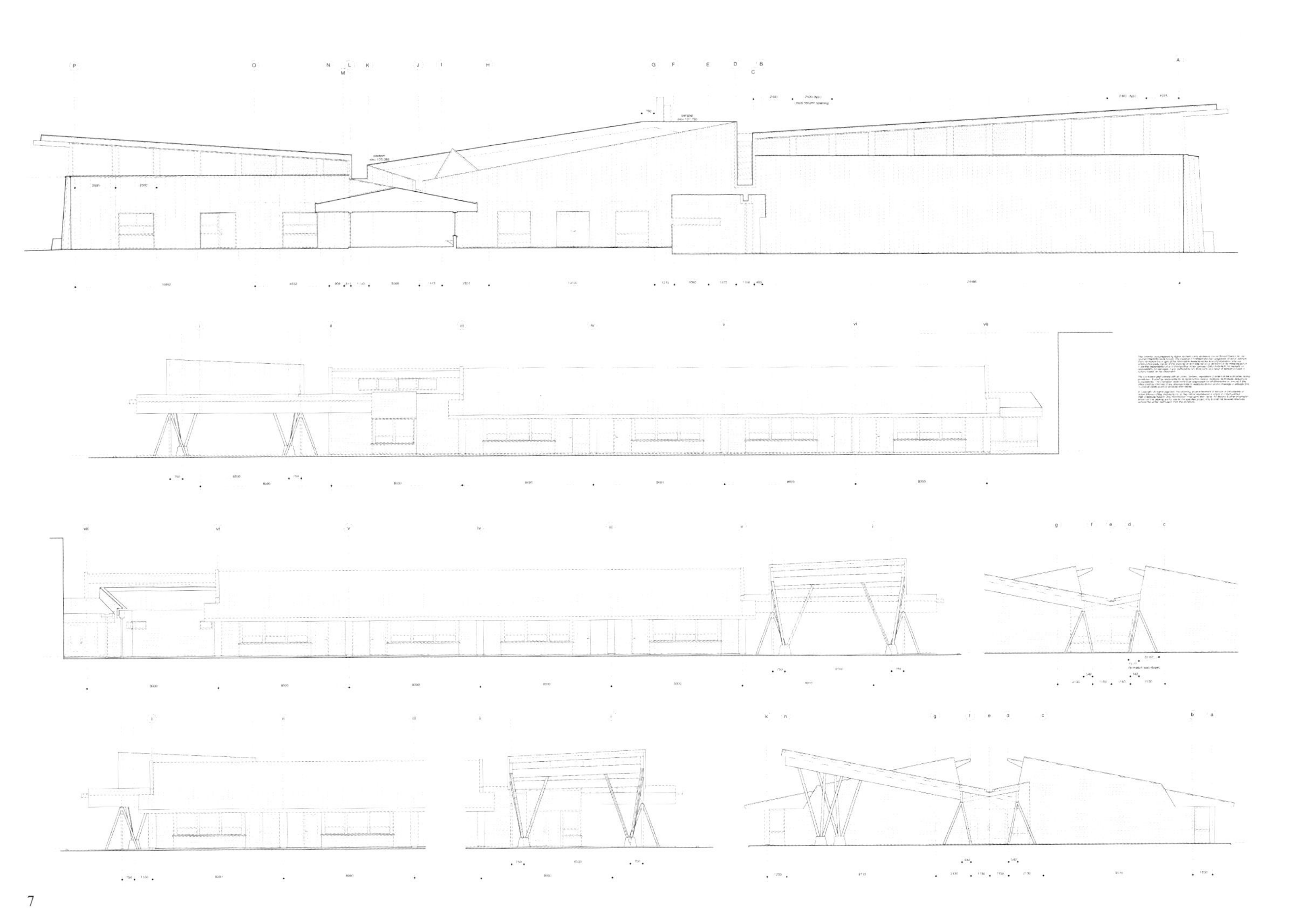

7

4　墙体细部
5　独特的走廊设计
6　平面图
7　建筑立面详图

8

9

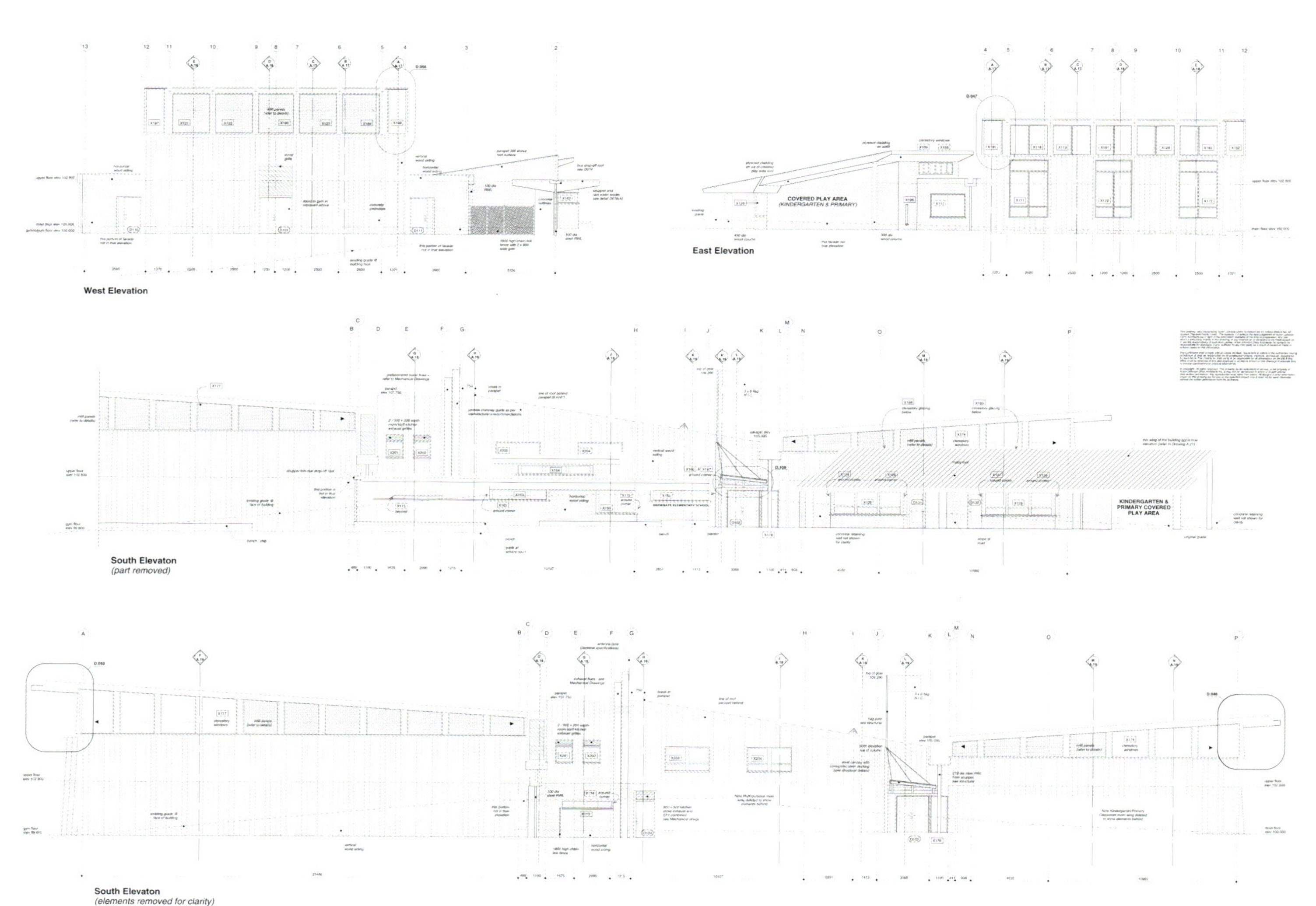
West Elevation
East Elevation
COVERED PLAY AREA
(KINDERGARTEN & PRIMARY)
KINDERGARTEN &
PRIMARY COVERED
PLAY AREA
South Elevaton
(part removed)
South Elevaton
(elements removed for clarity)

10

11

8 由室内窗户看室外建筑
9 学校院落细部
10 立面详图
11 图书馆细部
12 剖面详图

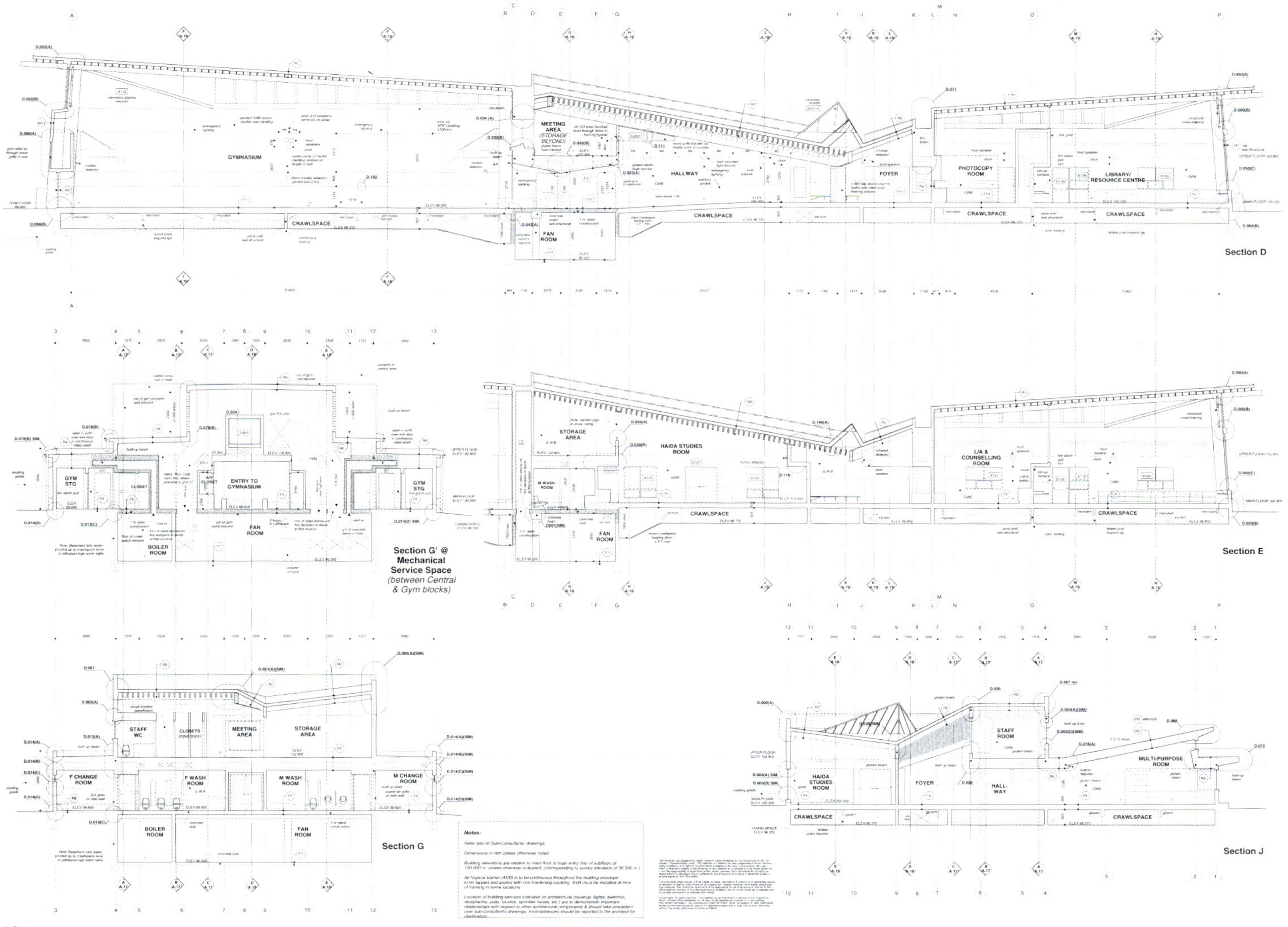

12

13 体育馆室内
14 立面详图
15 入口支架细部
16 细部
17 建筑剖面详图

13

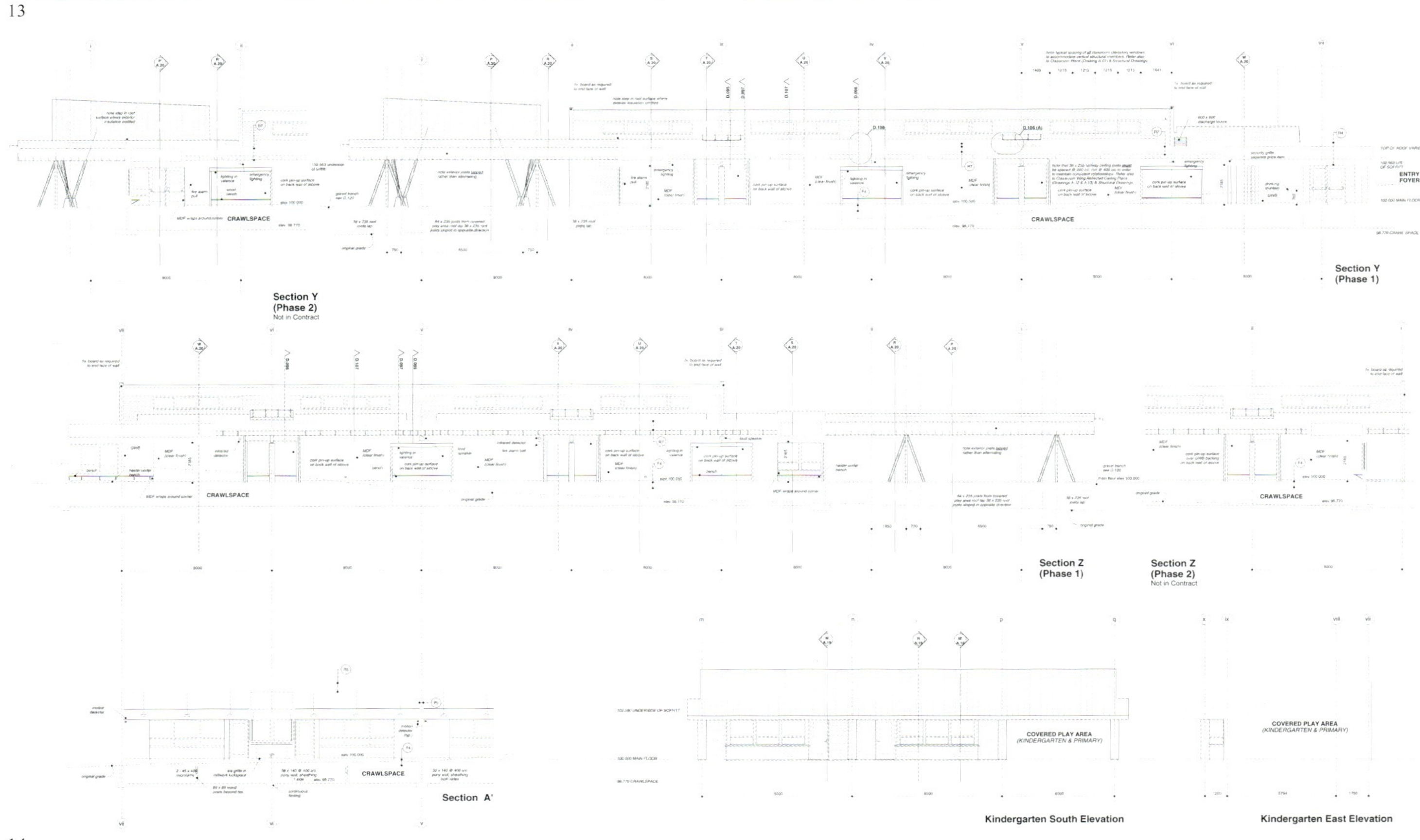

14

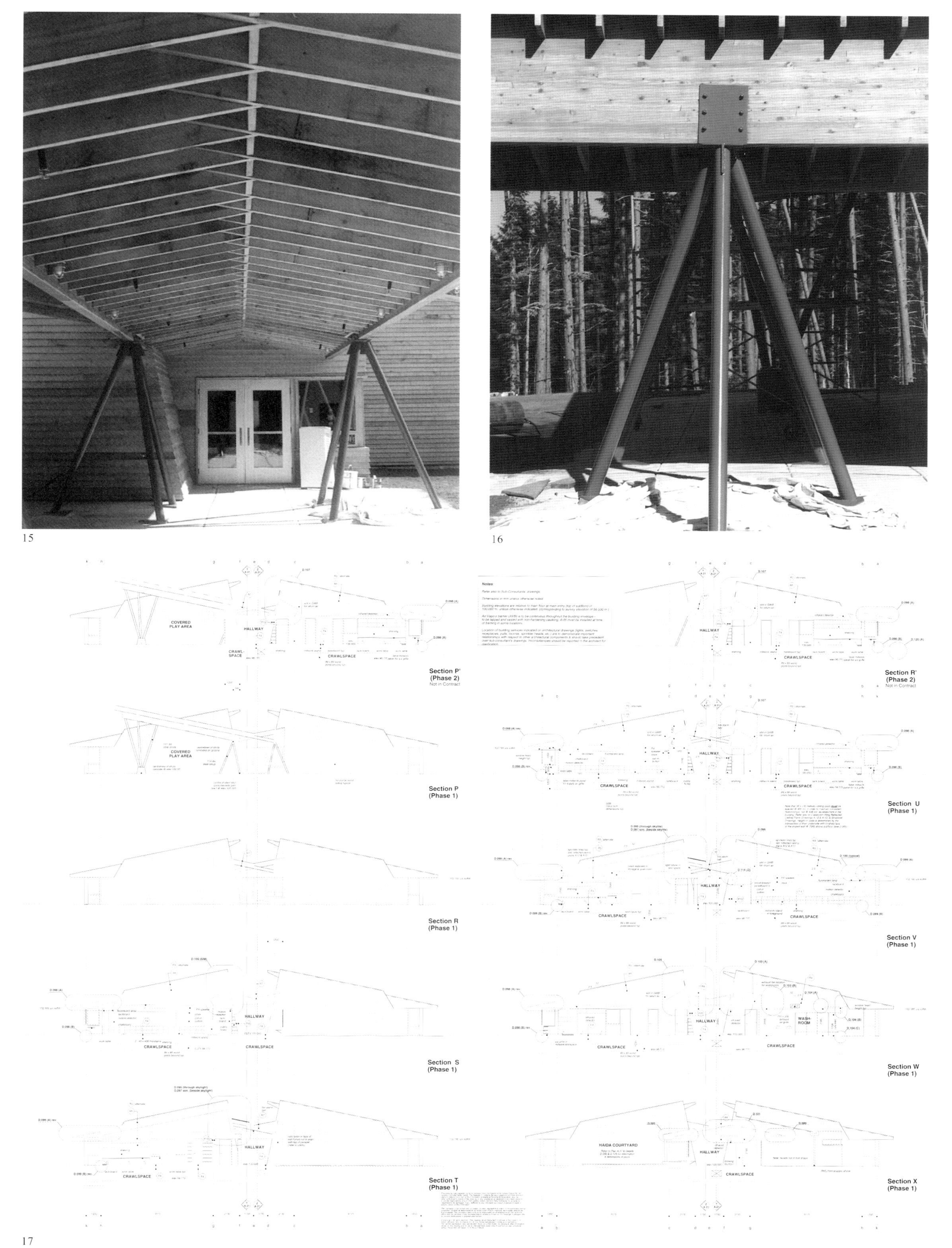
Section P'
(Phase 2)
Not in Contract
COVERED PLAY AREA
HALLWAY
CRAWLSPACE
Notes
Section R'
(Phase 2)
Not in Contract
Section P
(Phase 1)
Section U
(Phase 1)
Section R
(Phase 1)
Section V
(Phase 1)
Section S
(Phase 1)
WASH ROOM
Section W
(Phase 1)
HAIDA COURTYARD
Section T
(Phase 1)
Section X
(Phase 1)

15

16

17

Patkau Architects

帕特考建筑设计事务所

公司简介

Patkau Architects was founded by John and Patricia Patkau in Edmonton, Alberta in 1978. In 1984, the firm was relocated to Vancouver, British Columbia. Michael Cunningham became a principal in 1995. Since its founding, Patkau Architects has developed an international reputation for design excellence. Significant national and international awards have been received for a wide variety of building types, including ten Governor General's Medals, four Progressive Architecture Awards, ten Canadian Architect Awards of Excellence, and an RAIC Innovation in Architecture Award of Excellence.

Patkau Architects recently won the design competition for a major addition and renovation to the central branch of the Winnipeg Public Library. The firm has also won design competitions for the Nursing and Biomedical Sciences Facility for the University of Texas, Houston, College Housing for the University of Pennsylvania, and the Bibliothèque nationale du Québec, a new central library for the province of Québec. The work of Patkau Architects has been published and exhibited widely. Over 200 articles in books and professional journals and three books dedicated exclusively to the firm's work have been published. The work has also been exhibited in numerous exhibitions, including 20 solo exhibitions, in Canada, the United States, and Europe. In 1996, Patkau Architects was selected to represent Canada at the Venice Biennale.

In over 25 years of practice, both in Canada and in the United States, Patkau Architects has been responsible for the design of a wide variety of building types for a diverse range of clients. Projects have varied in scale from gallery installations to urban planning, and have included private residences, libraries, art galleries, schools, and university buildings. Many projects have involved facilities programming, management of detailed public processes, and design of complex buildings and sites. Comprehensive involvement throughout all phases of the design and construction process has consistently resulted in award-winning projects.

Patkau Architects has also been involved in research projects including an extensive investigation into issues of sustainable building and a detailed study of emerging educational technologies for the University of Texas Houston Health Science Center, as well as a student housing feasibility study for the University of Pennsylvania.

In addition to practice, the firm is also active in architectural education. John and Patricia have taught, lectured and been guest critics at a numerous universities in Canada, the United States, and Europe. Patricia is presently Professor of Architecture at the University of British Columbia. In 1995, Patricia and John were jointly Eliot Noyes Professor of Architecture at the Graduate School of Design, Harvard University.

帕特考建筑设计事务所由约翰·帕特考和帕特里夏·帕特考于1978年在阿尔伯达省的埃德蒙顿市创立，并于1984年迁至温哥华。Michael Cunningham于1995年成为总建筑师。从创立之日起，帕特考建筑设计事务所赢得了国际化的声誉。帕特考建筑设计事务所以不同类型的建筑获得了众多的奖项，包括：加拿大总督大奖，四次最先进建筑师奖，十次加拿大最佳建筑事务所奖，和一次加拿大皇家建筑师最佳创新奖。

最近，帕特考建筑设计事务所赢得温尼伯图书馆的扩建和改建工程。帕特考建筑设计事务所的设计还有：德克萨斯州立大学护理和生理学系馆，宾夕法尼亚大学住宅，魁北克省立图书馆等。其作品被广泛发表达200次，并参加美国、加拿大和欧洲多达20次的专题展览。1996年，作为加拿大的代表，帕特考建筑设计事务所参加了威尼斯建筑双年展。

在长达25年的实践中，帕特考建筑设计事务所在美国、加拿大等地参与了从展览安装到城市规划等各种类型的设计项目。帕特考建筑设计事务所的建筑经验集中在私人住宅、图书馆、艺术馆、学校、大学建筑。其中很多的项目涉及基础设施、公共咨询、细节管理、复杂的功能和场地设计。及时有效地参与设计全过程是帕特考建筑设计事务所获得奖项的原因。

帕特考建筑设计事务所同样涉及研究领域：包括可持续建筑研究、得克萨斯州立大学休斯顿健康科学中心教育科技合并细节研究和宾夕法尼亚大学学院住宅可持续性研究。

除了建筑设计工作之外，帕特考建筑设计事务所还涉及建筑领域。约翰·帕特考和帕特里夏·帕特考在美国、加拿大和欧洲的多所大学执教。现在帕特里夏·帕特考是不列颠哥伦比亚大学的建筑学院教授。1995年帕特考夫妇获得哈佛大学设计学院Eliot Noyes Professor of Architecture。

帕特考建筑设计事务所的设计工程照片由詹姆斯·道（James Dow）摄影。

Private residence
私人住宅

业　　主：私人
地　　点：温哥华
规　　模：285m^2
完工时间：2000年
摄 影 师：James Dow, Paul Warchol, Undine Prohl

The private residence is for a single person. The program includes typical living spaces, a single bedroom, a study, a music room and a lap pool.

The site is a small waterfront property, 33 feet wide by 155 feet deep, looking across English Bay to the North Shore Mountains, which dominate the skyline of Vancouver. Required side yard setbacks result in a plan, which is limited to 26.4 feet in width.

The house is organized simply with living spaces on grade, private spaces above grade and music room below grade. The dimensions of the site make it difficult to locate the lap pool on grade while retaining generous living spaces. Consequently the lap pool is located above grade, along the west side of the house, connected at either end to terraces associated with bedroom and study. Within the small narrow floor plates spatial expansion is only possible outward over the water and upward through the volume of the house. Small spaces are enlarged by virtue of generous ceiling heights, while the fully interiorized dining room rises through the floor above to a clerestory, made possible by the presence of the lap pool on the west side of the house, to bring both daylight, and light reflected from the pool, deep into the central area of the plan.

Vancouver is located in an area of high seismic risk. In this context, with the lap pool located above grade, a robust structure is required, one that is resistant to significant lateral forces. The house is constructed almost entirely of reinforced concrete as a result. Within this structural concrete shell the interior is insulated and clad with painted gypsum board. In areas where insulation is not required the concrete structure is exposed.

1

私人住宅是一个单身住宅建筑，包括：典型的起居室、一个卧室、一个学校空间、一个音乐室和一个戏水池。

住宅的用地是一个小的水边场地（10m × 47m），可以看见温哥华的主要景观——英吉利海湾和北岸山脉。因侧花园后退红线的需要，建筑物宽8m。

住宅的起居空间简单地安排于一层，私密空间位于上层，音乐室位于下层。由于场地的限制，戏水池和起居空间不能位于一层，因此，戏水池被设置于一层之上，建筑的西向末端与花园，卧室和学习间相连。因为用地狭窄，建筑只有向水面和上空伸展。室内的餐室抬高，高窗使西面的戏水池带来的日光和水池反射的光线可以进入室内的最深处。加之高大优雅屋顶，使狭小的空间显得十分宽大。

温哥华是地震高危险区，加之屋顶的戏水池，必须有一个结实的结构承担横向荷载。住宅由加强混凝土构成。室内的隔绝和装饰材料为粉刷的石膏板，而不需要保暖部分的混凝土则暴露出来。

2

1 住宅全景
2 住宅背立面
3 二层平面图
4 首层平面图
5 建筑模型
6 地下层平面图

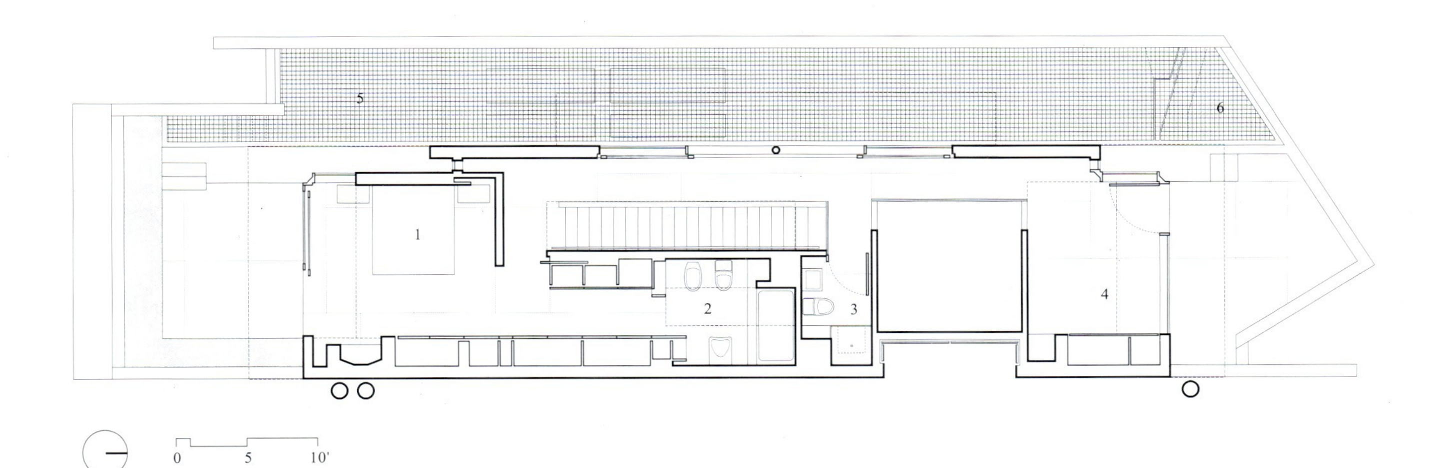

Upper Floor Plan
1 Bedroom
2 Bathroom
3 Guest Bathroom
4 Study
5 Lap Pool
6 Hot Pool

3

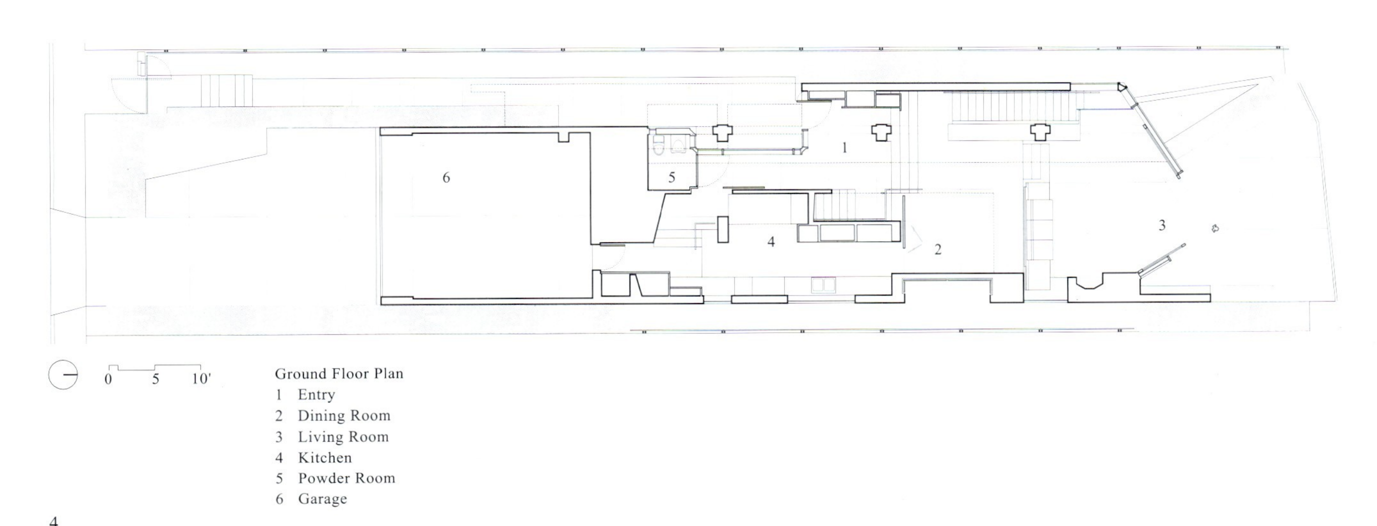

Ground Floor Plan
1 Entry
2 Dining Room
3 Living Room
4 Kitchen
5 Powder Room
6 Garage

4

5

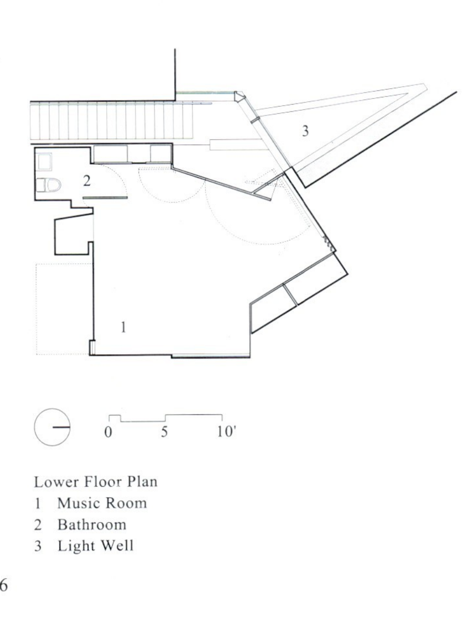

Lower Floor Plan
1 Music Room
2 Bathroom
3 Light Well

6

7 墙体细部
8 沿街立面
9 水池细部
10 水池细部
11 水池及独特的室内光线效果
12~14 剖面图
15~17 住宅室内细部

8

7

9

10

11

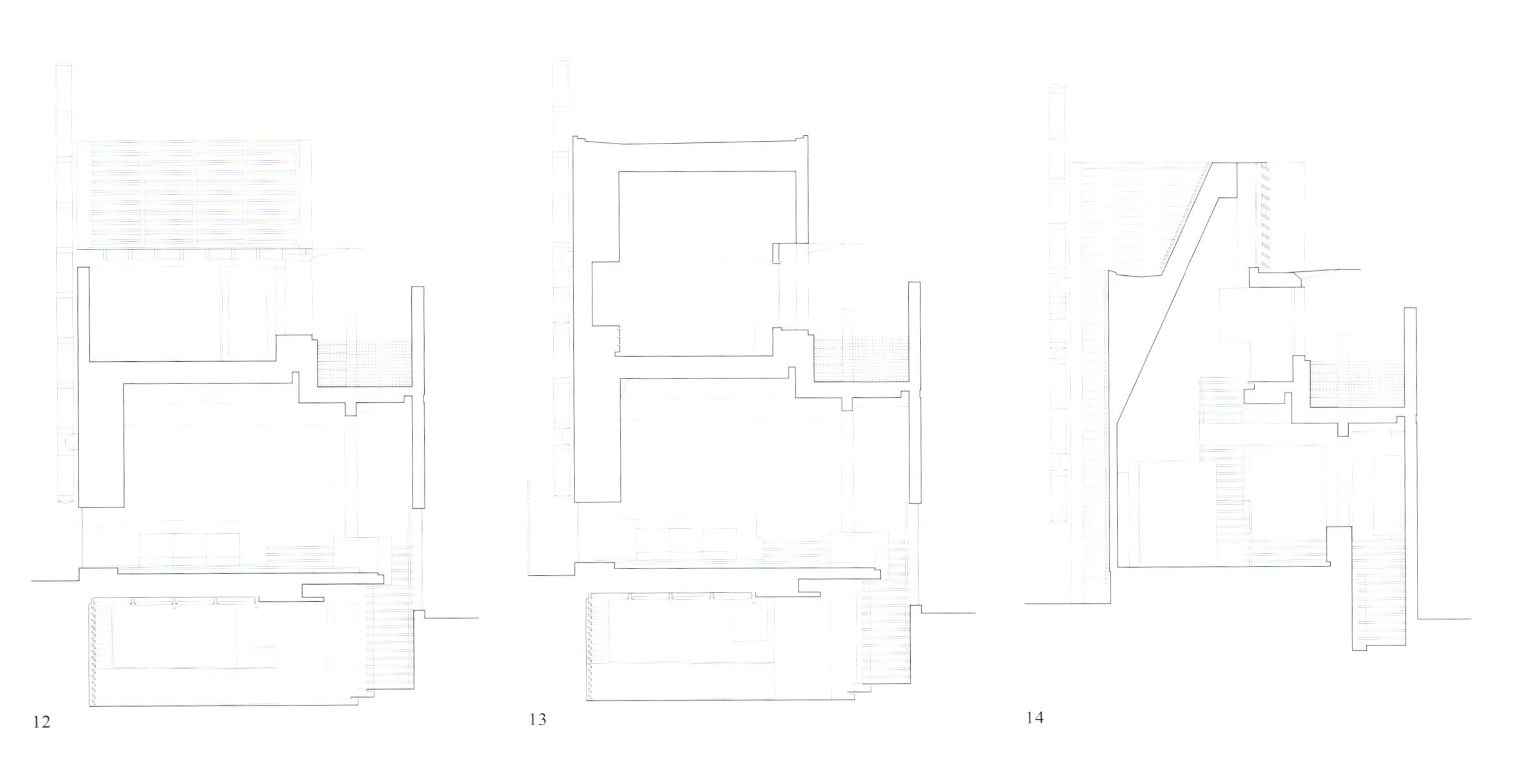

12　　13　　14

15

16

17

18~20 住宅室内细部
21~23 剖面图

18

19

20

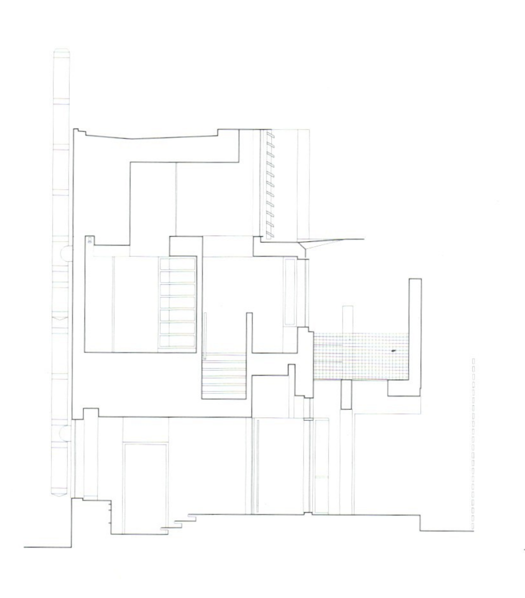
21

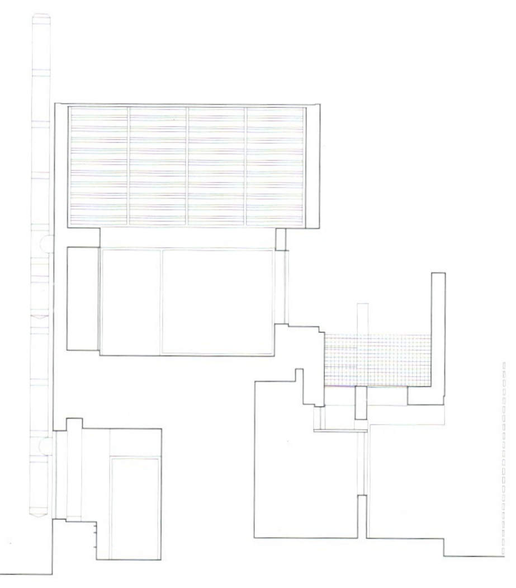
22

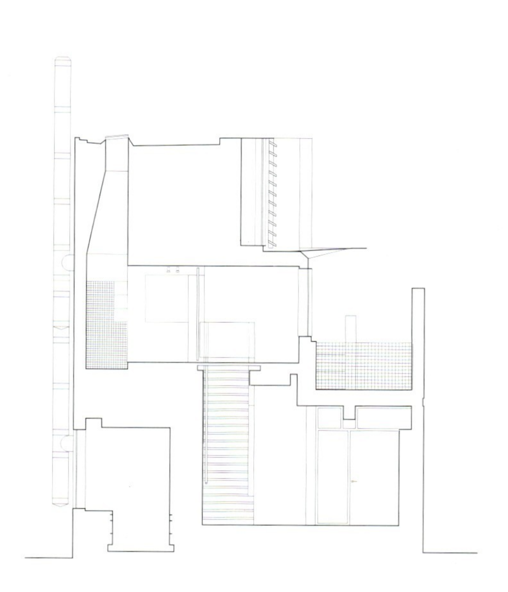
23

Ian MacDonald Architect Inc.

Ian MacDonald 建筑设计事务所

The firm of Ian MacDonald Architect Inc. was established in 1984. Their work includes projects of varied scale and mixed use with an emphasis on residential work. The focus of their work has been on developing legible, sitespecific architecture that articulates a clear idea and sense of place. What contribute to this focus are responsive sitting, spatial richness, and fine detail of expressive materials - all within a well-crafted, modern vocabulary.

Ian MacDonald 建筑设计事务所成立于1984年，其建筑作品主要是不同规模和用途的居住建筑。他们擅长创造适合不同场所的建筑物，清晰地反映用地的特殊性。所有精美的当代建筑设计手法——适宜的场地设计、丰富的空间和建筑材料的精美细节——都是Ian MacDonald 建筑设计事务在每一个设计项目里追求的目标。

House In Erin Township

Erin镇住宅

业主：私人
地点：Erin镇

The design challenge of this country retreat for two Toronto-based urban professionals: how to create a place with a strong connection to the land and the sense of the 10-acre site as its own world, in an area only 45 minutes from downtown Toronto with suburban encroachment all around.

The sitting strategy became particularly important in realizing this goal. Design began with a careful reading of the site for potential locations. Rather than choose the obvious vantage point up on the hill overlooking the landscape, yet vulnerable to "view pollution" from unpredictable future sprawl, architects chose to embed the house in the tree row adjacent to the road. This decision was the least invasive from an environmental point of view, minimizing the length of driveway. But more important, it also allowed the views from the house to focus on the undulating landscape of gentle hills and a wetland pond (which had originally drawn the couple to the site) and to ensure that whatever development might happen around them, the sense of retreat and views of landscape from within the house would not be jeopardized.

The virtues of this house lie in its overall modesty and lightness of sitting. Its design highlights the things architects feel matter in the making of good architecture: responsive sitting with fine detail of expressive materials in a well crafted, modern vocabulary, spatial richness within a modest program and budget, and the optimization of limited client budget on elements that matter most. Architects also think of it as something of a prototype of environmentally responsive design addressing an important social issue: how to build thoughtfully on an increasingly populated urban fringe, where all too often immodest and ostentatious construction dominates the landscape in a way that is both wasteful and insensitive to the hidden beauty of the landscape.

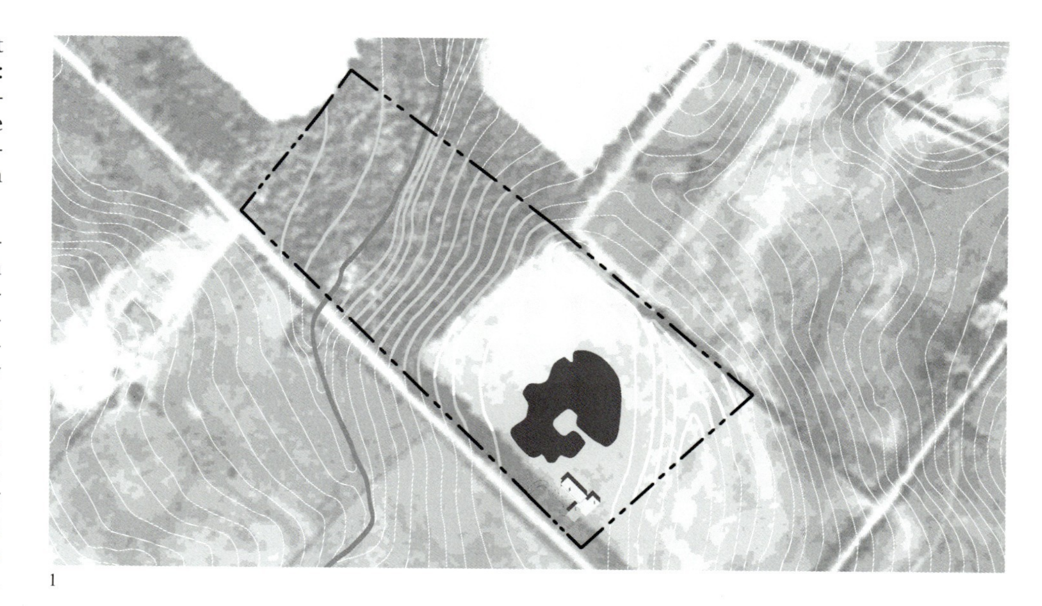

1

Erin镇住宅的业主是两个在多伦多工作的专业人士。住宅用地面积4hm²。Erin镇为距离多伦多市区45km的城市郊区。设计的挑战性在于如何寻求一种建筑形式以适应其用地的独特性。

总平面设计对于设计目标的达成尤为重要。设计以研究场地的各种可能性为开端。建筑师选择住宅坐落于林荫道旁，而不是选址于用地最显要的部分。其原因是避免未来城市扩张对住宅的"景观污染"，对场地内的环境最小的冲击和最短的行车道。但是最重要的是保留了从住宅内的面向山坡和湿地的景观（这正是最初吸引业主夫妇的场地特点），不论将来周围建造任何的建筑物，此住宅独有的景观不会受到威胁。

住宅的设计以整体的气度和轻松愉悦的姿态与场地结合。建筑师专注于理想中完美设计的最主要的部分：精致的建筑材料的使用（以反映现场地特色）、现代的建筑因素、满足现代生活和造价的丰富空间、在有限造价下，业主对建筑最重要部分的优先性策略。建筑师同样关注环境设计的原型——一个重要的社会问题：如何在人口增长的城市边缘开发建造，而不是以粗俗的和虚伪的建筑物肆意浪费景观资源和掩盖自然景观的美丽。

2

1 总平面图
2 由沼泽地远望住宅
3 南、北立面图
4 东、西立面图

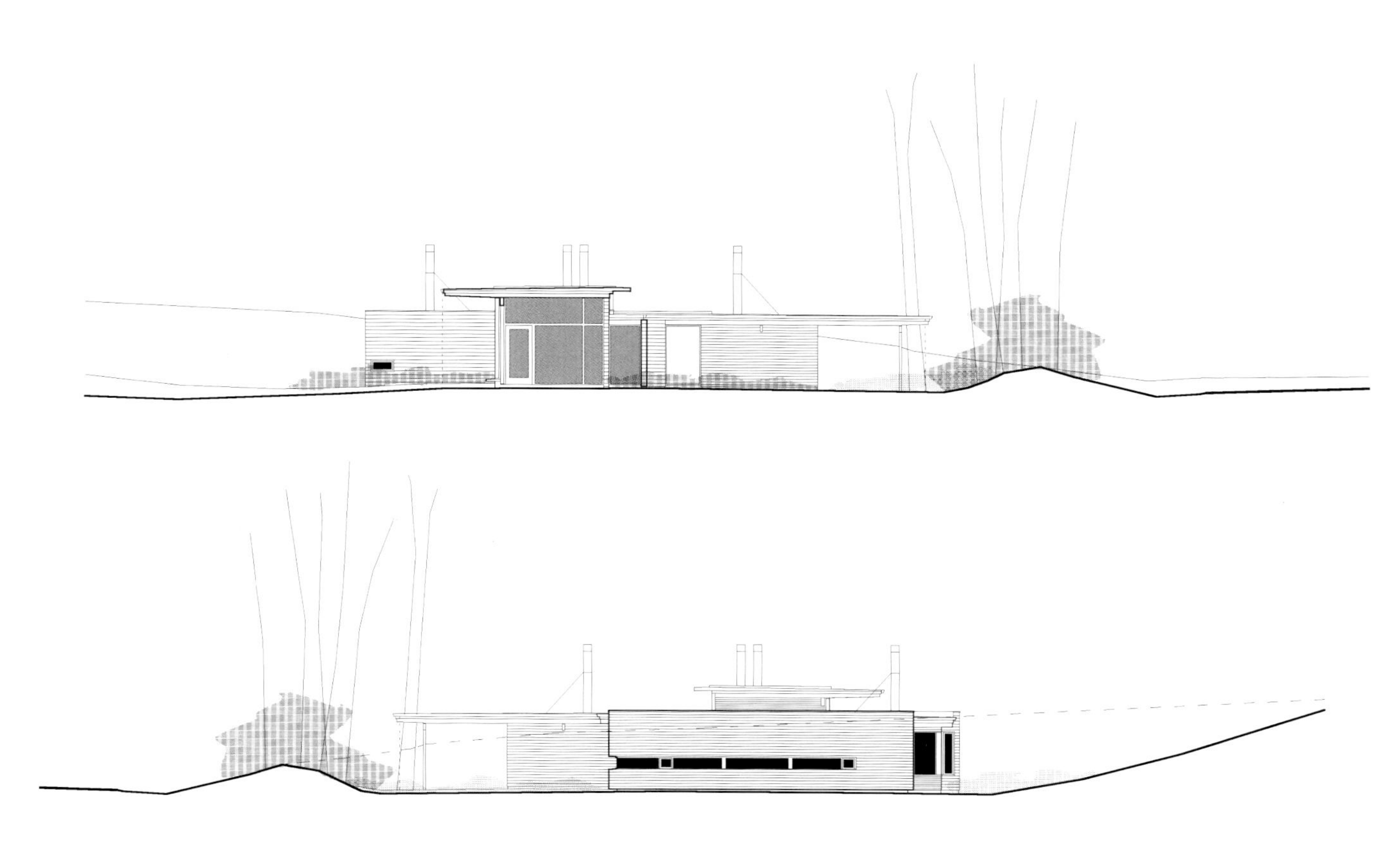

3

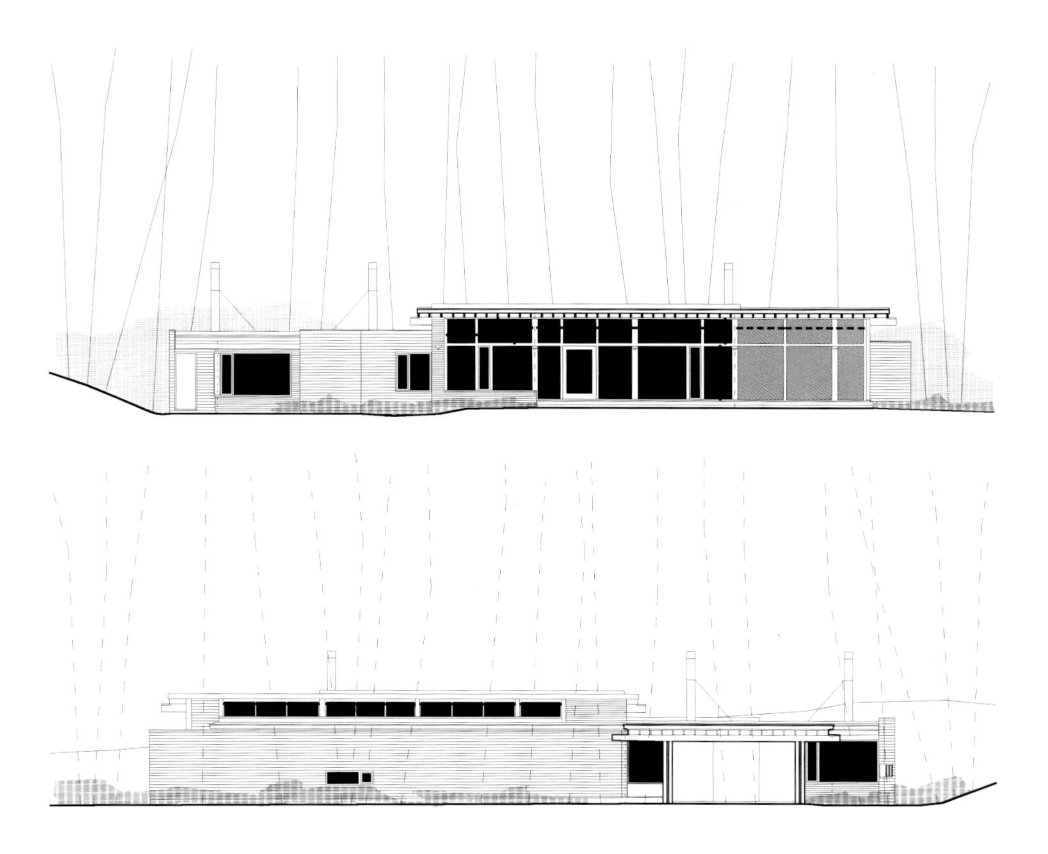

4

5

6

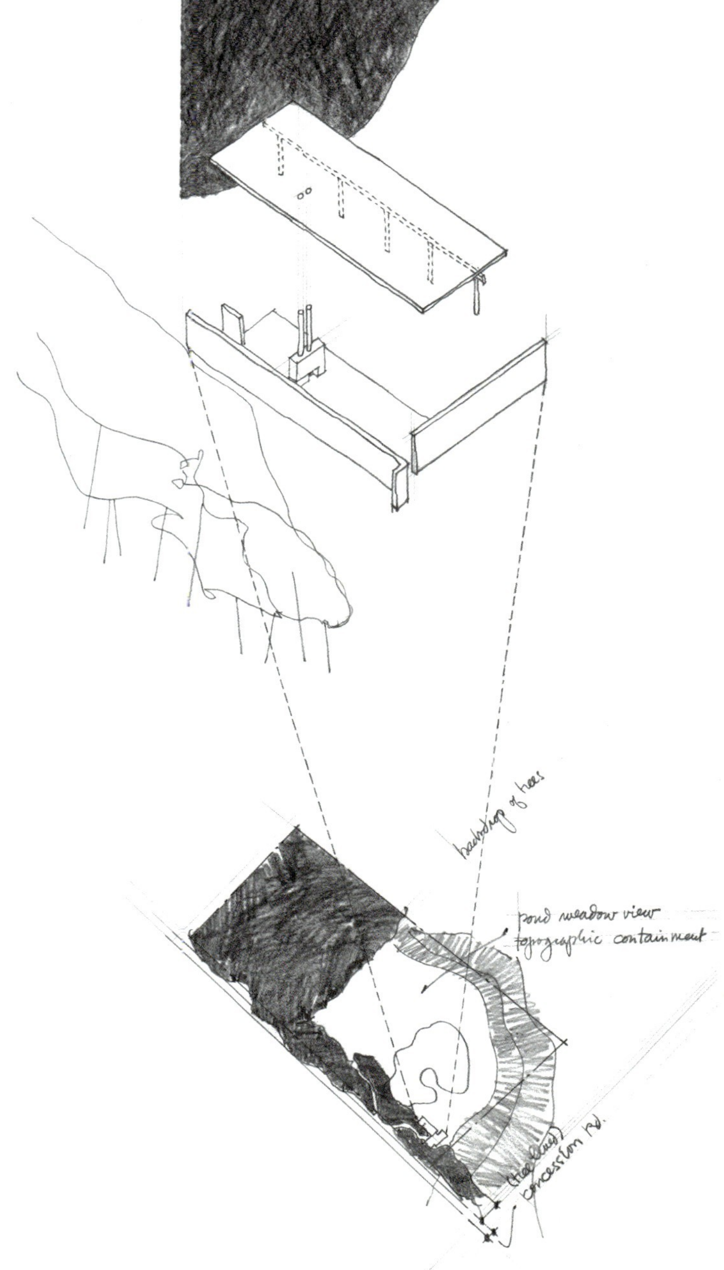

7

8

5 住宅黄昏景色
6 住宅夜景
7 总平面构思草图
8 由公路远望住宅
9 二层平面图
10 细部剖面图
11 走廊
12 室内外景观由大玻璃窗融为一体
13 厨房和餐厅

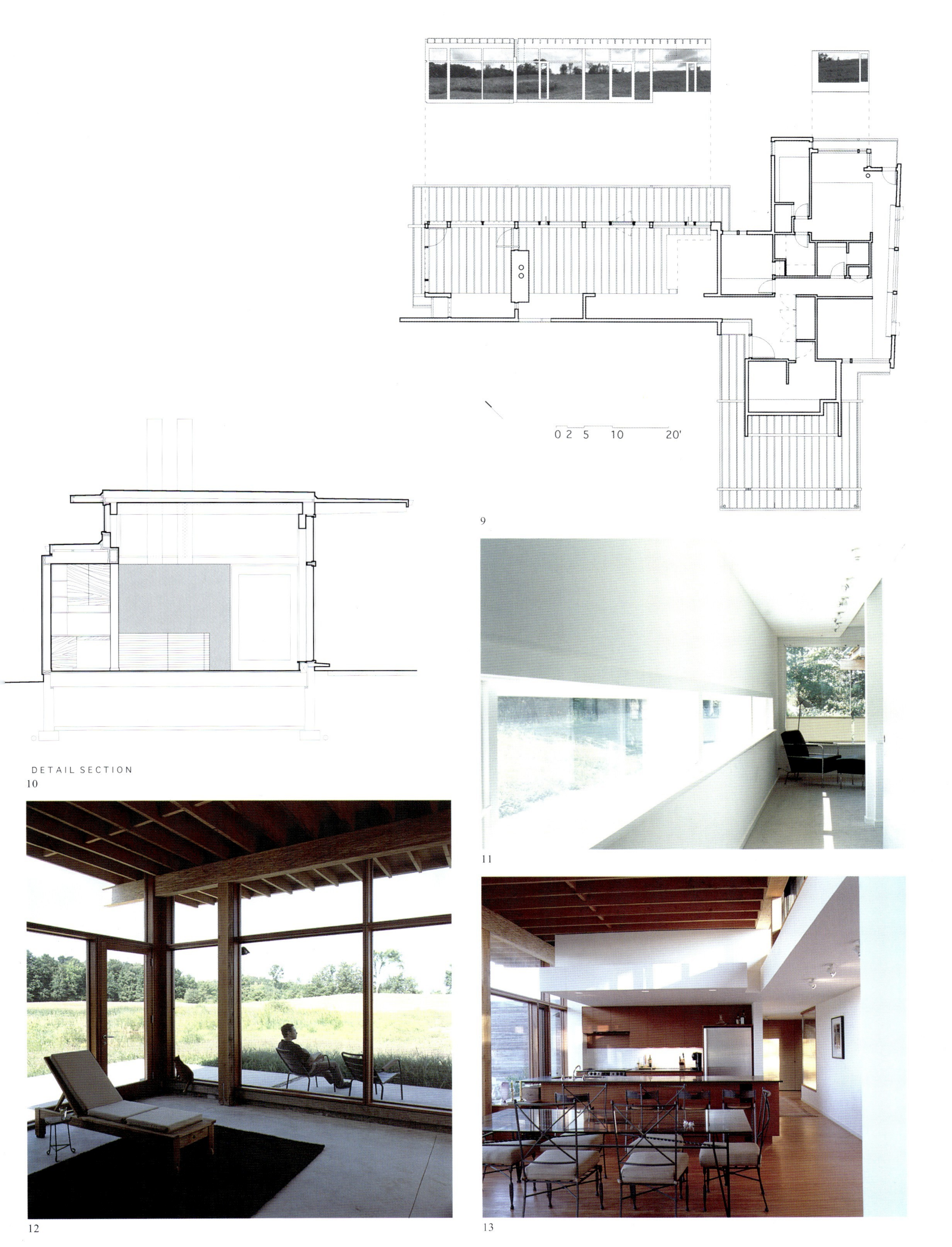

9

10

11

12

13

14　首层平面图
15　室内
16　室内

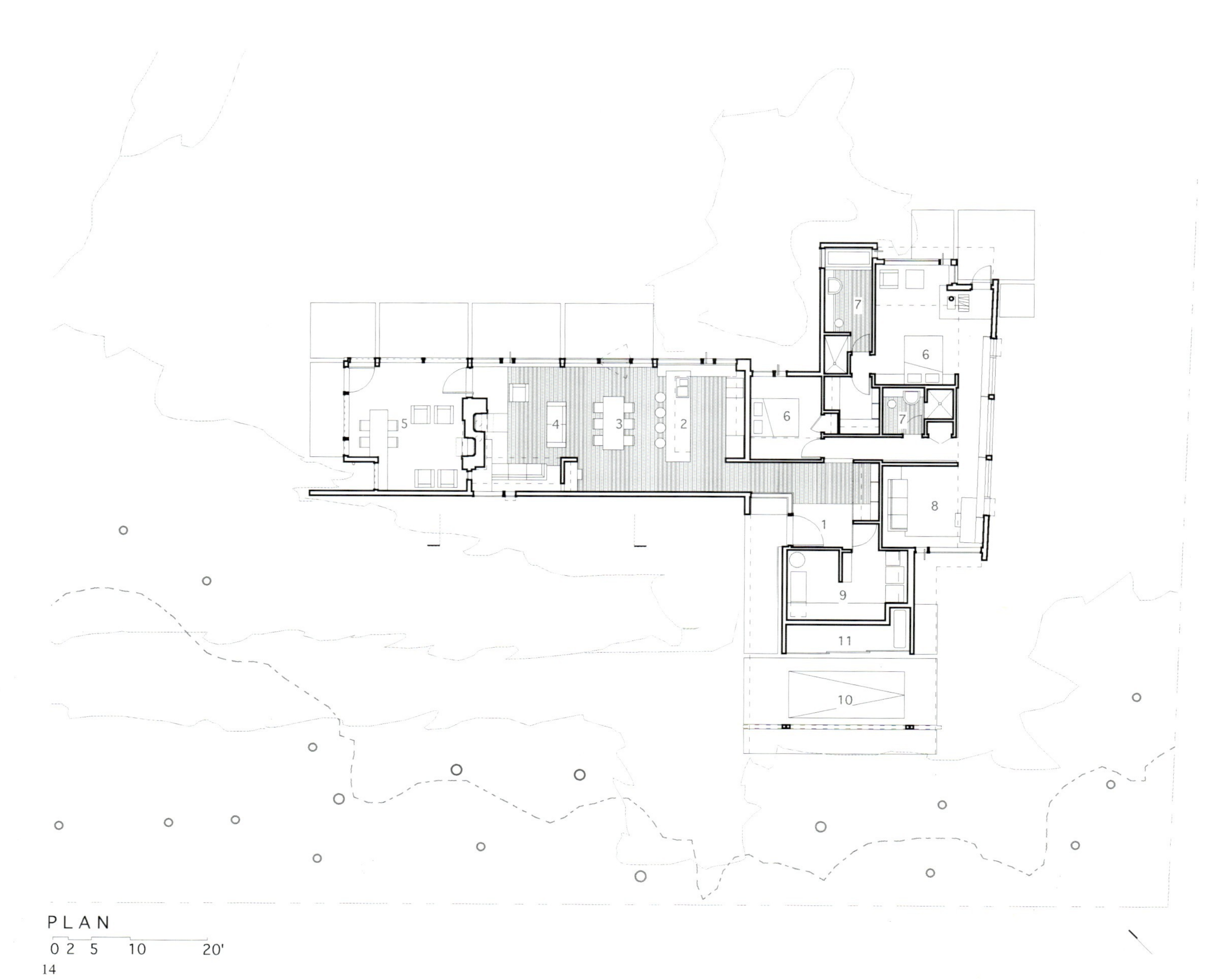

14

15

16

Nicolson Tamaki Architects Inc.

Nicolson Tamaki 建筑设计事务所

As the first Canadian architect firm, which received LEED certification, Nicolson Tamaki Architects Inc. was found in 1986, and offers a board range of consultant services in architectural design and project management, urban planning, zoning studies and applications, heritage-related projects, interior design, and development studies.

作为加拿大第一个得到绿色建筑设计（LEED）资质的建筑设计事务所，Nicolson Tamaki 建筑设计事务所成立于 1986 年。Nicolson Tamaki建筑设计事务所提供广泛的设计咨询服务：建筑设计、工程管理、城市设计、用地研究、历史建筑、室内设计和开发研究。

Northlands Golf Course
Northlands 高尔夫球场

地　　点：温哥华
完成时间：1998 年

This publicly run clubhouse is set amidst the south facing slopes of Vancouver's North Shore. It is a wood building set on a concrete base and utilizes large decks and extensive glazing to create a strong link to its beautiful natural setting. Granite acquired from the road construction is used in facing along the entry façade and all exterior soffits are clad in combed-faced clear Western Red Cedar.

The building is set up to receive future photovoltaic and per-heated solar water heating. The entire building is day lit with operable windows at the clerestory to provide cross-ventilation. The color palette is natural with few accent colors.

Northlands 高尔夫球场坐落于温哥华市的北岸山脉的南向山坡之上，背对着巨大体量的北岸山脉和茂密的北岸森林。如何以相对极小体量的球场建筑取得与其巨大的用地之间的协调是整个建筑设计的关键。

木结构的建筑位于混凝土的地基之上，宽敞的室外平台为使用者提供舒适的与优雅的自然环境对话的场所。室外建筑材料为花岗岩和西海岸红杉。整个建筑设有未来太阳电能和太阳能热水系统。可开启的窗户可以提供舒适的穿堂风。

1

2

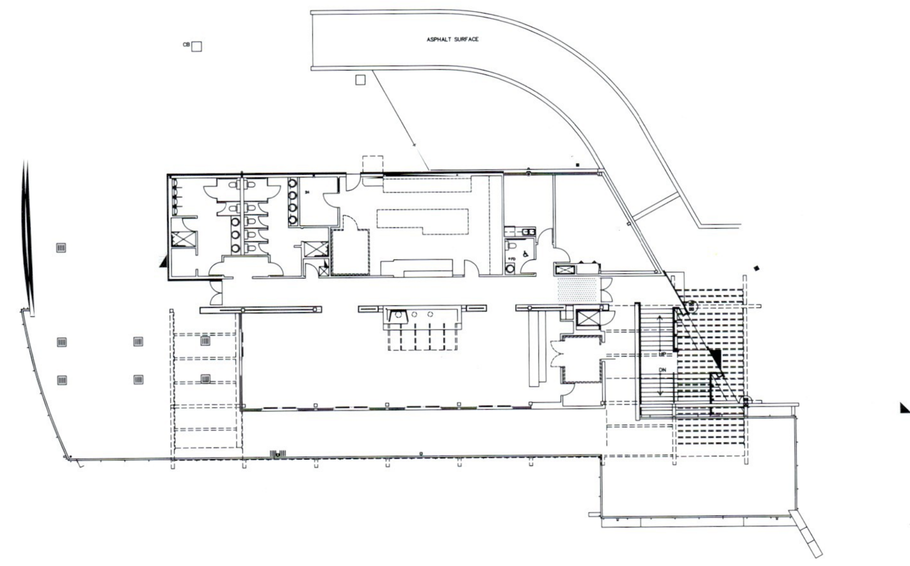

3

1 球场会所入口
2 会所外观
3 首层平面图

4

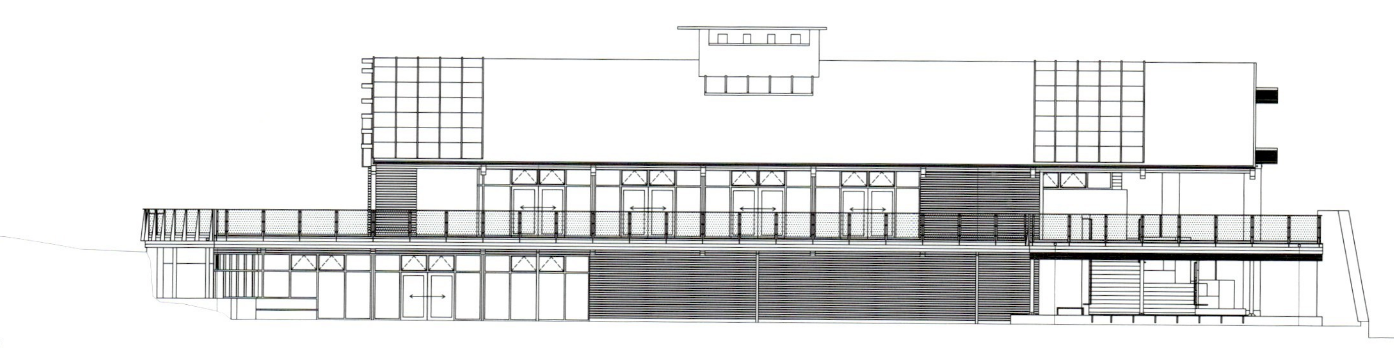

5

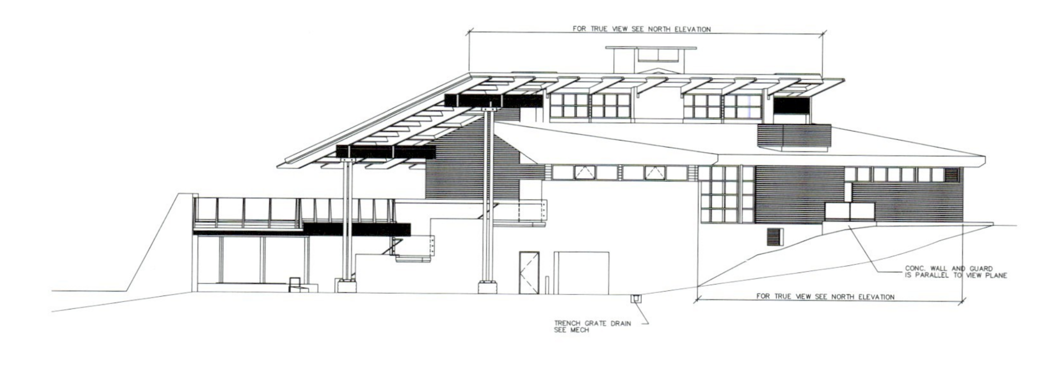
FOR TRUE VIEW SEE NORTH ELEVATION
CONC. WALL AND GUARD IS PARALLEL TO VIEW PLANE
FOR TRUE VIEW SEE NORTH ELEVATION
TRENCH GRATE DRAIN SEE MECH

6

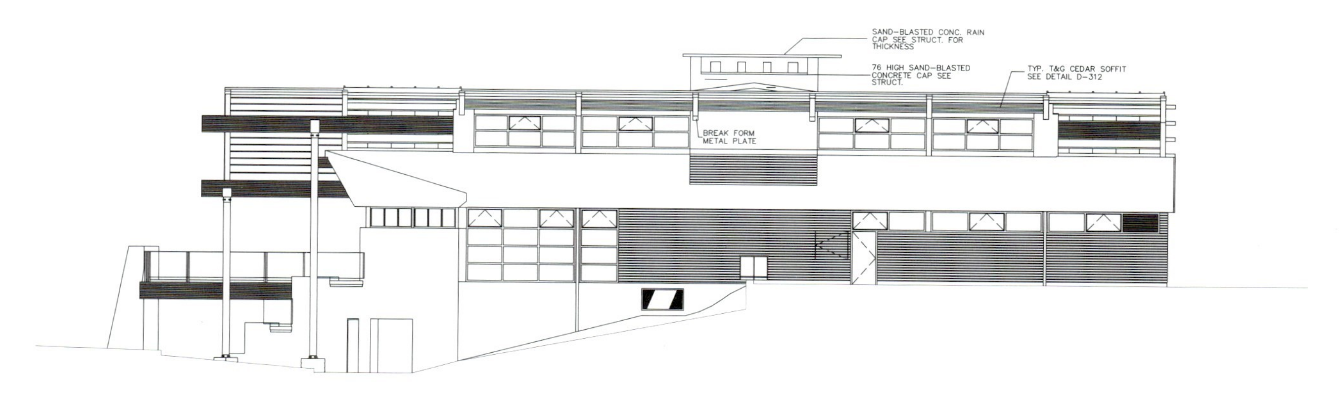
SAND-BLASTED CONC. RAIN CAP SEE STRUCT. FOR THICKNESS
76 HIGH SAND-BLASTED CONCRETE CAP SEE STRUCT.
TYP. T&G CEDAR SOFFIT SEE DETAIL D-312
BREAK FORM METAL PLATE

7

4 北面景观
5~8 立面图
9 剖面图
10~11 二层平台

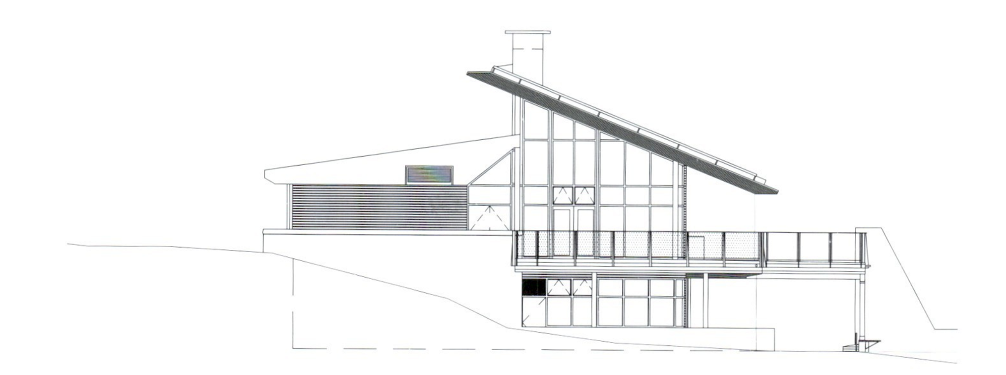

8

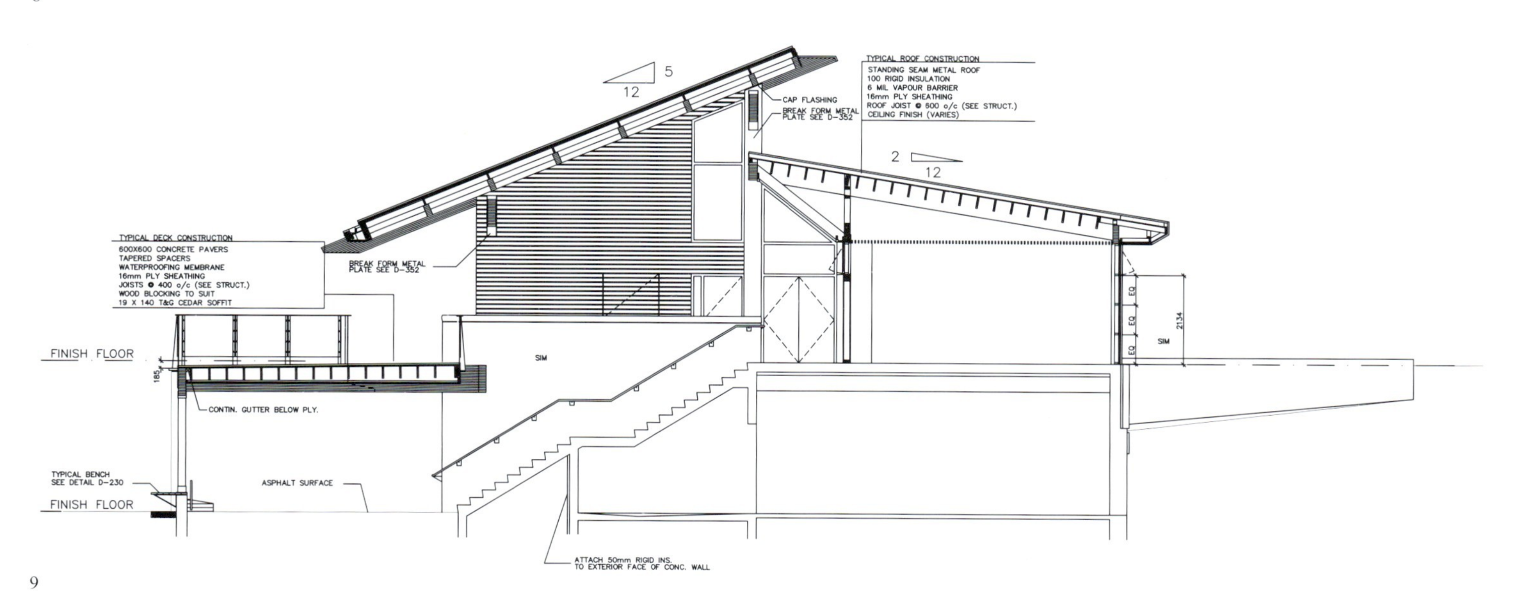

9

10

11

12~13,16　室内细部
14　剖面图
15　构造详图
17　墙体构造详图

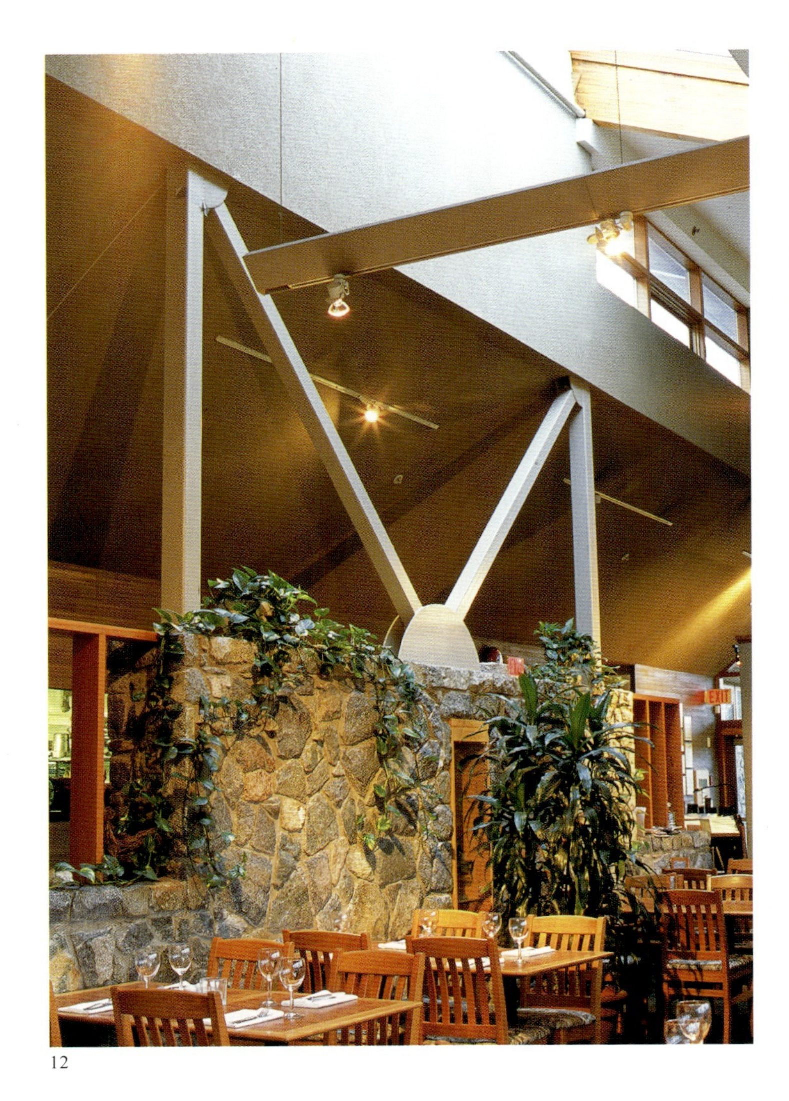

12

13

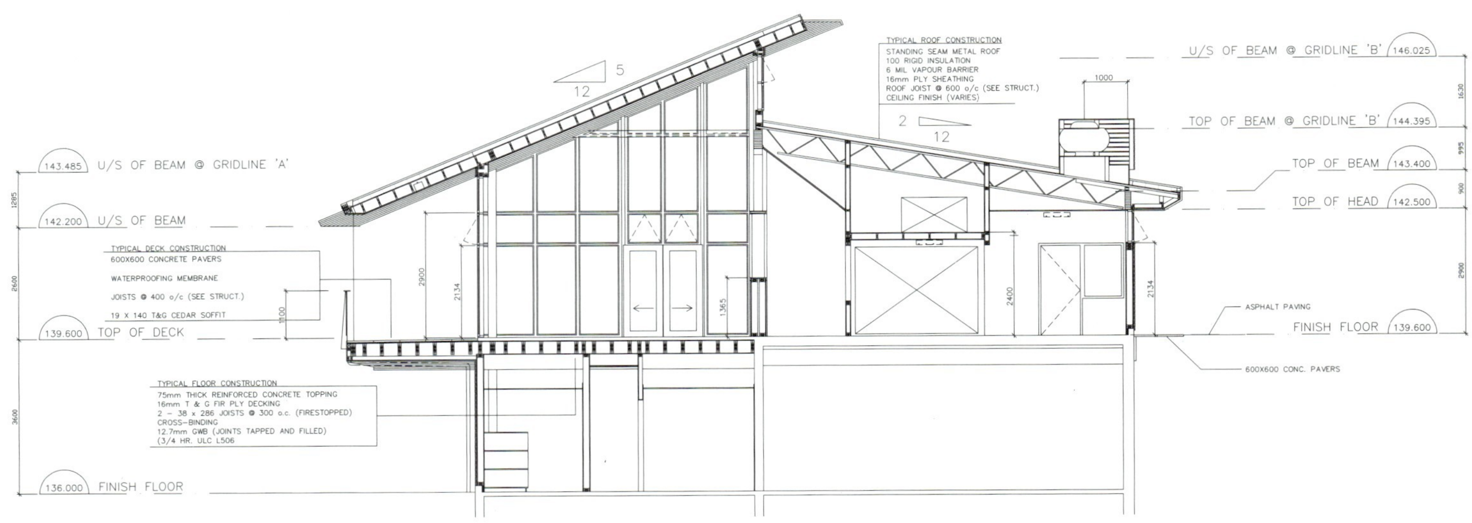

14

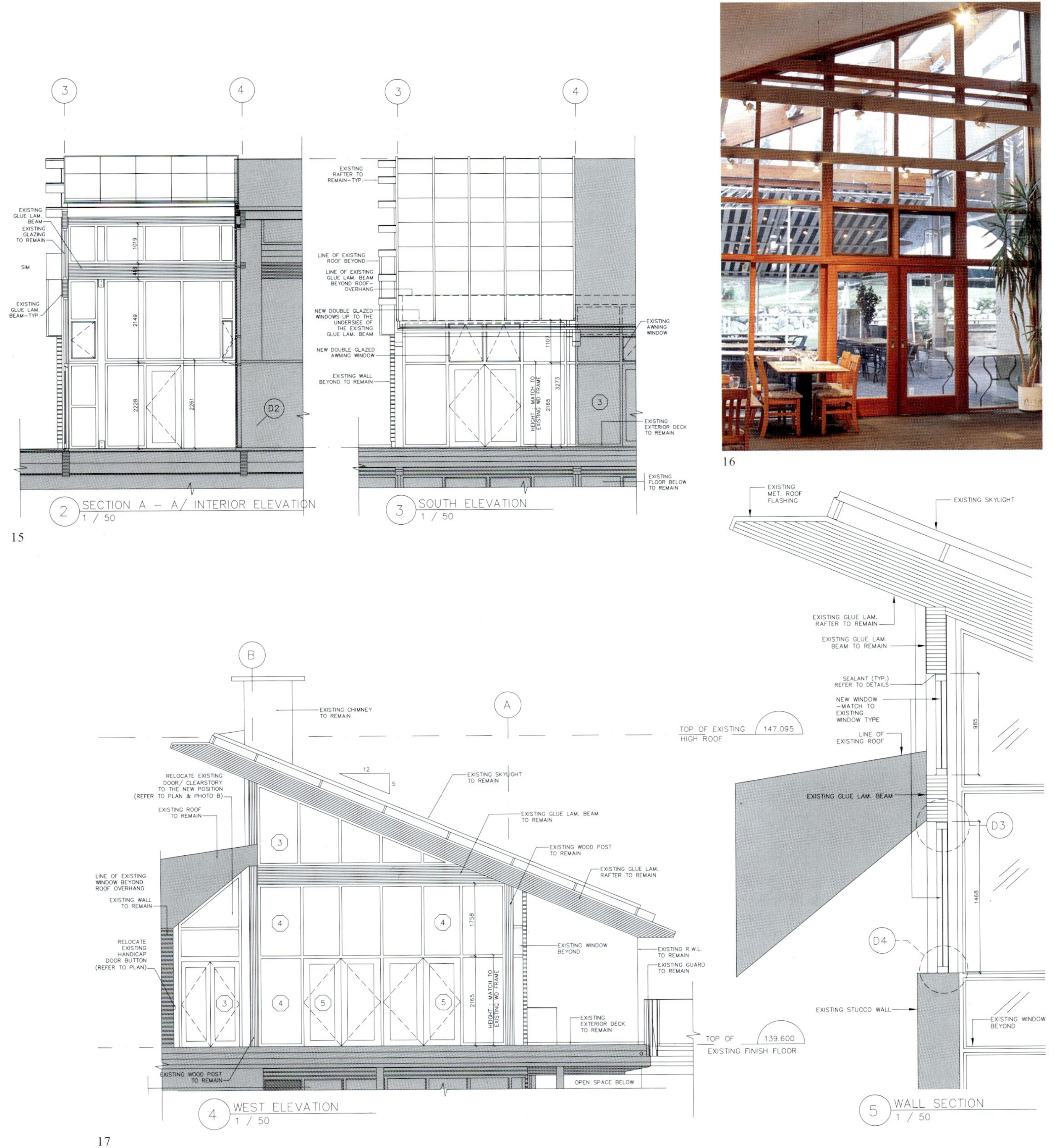

EXISTING GLUE LAM. BEAM
EXISTING GLAZING TO REMAIN
SIM
EXISTING GLUE LAM. BEAM-TYP.
D2
2 SECTION A – A/ INTERIOR ELEVATION
1 / 50
EXISTING RAFTER TO REMAIN-TYP.
LINE OF EXISTING ROOF BEYOND
LINE OF EXISTING GLUE LAM. BEAM BEYOND ROOF-OVERHANG
NEW DOUBLE GLAZED WINDOWS UP TO THE UNDERSIEE OF THE EXISTING GLUE LAM. BEAM
NEW DOUBLE GLAZED AWNING WINDOW
EXISTING WALL BEYOND TO REMAIN
HEIGHT : MATCH TO EXISTING WD FRAME
EXISTING AWNING WINDOW
EXISTING EXTERIOR DECK TO REMAIN
EXISTING FLOOR BELOW TO REMAIN
3 SOUTH ELEVATION
1 / 50
EXISTING CHIMNEY TO REMAIN
TOP OF EXISTING HIGH ROOF
147.095
RELOCATE EXISTING DOOR/ CLEARSTORY TO THE NEW POSITION (REFER TO PLAN & PHOTO B)
EXISTING ROOF TO REMAIN
EXISTING SKYLIGHT TO REMAIN
EXISTING GLUE LAM. BEAM TO REMAIN
EXISTING WOOD POST TO REMAIN
EXISTING GLUE LAM. RAFTER TO REMAIN
LINE OF EXISTING WINDOW BEYOND ROOF OVERHANG
EXISTING WALL TO REMAIN
RELOCATE EXISTING HANDICAP DOOR BUTTON (REFER TO PLAN)
EXISTING WINDOW BEYOND
EXISTING R.W.L. TO REMAIN
EXISTING GUARD TO REMAIN
EXISTING EXTERIOR DECK TO REMAIN
TOP OF EXISTING FINISH FLOOR
139.600
EXISTING WOOD POST TO REMAIN
OPEN SPACE BELOW
4 WEST ELEVATION
1 / 50
EXISTING MET. ROOF FLASHING
EXISTING SKYLIGHT
EXISTING GLUE LAM. RAFTER TO REMAIN
EXISTING GLUE LAM. BEAM TO REMAIN
SEALANT (TYP.) REFER TO DETAILS
NEW WINDOW -MATCH TO EXISTING WINDOW TYPE
LINE OF EXISTING ROOF
EXISTING GLUE LAM. BEAM
D3
D4
EXISTING STUCCO WALL
EXISTING WINDOW BEYOND
5 WALL SECTION
1 / 50

15

16

17

DGBK Architects

DGBK 建筑设计事务所

Based in Vancouver, DGBK Architects has achieved national and international recognition through the development of consistently thoughtful and sensitive solutions to complex architectural problems. Their practice focuses on building and interior design, master planning and preplanning and design for projects with large user-groups, new design requirements, challenging schedules and fixed capital and operating budgets.

DGBK Architects has practiced predominantly in British Columbia since 1972, but includes both national and international projects in countries such as; the U.S.A, China, Japan and Micronesia. R.J. Goodfellow Architects of Calgary, joined with DGBK in 2001 to form Goodfellow DGBK Architects. Partners in the firm hold professional registrations in British Columbia, Alberta and several U.S. states. DGBK's work, with the assistance of experienced and knowledgeable consultants, includes major hotels, resorts, recreation centers, health care projects, airports, schools, laboratories, corrections facilities, residential high-rise developments and ecclesiastic projects. The basis of their practice for the past thirty years has been a humanistic approach in the design of places for people, accomplished in a collaborative process with our clients.

DGBK 建筑设计事务所位于温哥华，以解决复杂的建筑问题而得到加拿大全国和国际上的认同。其工作范围包括：建筑和室内设计，总体规划和大型项目，新项目、短周期以及低预算项目的前期工作。

从1972年起，DGBK 建筑设计事务所的业务主要集中在加拿大不列颠哥伦比亚省，但是其设计项目遍布世界各地：美国、中国、日本和密克罗尼西亚。位于加拿大卡尔加里市的 R.J. Goodfellow 建筑设计事务所于2001年加入DGBK。DGBK 建筑设计事务所拥有不列颠哥伦比亚省、阿尔伯达省和美国多个州的建筑专业设计执照。

DGBK 建筑设计事务所的工程项目集中在大型酒店、度假村、休息中心、医疗建筑、机场、学校、实验室、高层居住建筑和宗教建筑。三十年来，DGBK 建筑设计事务所与业主一起设计以人为本的建筑。

Kitimat Hospital and Health Centre

Kitimat 医院及健康中心

地点：Kitimat
规模：11 000m²

Kitimat Hospital & Health Center's stated mandate is to help local communities achieve optimal physical, psychological and spiritual health. Formally, the design reflects the identity of the community it serves, and is at once familiar and provocative. Operationally, it represents a completely new approach to the provision of healthcare services in British Columbia through its integration of acute, multi-level, and community healthcare into a single comprehensive facility.

The Center's two primary program areas are expressed in complementary, but distinct, formal and material languages: Within the acute care wing, hospital services are grouped around a glazed atrium that acts as a public medical "mall" from which users have easy access to a variety of services. The dramatic proportions of this atrium, with views to the landscape beyond a double-height curtain wall and v-shaped steel columns, are reminiscent of the area's industrial heritage; By contrast, the more private, residential nature of the multi-level care wing is expressed in its smaller scale, gently sloping roofs, and extensive use of wood. These two wings are tenuously connected by a glazed walkway that is simultaneously a point of connection, and also of separation.

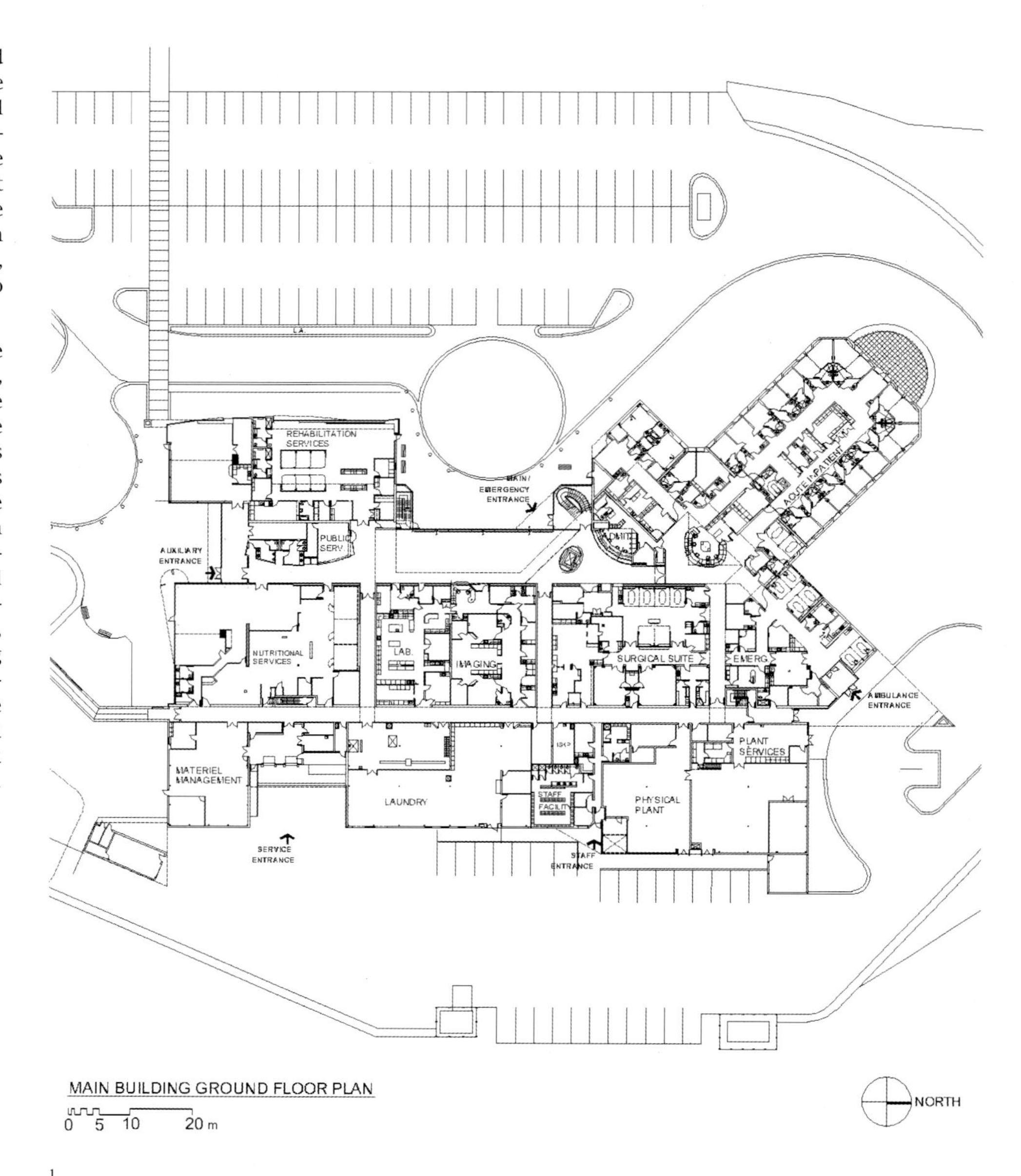

1

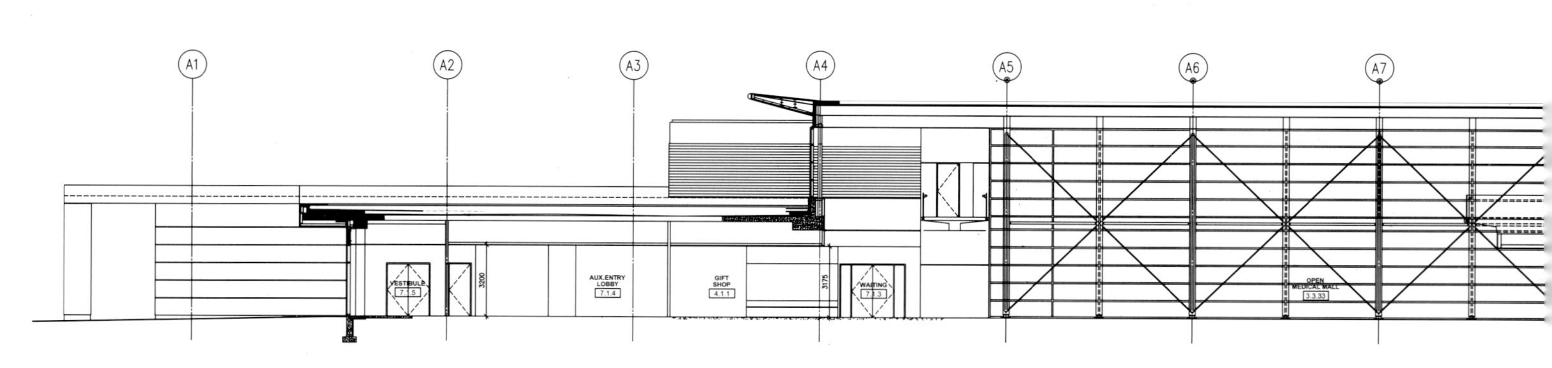

2

3

Kitimat医院和健康中心建筑帮助当地社区实现最佳生理、心理和精神健康。从表面上看，建筑设计反映了所服务的社区的特点，并且立刻被人们熟悉；从实际操作的层面上讲，新的建筑将不列颠哥伦比亚省的医疗服务设施要求的精确治疗、多层次和社区的医疗置于同一建筑之内。

中心设有两个主要功能区域，在形态和材料的语言上迥异：在精确治疗区，医疗设施围绕着一个玻璃中庭，并以其作为公共的医疗"商场"，以方便使用者进入不同的服务区。夸张的中庭尺度，包括可以远眺自然景致的两层高的玻璃幕墙和V字形的钢架，反映出本区域的工业建筑的历史。作为对比，私密性较强的多层次医疗区有小一些的尺度、缓坡屋顶和木材的恰当使用。这两个区域由玻璃通道连接，既是一种连接，又是一种分割。

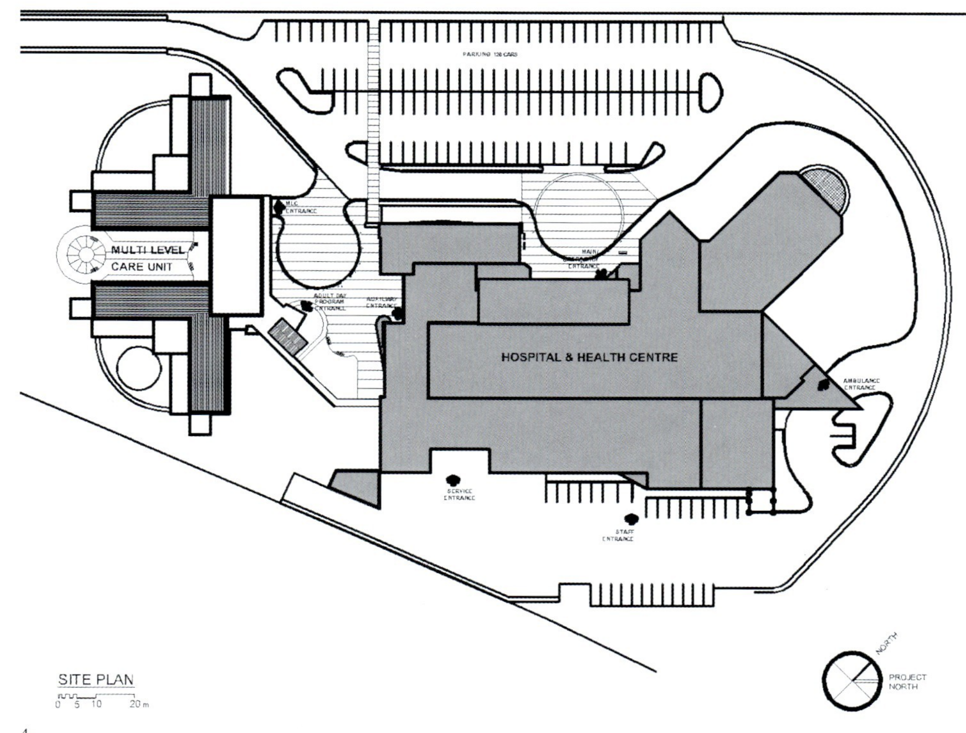

4

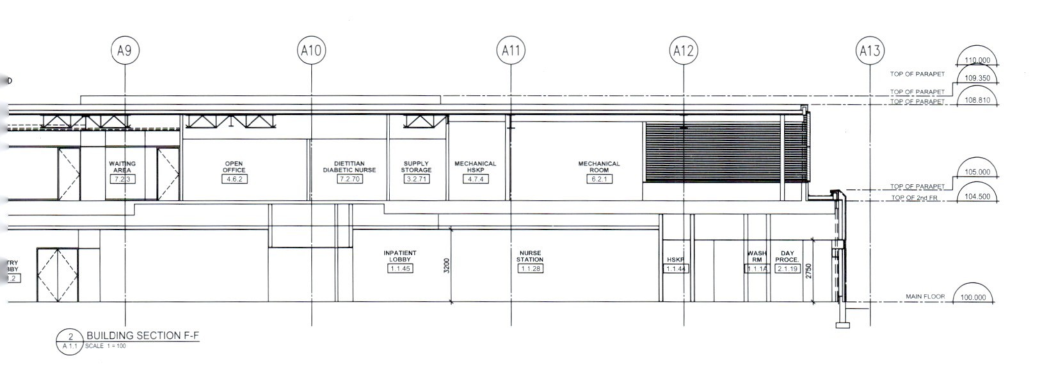

1　平面图
2　主入口及立面图
3　入口
4　总平面图

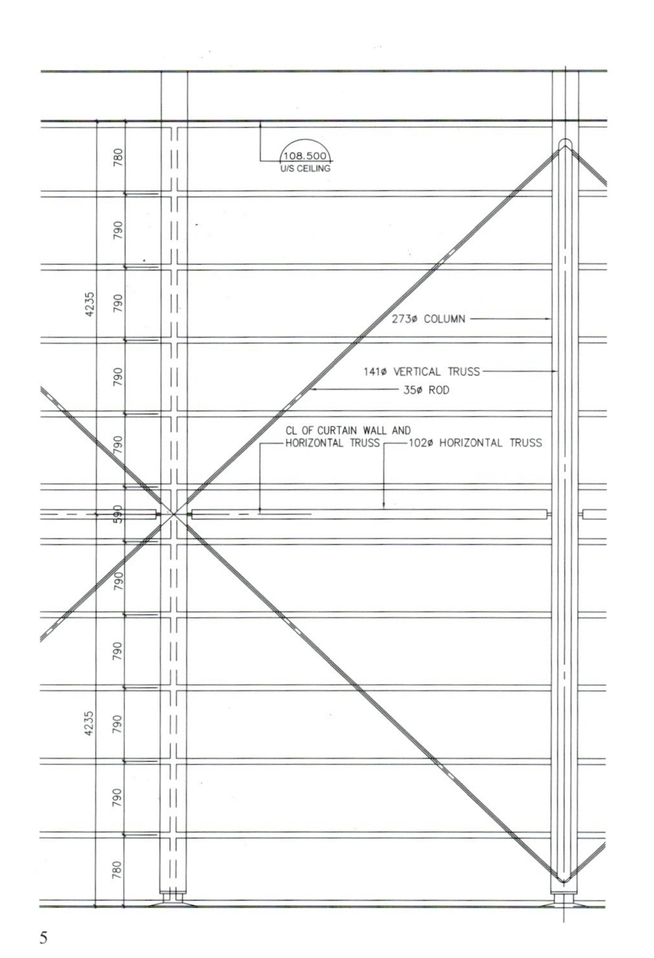
108.500
U/S CEILING
273ø COLUMN
141ø VERTICAL TRUSS
35ø ROD
CL OF CURTAIN WALL AND
HORIZONTAL TRUSS
102ø HORIZONTAL TRUSS
4235
4235
780
790
790
790
790
590
790
790
790
790
780

5

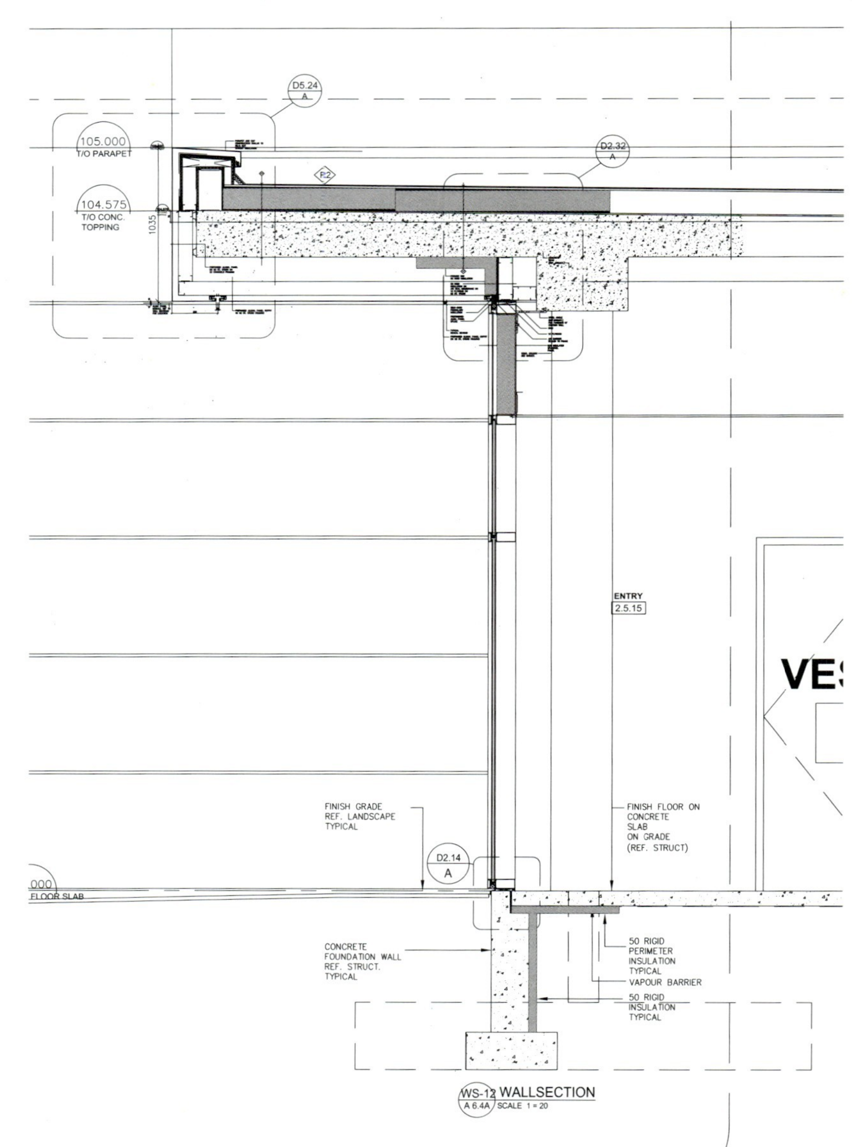
D5.24
A
105.000
T/O PARAPET
104.575
T/O CONC.
TOPPING
D2.32
A
ENTRY
2.5.15
VES
FINISH GRADE
REF. LANDSCAPE
TYPICAL
D2.14
A
FINISH FLOOR ON
CONCRETE
SLAB
ON GRADE
(REF. STRUCT)
FLOOR SLAB
CONCRETE
FOUNDATION WALL
REF. STRUCT.
TYPICAL
50 RIGID
PERIMETER
INSULATION
TYPICAL
VAPOUR BARRIER
50 RIGID
INSULATION
TYPICAL
WS-12
A 6.4A
WALLSECTION
SCALE 1 = 20

6

7

8

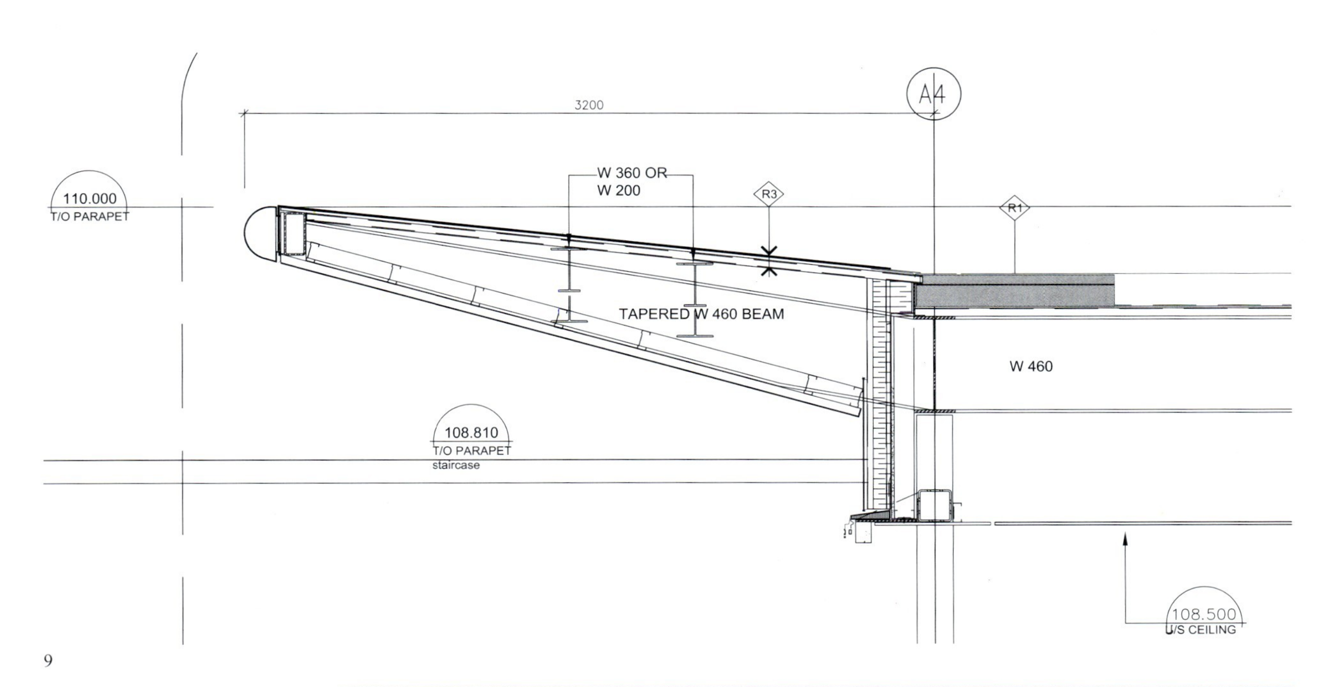

9

5　玻璃幕墙详图
6　墙体剖面详图
7~8　室内细部
9　挑檐详图
10　室外挑檐
11~12　入口细部

10

11

12

Saucier + Perrotte Architects

Saucier + Perrotte 建筑设计事务所

Since their inception in 1988, Saucier + Perrotte Architects have quickly gained international renown for their cultural and institutional buildings. The office is implicated in institutional, university and residential project at national and international levels.

At the year of 2004, Saucier+Perrotte Architects was the Canadian representatives at the Venice Biennale in Architecture.

Saucier + Perrotte 建筑设计事务所成立于1988年，并且迅速地在文化和教育研究建筑设计领域赢得了国际声誉。现今，Saucier + Perrotte 建筑设计事务所的特长集中在文化建筑、大学建筑和居住建筑。

2004年，Saucier + Perrotte 建筑设计事务所作为加拿大的代表参加了威尼斯建筑双年展。

First Nations Garden Pavilion

土著公园展厅

业　　主：蒙特利尔植物园
地　　址：加拿大魁北克省蒙特利尔市
完工时间：2001年9月
规　　模：新建筑185m²
建筑功能：展览/礼品商店/服务用房/办公空间

The First Nations Garden is a permanent commemoration of the great peace of Montreal of 1701. It is a crossroads of cultures, designed to help visitors to discover the culture of the first inhabitants of North America. It also offers an opportunity for the First Nations to share their traditions, wisdom and knowledge. The pavilion is a museum within the garden. Sheltering less than 2% of the garden grounds, the pavilion is mostly outdoor space. Built along the garden's main pathway, the pavilion metaphorically raises the path to reveal the cultural memory of the place. The undulating roof recalls a wisp of smoke through the trees. Outdoor displays sheltered by the roof are framed by two indoor spaces at opposite ends of the pavilion - exhibition and orientation spaces at one end, public washrooms and a meeting space at the other. The pavilion also houses a boutique and offices.

The relationship between building and site, and the environmental sensitivity needed to maintain the spirit of the garden, was critical to the design of the pavilion. The new building acts as both a filter and a link between two garden environments: an area of spruces and a maple forest.

Wherever possible, the pavilion's exhibition was planned outdoors. These exterior spaces orient the visitor and help to reduce the apparent size of the building by integrating the exhibition with the wider environment. Vertical surfaces are minimized so as to limit the visual impact of the building on the environment, and half of the built spaces are located underground to further reduce the influence of the new building on the existing setting. The new building was sited to retain all existing trees and maintain a relatively open terrain in an attempt to integrate the building and the site.

1

2

土著公园是纪念蒙特利尔市1701年和平的场所。公园的设计涉及多元文化的内容，其目标是帮助参观者发掘北美印第安人的文化传统。同样地，公园也提供给印第安人一个共享其传统、智慧和知识的机会。土著公园展厅是公园中的一个博物馆，其遮盖面积少于公园总面积的2%，而且几乎是完全室外的。展厅沿公园的主要道路布置，寓意着道路是展示文化传统的场所。屋顶起伏形状的灵感来自一束由树林中升起的烟。室外的有上盖的空间位于展厅两端的围合空间之间。一端是展览和公园引导室，而另一端是公共卫生间和一个集会空间。展厅同样设有一个礼品商店和办公室。

对土著公园展厅设计的限制不仅有来自如何处理建筑物本身和场地敏感关系的要求，而且有如何保护公园的精神功能的限制。新的展厅建筑对公园内的两个区域——云杉树林和枫树林——是一种过滤和沟通。

只要条件允许，展览活动都在室外举行。这些室外展览空间不仅可以引导参观者，而且通过与大环境的融合，在视觉上减小其体量。展厅的层高降至最小，以限制视觉上对环境的冲击。一半的新建建筑空间位于地下，进一步降低建筑物对场地的冲击。土著公园展厅保留了原址上全部的树木，并设有一块相对开阔的区域，以加强建筑物和场地之间的融合。

3

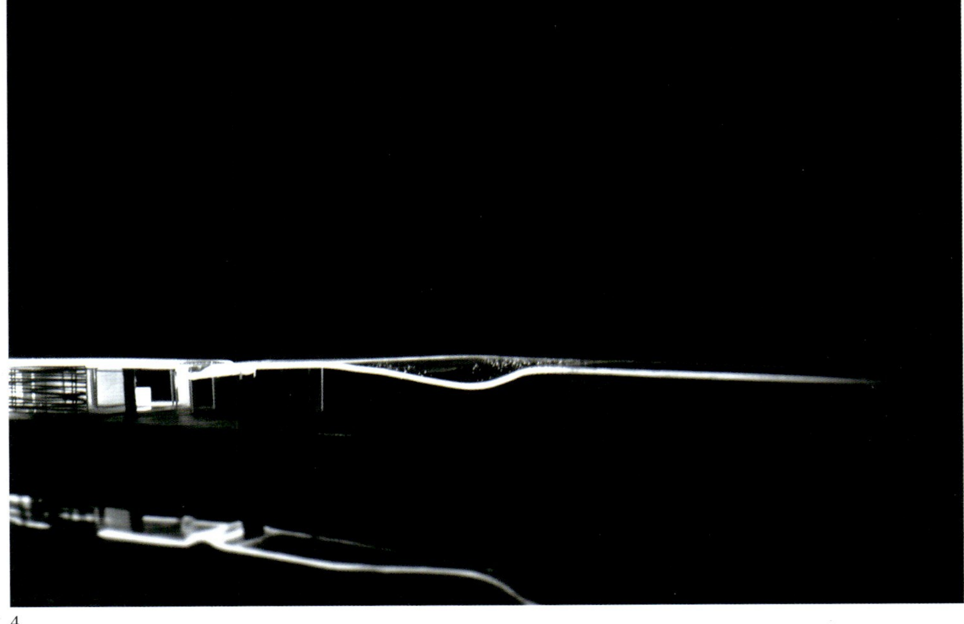

4

1 由道路远观展厅
2 构思草图
3 展厅的开敞空间
4 建筑模型

5
6
7
8
9

5　建筑模型
6　首层平面图
7　地下层平面图
8　半地下的展示空间
9　室外的展示空间
10　屋顶材质的细部质感
11　墙体详图
12　屋顶的自由曲线
13　屋顶细部

1　Wall construction

random size tree branches
galvanized steel supporting structure
curtain wall assembly

2　Roof construction

SBS waterproofing membrane
13mm wood fiber board
76mm rigid insulation
19mm pressure treated plywood
38mm steel deck
steel structure
ceiling assembly

3　foundation construction

hot rubberrized waterproofing membrane
sitecast concrete
76mm rigid insulation
vapour barrier
22m furring
16mm wallboard

10

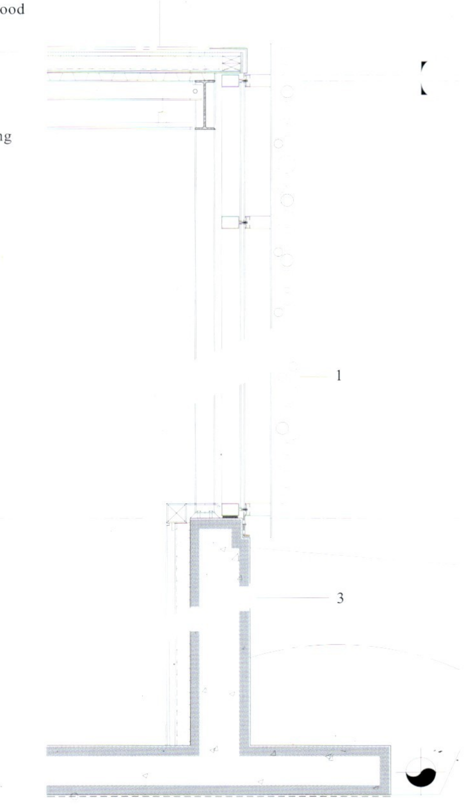

11

12

13

NEW COLLEGE RESIDENCE

新学院住宅

业　　主：多伦多大学
地　　址：加拿大安大略省多伦多市
完工时间：2003年9月
规　　模：新建筑11 600m²
建筑功能：学生住宅（214间－277个床位）/多功能演讲厅/行政办公空间

Two hanging gardens assure the well-being of the student population. Standing on Spadina Avenue just below Willcocks Street, a wall of large perforated masonry plates floats above the transparent mezzanine level. This brick volume is matched by a zinc and glass clad volume which faces the St. George Campus and contains the East Garden. This three-storey space is a place of repose as well as an exterior community room available to all residents. Linked by an interior stairwell, the West Garden is carved out of the brick facade at the northern end from the 5th through the 8th floors. This allows the evening sun to creep through the glass façade at the back of the garden and into the common areas behind.

The organization of the architectural elements reflects the conceptual relationship between the student community and its environment. The seven floors of residences form a more contained community intended to provide both a home and opportunity for social interaction. All common areas are co-ed including washrooms and are contained in a central circulation area sandwiched between the two main volumes.

The lower levels become a gathering place for the community at large. Raised above the street, the mezzanine level contains the administration offices and reception, as well as access to both the residences above and the ground floor below. The versatile ground floor provides a theatre for public performances along with quiet study halls for the large student population. The New College is intended to be a meeting place, a venue for the exchanging of ideas within both an academic setting and an urban environment.

1

1　住宅的造型由砖饰面和锌板、玻璃饰面两部分组成
2　住宅夜景
3　东花园夜景

新学院住宅的两个空中花园提供给学生良好的休息场所。新学院住宅建筑位于Spadina大道上，Willcocks街的下方，在通透的夹层空间之上，像是一片巨大的墙承托着建筑物的基座。建筑物由以大面积砖为立面材料和以锌板及玻璃为立面材料的部分组成。面对圣乔治校区，并环绕着东花园的建筑立面以锌板及玻璃装饰。东花园是一个3层高的安静区域，也是对所有住客开放的室外社交场所。西花园位于砖立面建筑一侧的北端，由室内的楼梯相连，而且占据了五至八层的空间。这使得黄昏的阳光可以穿过花园背面的玻璃，射入下层的公共空间。

新学院住宅建筑元素的组合反映出学生之间交流的需求和其所处的环境之间的关系。七层的居住空间不仅提供了一个家的空间，而且提供了一个良好的社交场所。所有的公共空间都是男女共用的，包括卫生间。一个集中的交通区域位于两个主要的建筑物之间。

广义上来说，位于低层的空间是整个住宅居住者的集会场所。高于街道层之上的夹层空间是办公室、接待空间和进入上层居住空间以及下面首层的入口。首层空间为公共表演剧场和供学生使用的安静的学习区域。新学院住宅建筑的设计不仅在学术设施上提供了一个集会的场所，一个交换思想的空间，而且在建筑本身与城市环境的关系上也是如此。

2

3

4

5

6

7

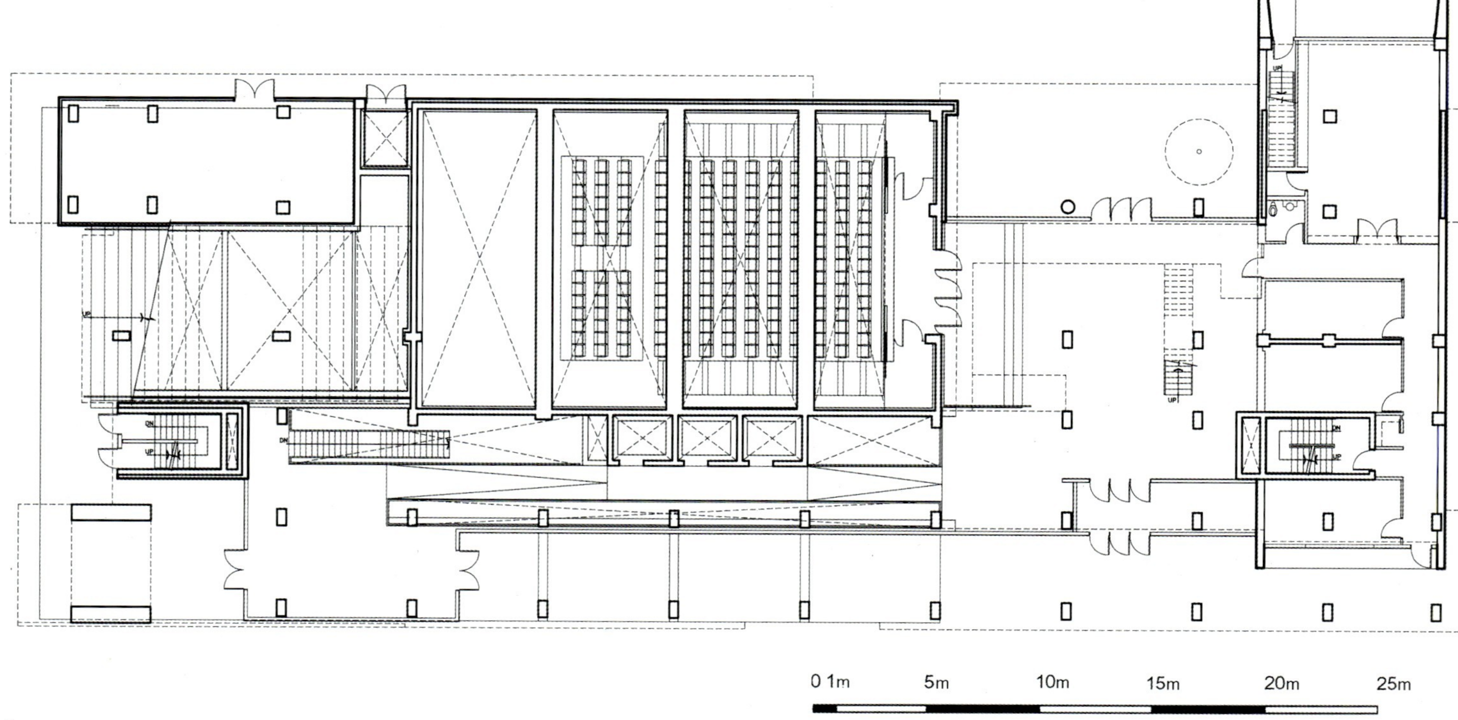

8

4 砖立面
5 砖立面细部
6 入口细部
7 窗细部
8 首层平面图
9 锌板和玻璃构成的立面
10 三维立体模型
11 二层平面图

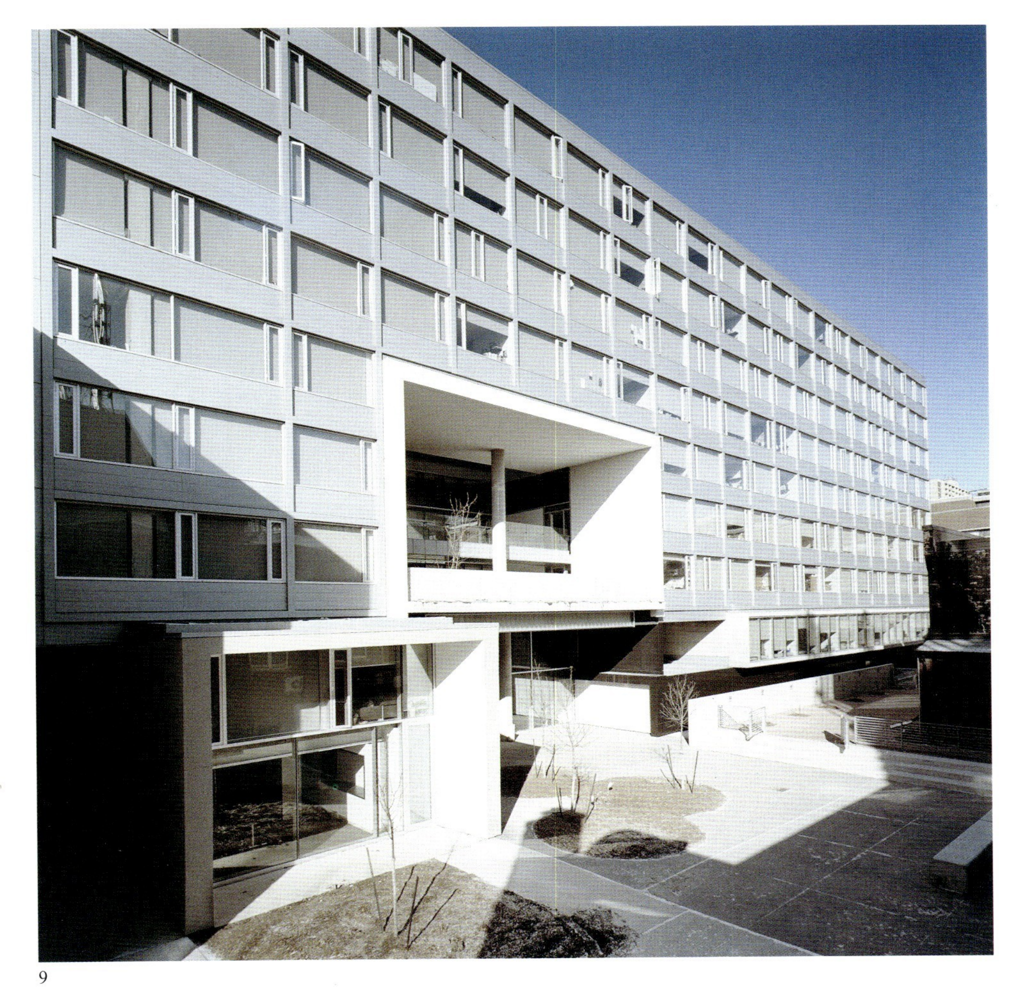

9

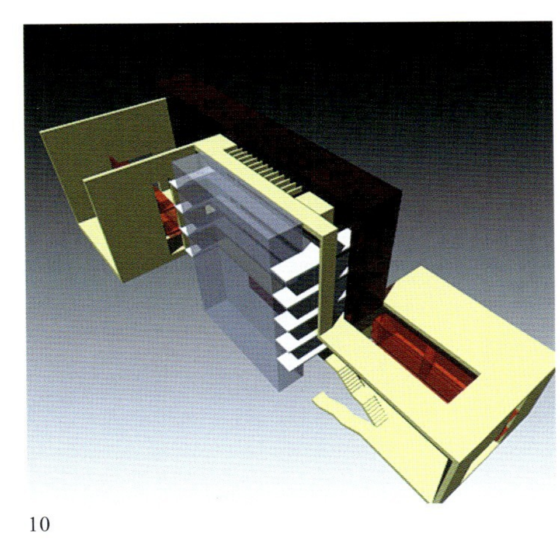

10

0 1m 5m 10m 15m 20m 25m

11

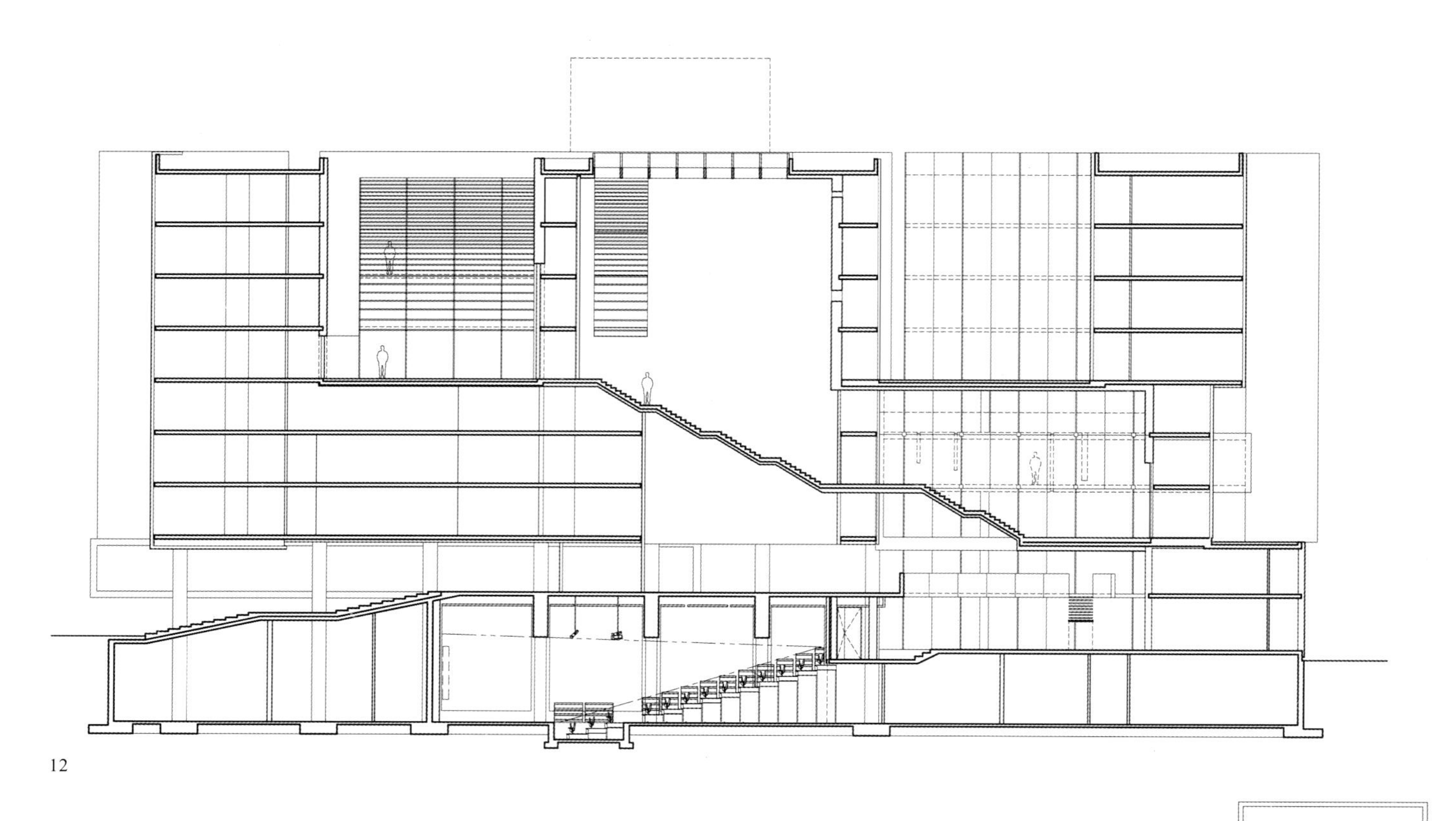

12

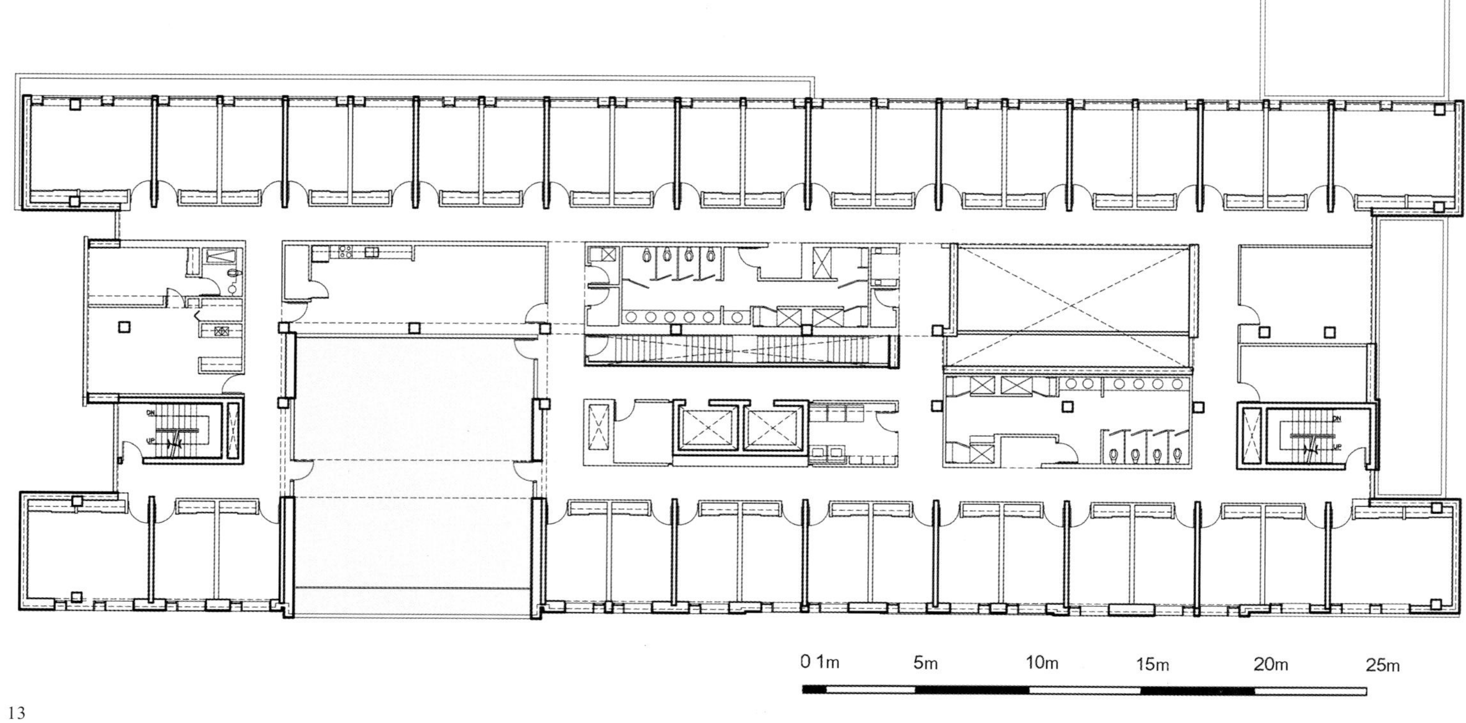

13

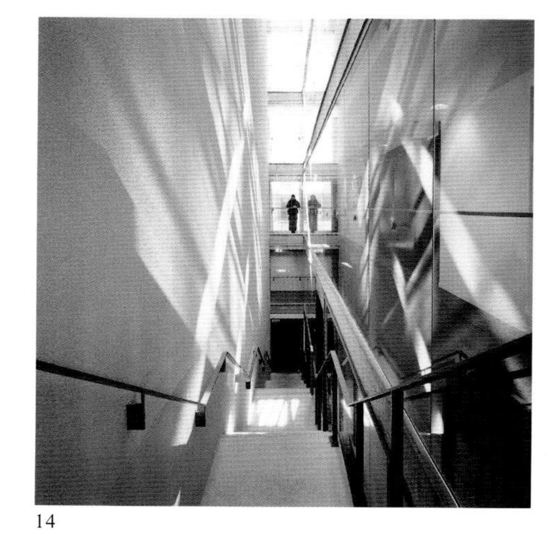

14

15

16

12 剖面图
13 五层平面图
14 室内通道
15 室内的公共空间
16 室内通道
17 玻璃窗及墙体详图
18 室内通道
19 东花园内部
20 建筑细部

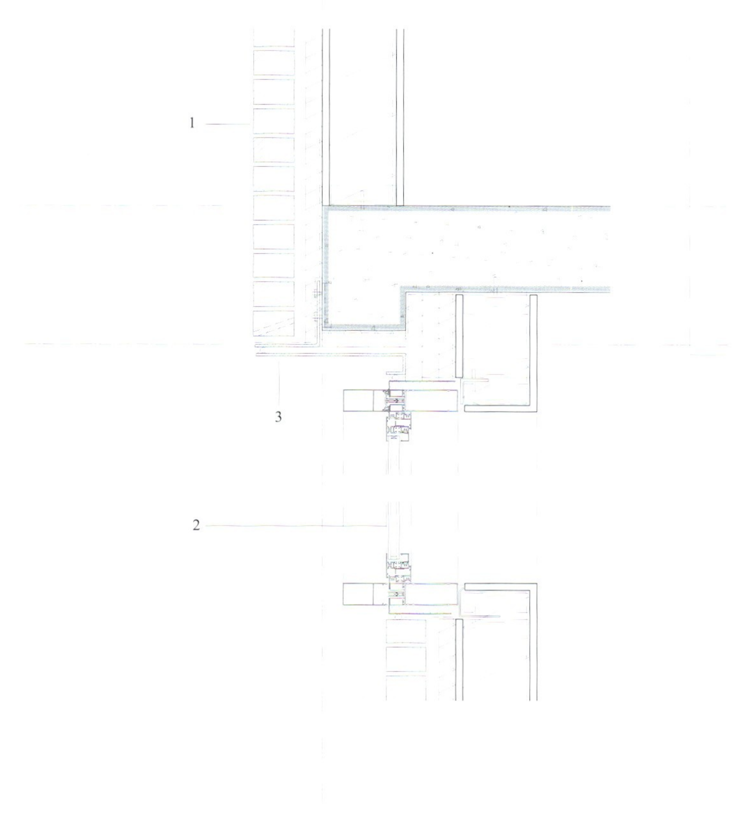

17

18

19

20

Perimeter Institute For Research In Theoretical Physics

Perimeter理论物理研究所

业　　主：Perimeter研究所
地　　址：加拿大安大略省滑铁卢市
完工时间：2004年10月
规　　模：6 000m²
建筑功能：210座演讲厅/研究图书馆/研究员办公室/集会区域/教室和会议室/行政办公室/快餐区/健身设施

Riding the controversial line between public and private space, this private research institute attempts to subvert the usual hard thresholds established by private enterprise in the public realm. The site is on the shore of Silver Lake, at the northern edge of Waterloo's downtown core and the southern edge of the city's central park. Adjacent to the primary pedestrian access between the University campus and the city center, the site is an urban wilderness between clearly defined worlds.

The design is also inspired by the nebulous spaces occupied by the subjects of theoretical physics, at once micro- and macrocosmic, rich in information and of indeterminate form and substance. Between city and park, the Perimeter Institute expands and inhabits the improbable space of the line that separates the two. The building defines the secure zones of the institute's facilities within a series of parallel glass walls, embedded in an erupting ground plane that reveals a large reflecting pool. The north façade, facing the park across this pool, reveals the institute as an organism, a microcosm of discrete elements. The south façade, facing the city across train tracks and the city's main arterial road, presents the institute as a unified but transforming entity, of enigmatic scale and content. Entry to the institute is possible from both the north, along the reflecting pool, and south, under the new ground.

The interior of the institute is organized around two central spaces, the main hall on the ground floor and the garden on the first. Spaces for administration, meeting and seminar rooms, leisure and fitness spaces, and a multipurpose theatre for symposia and public presentations, have direct access to the main hall. The circulation corridors running east-west are sandwiched between the opalescent glass planes, which are occasionally punctured or shifted to reveal views across the interior space of the hall. Vertical circulation climbs these walls, tendrils of ground that run from the garden through the building. The garden - nature emerging from the vacuum - is crossed by three bridges that puncture all the planes, as well as the north and south façades. The bridges are conduits for quick access to facilities, information and colleagues, routes crossing the improbable space between theoretical physics and everyday life.

1

很难限定私营的Perimeter理论物理研究所是属于公共的空间，还是私家的场所。Perimeter研究所的建筑设计试图打破在公共区域内的私营企业建筑固有的和强烈的规则。Perimeter研究所的建筑位于银湖的岸边，滑铁卢市中心的北侧边缘，城市中心公园的南侧，并与通向大学校园和城市中心的主要步行通道相邻。其用地是一块位于城市中心和中心公园之间的生态保护树林。

建筑设计造型借鉴了太空的星云图案。在微观和宏观的宇宙中，理论物理学所展示的丰富信息和具有不确定形态的物质也同样启发了建筑师的灵感。在城市和公园之间，Perimeter理论物理研究所的建筑以一个不真实的空间形态区分两者。由倒影水池之中伸出的一组玻璃墙界定了研究所的安全保护区域。建筑的北面为水池以及更远处的中心公园，北立面的设计寓意着一个有机体，一个微观的抽象元素。建筑物的南侧是火车道和城市主要交通支线，以及更远处的城市中心。南立面的设计则是通过迷一般的尺度和细节，表达出建筑物统一的但是变化的实体。建筑物的入口位于沿倒影水池的北面和首层南面的下方。

建筑室内空间围绕着两个核心区域组织：一个是首层的中庭，另一个则是二层的花园。行政办公区、会议室、教室、演讲室、健身室以及多功能座谈和公共演讲剧场都有直接面对中庭的出入口。室内交通沿东西向布置，且位于彩色玻璃板之间。玻璃板图案的变化——或贯穿或移动——改变了室内中庭的景观。竖向楼梯像是植物的卷须一样由花园内的墙攀爬向上。在作为天然的吸尘器的花园内，3座桥贯穿所有楼层平面，以及南北立面。这些桥不仅方便了设施的使用、信息的传播、同事之间的交流，而且加快了使用者在虚幻的理论物理和现实生活之间的身份转变。

2

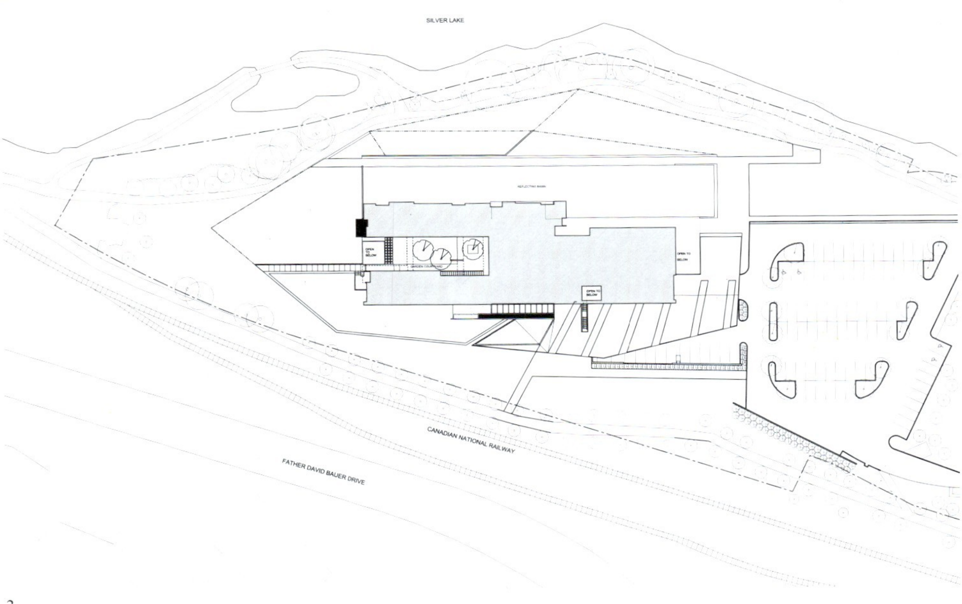

3

1 夕阳下的倒影水池及建筑北立面
2 入口夜景
3 总平面图

4

5

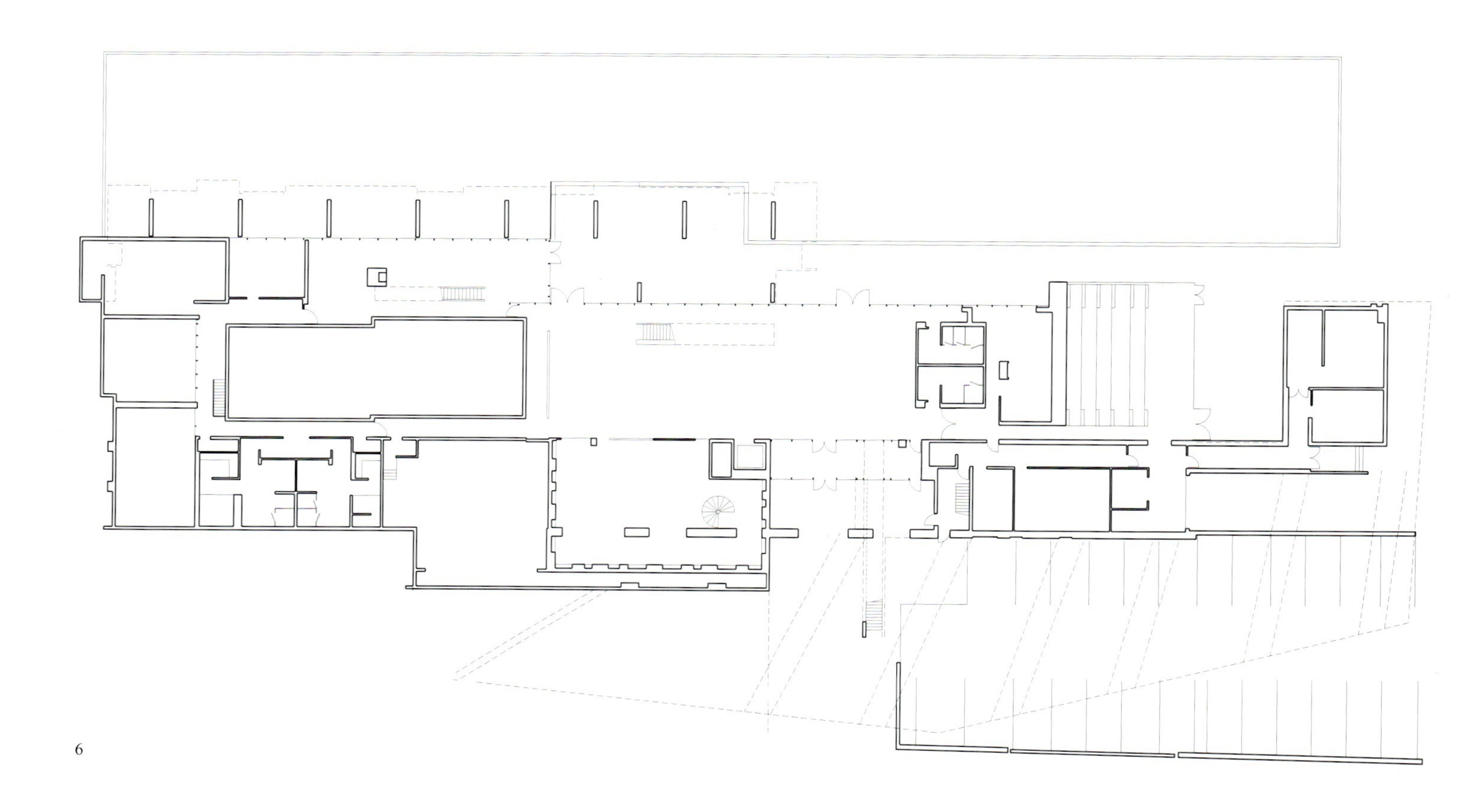

6

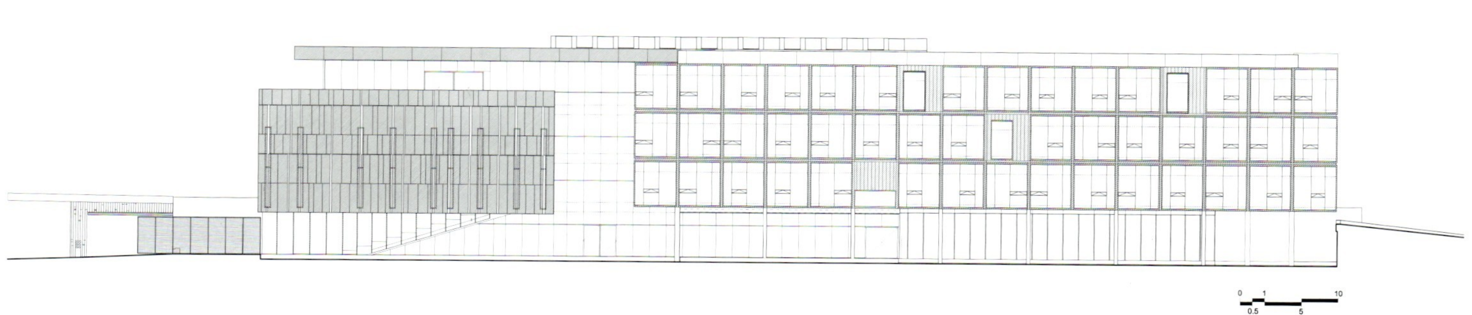

7

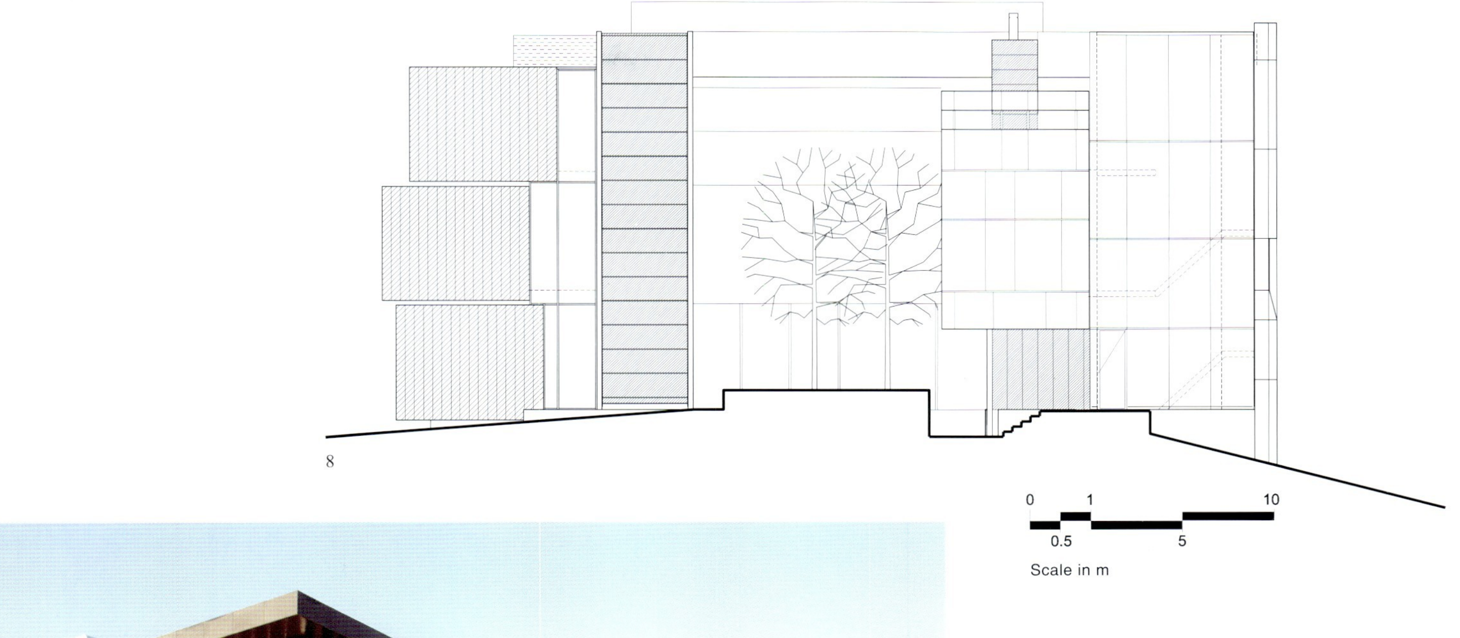

8

9

4 南立面
5 北立面的独特造型
6 首层平面图
7 北立面图
8 西立面图
9 建筑西北角
10 东立面图

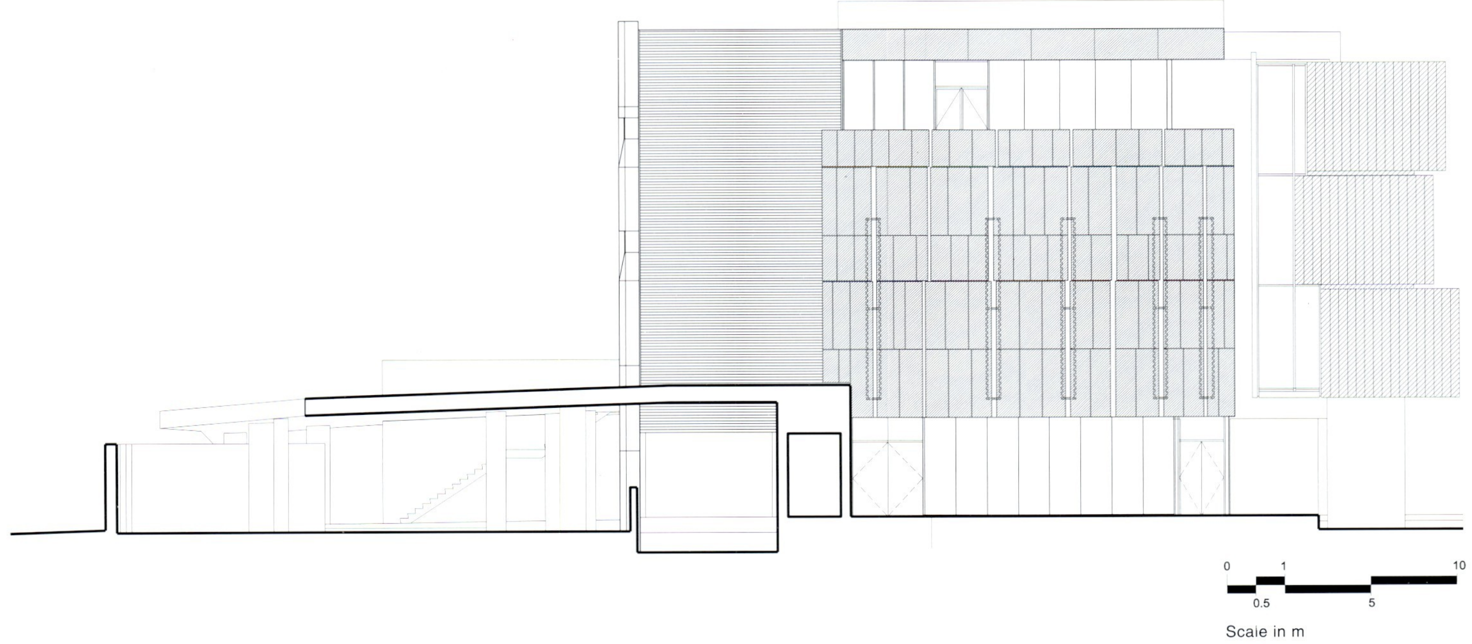

10

11

12

11 南立面夜景
12 南立面细部
13 室内天窗仰视南立面细部
14 南立面细部
15 仰视南立面细部
16 西立面细部
17 南立面图

13

14

15

16

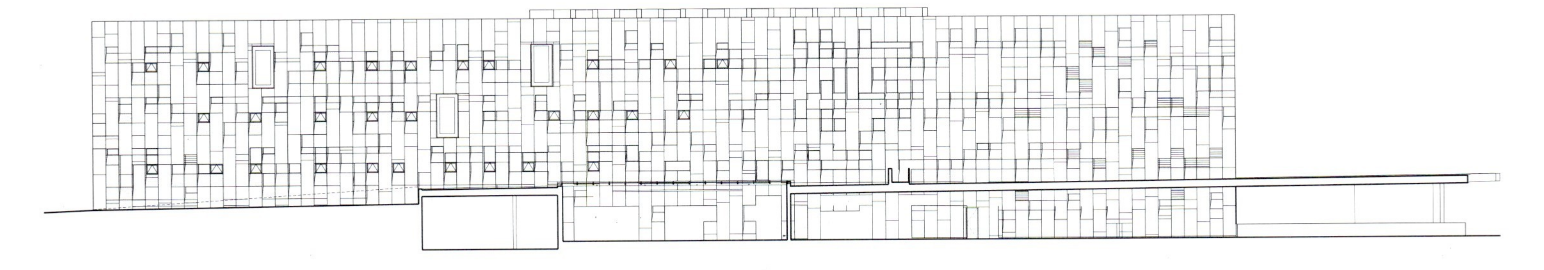

17

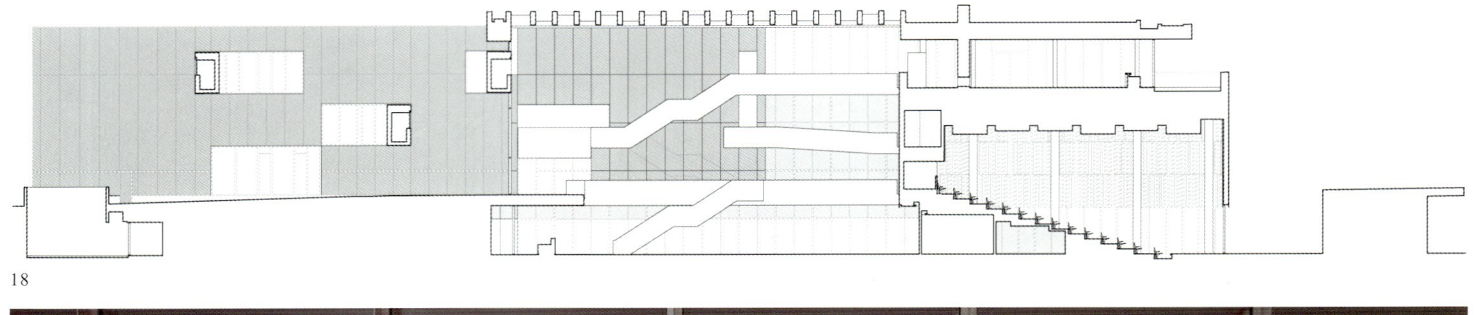
18

19

20

21

22

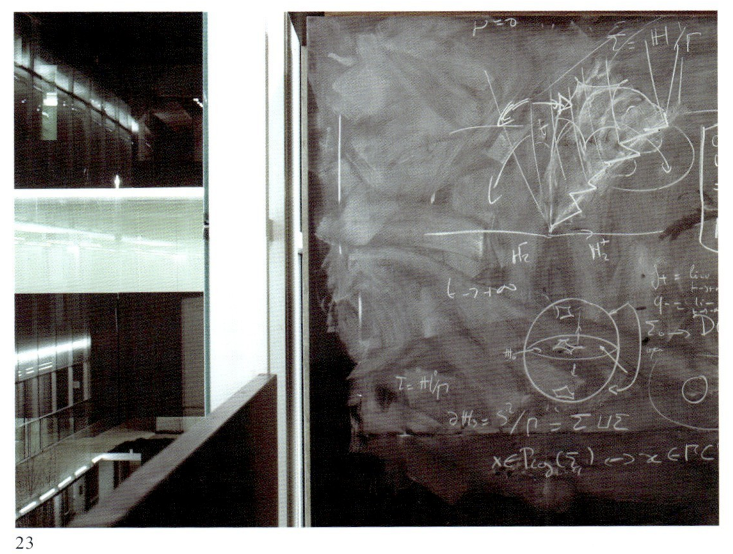
23

24

25

18 东西向剖面图
19 中庭细部
20 中庭细部
21 中庭天窗细部
22 中庭的玻璃天窗改变室内的光环境
23 室内细部
24 墙体细部
25 中庭的玻璃墙

26

27

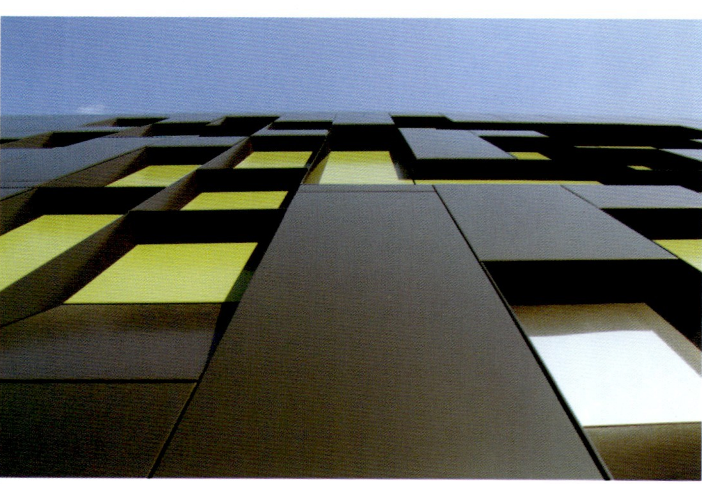

28

29

1 Roof Composition

zinc sheet coping
water shield
16mm treated plywood
variable air space
40mm rigid insulation
reinforced asphalt membrane
sitecast concrete

2 Wall Composition

25mm interlock zinc panel
100mm adjustable z bars
10mm air space
50mm polyurethane insulation
blueskin membrane
sitecast concrete

3 Curtain wall

4 Floor Composition

sitecast concrete
blueskin membrane
75mm polyurethane insulation
25mm air space
100mm adjustable double steel angles
zinc interlock panels

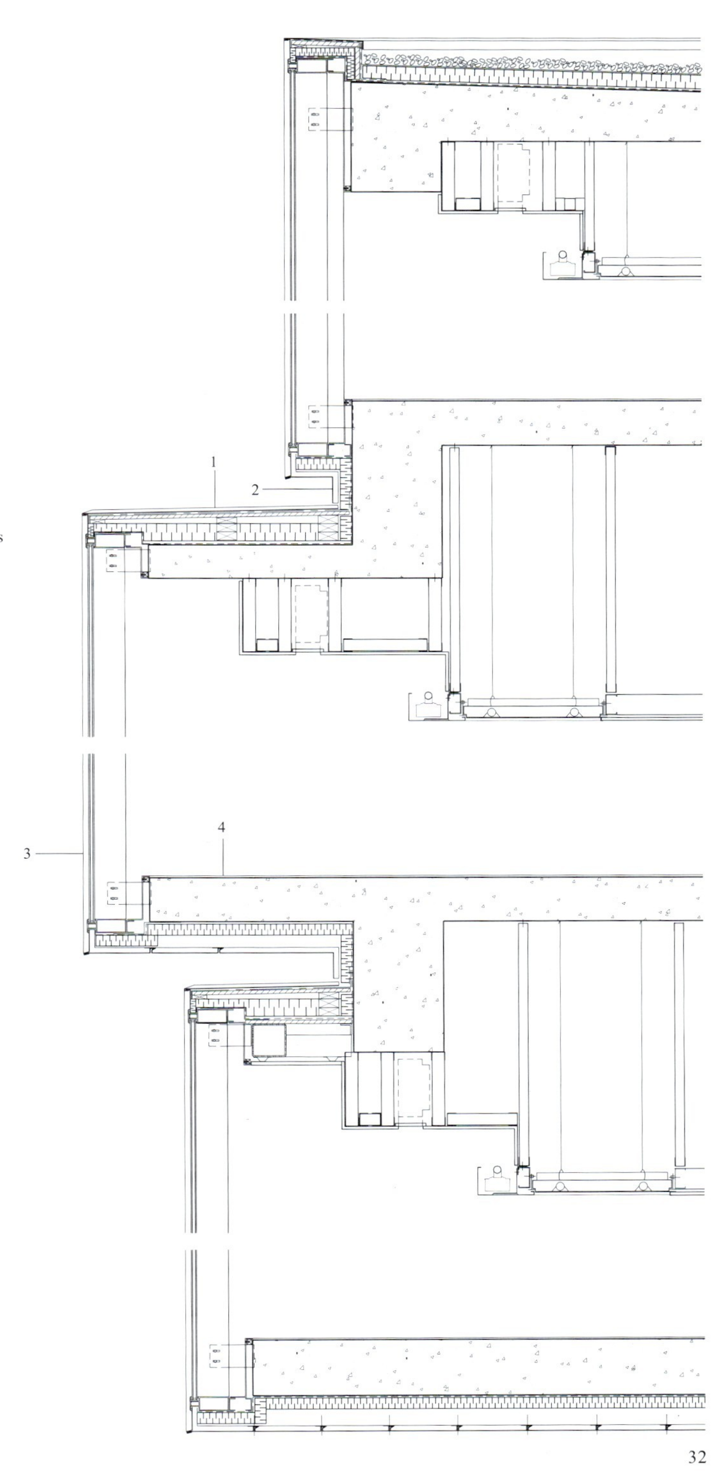

32

30

31

26 北立面夜景
27 玻璃与墙体细部
28 玻璃窗细部
29 细部
30~31 建筑墙体细部
32 墙体细部

特别鸣谢：

- Canada Mortgage and Housing Corporation
  Alvin Singh, Nellie Cheng (程乃立)
- 章捷 Maggie Zhang
- 加拿大高地建筑咨询 （上海）
  贾建南　王笑强　许宁　王承富　王钢　胡淑荣
- 上海哲创广告
  孙维　顾逸凡　沈磊
- Stantec
  Peter Buchanan, William Vo
- Arthur Erickson with Nick Milkovich Architects Inc.
  Arthur Erichson, Merle Ginsburg, Nicole Milkovich, David Liang
- Bing Thom Architects Inc. ——谭秉荣建筑设计事务所
  Helen Ritts, Ying lee
- Teeple Architects Inc. —— Teeple 建筑设计事务所
  Tom Arban, Sonia Syrko
- VIA Architecture —— VIA 建筑设计事务所
  Lindsay Steenberg, Leyla Roshanshad
- Downs/Archambault & Partners —— Downs/Archambault 及合伙人建筑设计事务所
  Mark Ehman
- Patkau Architects —— Patkau 建筑设计事务所
  Kayna Merchant
- DGBK Architects —— DGBK 建筑设计事务所
  Wade Comer
- Walter Francal Architects —— Walter Francal 建筑设计事务所
  Walter Francal, Mark Ashby
- Acton Ostry Architects INC —— Acton Ostry 建筑设计事务所
  Monica Feldman
- Nicolson Tamaki Architects —— Nicolson Tamaki 建筑设计事务所
  Don Nicolson, Anna Teng
- Hancock Bruckner Eng + Wright Architects —— HBEW 建筑设计事务所
  Gwyn Vose
- Saucier+Perrotte 建筑设计事务所
  Joan Renaud
- BCIT 不列颠哥伦比亚理工学院
  Gordon Kennedy
- Cameron Clark
- Douglass Weinberg
- Ryan Huang 黄建容（台北）
- Michael Appel, Katarina Adami 夫妇（德国 慕尼黑）